Education for Sustainability

While the Decade of Education for Sustainable Development (DESD) (2005–2014) has been completed, the status and advocacy of Education for Sustainable Development (ESD) remains prominent. The United Nations (UN) goals of Education for All (EFA) and the Millennium Development Goals (MDGs 2000–2015) were complementary and provided a rationale for the importance of environmental education (EE) and Education for Sustainable Development (ESD). The United Nations Educational, Scientific and Cultural Organization's (UNESCO) Muscat Agreement in 2014 advocated seven global education targets, one of which was to cultivate skills for global citizenship and environmental sustainability. As part of the UN Sustainable Development Goals (SDGs 2015–2030) and as echoed by the Aichi-Nagoya Declaration on Education for Sustainable Development, education is embedded in goals which pertain to biodiversity, sustainable consumption and production, and climate change. Supporting these goals, there is a call for research and development as well as coordinated actions with an emphasis on the principles of human rights, gender equality, democracy, and social justice. There is also a call for attention to the importance and relevance of traditional knowledge and indigenous wisdom in various geographical, socio-cultural, and educational contexts.

With this background, and in light of UNESCO's Education 2030 Agenda (2017), this *Education for Sustainability* Book Series has been launched. Its purpose is to echo and enhance the global importance of education for a sustainable future as an educational vision. The series provides insights on a broad range of issues related to the intersection of, and interaction between, sustainability and education. The series showcases updated and innovative practice, discusses salient theoretical topics, and uses cases as examples.

The Series adopts international, environmental education, and lifelong learning perspectives and explores connections with the agenda of education for sustainability and of Education for Sustainable Development. The intended audience includes university academics in educational studies, environmental education, geographical education, science education, curriculum studies, comparative education, educational leadership, and teacher education; the staff of international agencies with responsibilities for education; and school teachers in primary and secondary schools.

Supported by the expertise of a distinguished and diverse International Advisory Board, this series features authoritative and comprehensive global coverage, as well as diversified local, regional, national, and transnational perspectives. As a complement to the Schooling for Sustainable Development Book Series, it explores issues that go beyond primary and secondary schooling into university, vocational, and community education settings. These educational issues involve multiple stakeholders ranging from international agencies, governmental and non-governmental organizations, educational and business leaders to teachers, students, and parents. The research topics covered include global themes related to environment such as climate change education, disaster prevention and risk reduction, biodiversity education, and ecological education. They also include human ecological issues such as global citizenship, peace education, childhood development, intergenerational equity, gender studies, and human rights education. Further, they include society-oriented issues such as governance, green skills for sustainable development, sustainability leadership, and applied learning.

Researchers interested in authoring or editing a book for this series are invited to contact the Series Publishing Editor: Lawrence.Liu@springer.com

All proposals will be reviewed by the Series Editors and editorial advisors.

More information about this series at http://www.springer.com/series/15237

Stephanie Leder

Transformative Pedagogic Practice

Education for Sustainable Development and Water Conflicts in Indian Geography Education

Stephanie Leder
University of Cologne
Cologne, Germany

ISSN 2367-1769 ISSN 2367-1777 (electronic)
Education for Sustainability
ISBN 978-981-13-2368-3 ISBN 978-981-13-2369-0 (eBook)
https://doi.org/10.1007/978-981-13-2369-0

Library of Congress Control Number: 2018955927

This Springer imprint is published by the registered company Springer Nature Singapore Pte Ltd.
The registered company address is: 152 Beach Road, #21-01/04 Gateway East, Singapore 189721, Singapore

Preface

Keywords Geographical educational research · Transformative pedagogic practice
Education for sustainable development · Geography education
Teaching methodology · Schooling · India · Translation of educational policies
Power · Empowerment · Cultural values of learning and teaching
Basil Bernstein · Paolo Freire · Critical consciousness · Environmental education
Knowledge · Skill development · Argumentation · Learner-centered pedagogy
Textbook · Water conflicts · Natural resource management

Transnational educational policies such as *Education for Sustainable Development (ESD)* stress the role of education for critical environmental consciousness, sustainable environmental action, and societal participation. Approaches such as the promotion of critical thinking and argumentation skill development on controversial human–environment relations are relevant to participate in decision-making on sustainable development. The transformative potential of ESD is based on these approaches.

This book empirically examines the following research question: "Which challenges exist for the translation of the transnational educational policy Education for Sustainable Development (ESD) in pedagogic practice in geography teaching at English-medium secondary schools in Pune, India?" This study investigates pedagogic practice and the transformative potential of ESD within the setting of school education in India's heterogeneous educational system. The analysis at five English-medium secondary schools in the emerging megacity of Pune focuses on the institutional regulations, power relations, and cultural values that structure Indian geography education with the example of the topic of water. To analyze the challenges that exist for the implementation of ESD in pedagogic practice, the study follows a theoretically anchored and didactically oriented analysis.

At the conceptual level, this study pursues an interdisciplinary synthesis with elements of geographical developmental research, geographical education research, and sociology of education research. The theoretical framework for transformative pedagogic practice links concepts of Basil Bernstein's Sociological Theory of Education (1975–1990), Paolo Freire's Critical Pedagogy (1996), and the didactic approach of argumentation skill development (Budke 2010, 2012). This conceptual approach offers an integrative multi-level analysis, which reflects the status quo of pedagogic relations and shows opportunities to prepare and encourage students to become vanguards for social and environmental transformation.

The study is based on nine months of fieldwork at five English-medium secondary schools in Pune between 2011 and 2013. The methodological framework combines qualitative social science methods such as qualitative interviews, document analyses, and classroom observations with an intervention study in geography lessons. The analysis of transformative pedagogic practice is differentiated into three interrelated levels: document analyses, field work, and action research.

Firstly, the thematic and methodological analysis of educational policies, curricula, syllabi, and geography textbooks examines how academic frameworks for formal school education in India relate to the principles of ESD. In contrast to the National Curriculum Framework (2005), which promotes pedagogic principles similar to ESD, the contents and methods in geography syllabi and textbooks display a fragmented, fact-oriented, and definition-oriented approach to the topic of water. The resource of water is presented as a fixed commodity, and the access to water is not depicted as socially constructed. The controversy of differing perspectives on water access in the students' urban environment is not presented. This contradicts ESD principles, which favor an integrated, skill-oriented, and problem-based approach to topics at the human–environment interface.

Secondly, the study examines how power relations and cultural values of teaching and learning shape pedagogic practice, and how these link to ESD principles. Teaching methodology in observed geography lessons depicts students as reproducers of knowledge, as they are expected to repeat teaching contents spelled out by the teacher and in textbooks. Strong framing and classification of classroom communication shape the teacher–student interaction. Current pedagogic practice in India transmits norms and values of respect and authority, rather than promoting questioning and critical thinking. The textbook governs classroom interaction, as the role of the teacher is to transmit a pre-structured selection of knowledge as depicted in the textbooks. The prescriptions in syllabi and textbooks barely leave enough time and space for students to develop skills in geography lessons and constrain teacher's agency and control over the selection, sequence, pacing, and evaluation of knowledge and skills. These norms represent a performance mode of pedagogy, which contrasts with the competence mode of pedagogic practice in ESD. As a democratizing teaching approach, ESD principles are in strong juxtaposition to the traditional hierarchical structures that occur and are reproduced in the country's myriad of educational contexts.

Lastly, an intervention study identifies institutional, structural, and socio-cultural challenges and opportunities to translate ESD principles into geography teaching in India. To examine how ESD principles can be interpreted through argumentation on urban water conflicts, three ESD teaching modules "Visual Network" (Leder, 2014), "Position Bar" (Mayenfels and Lücke 2012), and "Rainbow Discussion" (Kreuzberger 2012) were adapted to the topic and context of this study. The implementation process demonstrates how strong classification and strong framing in Indian geography education can be weakened. While the use of classroom space and teaching resources is changing and students actively participate, the focus on presentation, sequence, and formal teacher–student interaction is sustained. The latter shapes teachers and students' reinterpretation of the ESD teaching modules. This implies that ESD and the promotion of argumentation skills only partly intervene in prevalent principles of pedagogic practice.

The results demonstrate how the educational discourse of ESD fundamentally challenges the reproductive mode of pedagogic practice in the case of geography education in India, as it subverts cultural values, norms, and constructions of teaching and learning. Despite this, ESD as a transformative pedagogic practice can contribute to gradually revising current geography teaching contents and methods toward promoting learner-centered teaching, critical thinking, and argumentation skill development. A contextualized understanding of how power relations shape and are reproduced in pedagogic practice can better link educational reforms to social reality. The study emphasizes the need for researchers and policy makers to demonstrate how principles of schooling can be altered for empowering students to obtain skills and gain knowledge to participate in decision-making, for example, concerning water resource conflicts, and to espouse sustainable development as conscious and critical citizens.

Cologne, Germany Stephanie Leder

References

Budke, A. (2012). Argumentationen im Geographieunterricht. *Geographie und ihre Didaktik*, (1), 23–34.

Budke, A., Schiefele, U., & Uhlenwinkel, A. (2010). Entwicklung eines Argumentationskompetenzmodells für den Geographieunterricht. *Geographie und ihre Didaktik*, 3, 180–190.

Freire, P. (1996). Pedagogy of the Oppressed. London: Penguin Books Ltd.

Kreuzberger, C. (2012). Regenbogen-Vierer—Diskussion mit Redekarten In A. Budke (Ed.), *Kommunikation und Argumentation*. Braunschweig: Westermann.

Leder, S. (2014). Barrieren und Möglichkeiten einer Bildung für nachhaltige Entwicklung in Indien. In I. Hemmer, M. Müller, & M. Trappe (Eds.), *Nachhaltigkeit neu denken. Rio + X. Impulse für Bildung und Wissenschaft*. München: Oekom-Verlag.

Mayenfels, J., & Lücke, C. (2012). Einen Standpunkt "verorten"—der Meinungsstrahl als Argumentationshilfe. In A. Budke (Ed.), *Kommunikation und Argumentation* (pp. 64–68). Braunschweig: Westermann.

National Council of Educational Research and Training. (2005). National Curriculum Framework 2005. New Delhi: NCERT.

Acknowledgements

This book has been developed, revised, and thought through with the support of many people and institutions. I would like to express my sincere thanks to everyone who has been part of this journey.

First of all, I would like to thank my supervisors for their continuous support. Prof. Dr. Frauke Kraas awoke my interest in sustainable urban development in India and beyond and has always been supportive of my Ph.D. research approach. I am much obliged for her support which began during my master's thesis research on access to healthcare services for children in Mumbai, where I was first exposed to pedagogic practice in a slum and an international private school. Prof. Dr. Alexandra Budke comprehensively supported this rather unusual international Ph.D. research thesis in the field of geography education from the beginning. My sincere thanks for engaging in my research and providing insightful comments and input.

This study was only possible with the support of the principals, teachers, and students at the schools in Pune where I conducted numerous classroom observations and interventions over a period of two years. A heartfelt thanks to Gayatri, Ujala, Sana, Shital, Vaishali, Umar, Amita, Kavita, Monika, Neleema, Anitha, and the many sincere students who were willing to share their perceptions of their schooling with me.

This research project in Pune was developed in close cooperation with the Bharati Vidyapeeth Institute of Environment Education and Research (BVIEER) with the support of Prof. Dr. Erach Bharucha, Dr. Shamita Kumar, and Dr. Kranti Yardi. I benefitted from their experience in implementing environmental education in schools, as well as their national and local networks of textbook authors, schools, and teachers. Only through many interviews and discussions with a range of educational stakeholders, I could gain a greater insight into current issues in pedagogic practice and the role of Education for Sustainable Development in the Indian educational system. At BVIEER, I would like to further thank my colleagues and friends Binita Shah, Laxmikant Kumar, Sayani Dutta, Shivam Trivedi, Pallavi Gunju, Shalini Nair, and Anand Shinde for Pune water excursions and water samples, the teacher and textbook workshop organization, map development, and all other valuable support for my research.

I would like to thank Dr. Sonja Frenzel and Prof. Dr. Becker-Mrotzek at the Cologne Graduate School for Teaching Methodology which supported and funded my research through a three-year Ph.D. scholarship. Further, I would like to thank the DAAD for funding one of my four field trips to India. Thank you to Dr. Adejoke Adesokan, Dr. Carmen Carossella, Dr. Carsten Roeger, Gesa Krebber, Julia Fischbach, Dr. Lea-Kristin Behrens, Sebastian Mendel, and Sylvia Hudenborn for sharing these formative years as both colleagues and friends.

Many thanks to my wonderful colleagues at the Institute of Geography at the University of Cologne, Johanna Mäsgen, Dr. Stephan Langer, Harald Sterly, Tine Trumpp, Benjamin Casper, Gerrit Peters, Franziska Krachten, Dr. Lutz Meyer-Ohlendorf, Dr. Zinmar Than, Susanna Albrecht, Dr. Megha Sud, Dr. Marie Pahl, Dr. Alexander Follmann, Dr. Mareike Kroll, Dr. Carsten Butsch, Dr. Veronika Selbach, Dr. Regine Spohner, and Dr. Pamela Hartmann. Many have read, discussed, and commented on my work, and I am grateful for all thoughts, suggestions, and fruitful efforts in making it better. A special warm thanks to my office mate Dr. Birte Rafflenbeul and my virtual office mate Dr. Judith Bopp in Bangkok, with whom I closely shared both joyful and demanding phases of research.

For valuable feedback in doctoral research workshops, I would like to express special thanks to my colleagues at the Institute of Geography Didactics, Jun.-Prof. Dr. Miriam Kuckuck, Dr. Nils Thönnessen, Dr. Andreas Hoogen, Veit Meier, Jacqueline Jugl, Frederic von Reumont, Dr. Sabrina Dittrich, Dr. Beatrice Müller, and Cristal Schult.

At the Centre de Science Humaine (CSH) in Delhi, I would like to thank the director Dr. Leïla Choukroune as well as Dr. Roland Lardinois, Dr. Somsakun Maneerat, Dr. Mélissa Levaillant, Dr. Jules Naudet, and Dr. Rémi de Bercegol. During my research stay at CSH, I received both valuable theoretical input and precious contacts to key informants in Delhi.

I would like to thank Prof. Dr. Shimrey at NCERT, Dr. Ravi Jhadav at MSCERT, Mangesh Dighe at PMC, and Sharmila Sinha at CSE for sharing their deep insights into the role of ESD in India. Furthermore, I would like to thank Father Savio from Don Bosco for appointing me as a judge for the green school program Greenline in Mumbai, through which I could visit and interview students at some of the most engaged schools in environmental education. For great insights into rural government schools in Uttarakhand, I thank the NGO Aarohi and the support of Rahul Nainiwal, Pankaj Wadhwa, and Dr. Sushil Sharma.

I would like to thank my research assistants Priyanka Shah, Meha Sodhani, Rohan Jayasuriya, and Dr. M. M. Shankare Gowda for transcribing my interviews. For proofreading the manuscript, I would like to thank Karen Schneider and Dr. Sigrid Newman. For critical inputs on abstracts and meaningful syntheses to a broader audience, I thank Dr. Harry Fischer.

For elaborate feedback on the initial draft of this thesis, I owe warm thanks to my friend Dr. Anna Zimmer. For reviewing chapters and encouragement, I would like to thank my friends Dr. Dana Schmalz, Vanessa Willems, Ditte Broegger, and

Rebecca Neumann. I particularly would like to thank Caroline Saam for supporting my research with valuable exchange during field visits for her master's thesis research on student perceptions on water pollution at the same five English-medium schools in Pune where I conducted the intervention study.

At the International Water Management Institute (IWMI) in Nepal and Sri Lanka, and the Swedish University of Agricultural Sciences (SLU), I have found supportive work environments with many inspiring colleagues which helped me finalize this book.

My special thanks to Saurabh Mehta for the figures and drawings, and his continuous presence, energy, and support during the whole journey.

To my family and my friends, I sincerely thank you for your patience with me throughout this thesis writing process in which you daily revealed the wonderful things there are to learn about life.

To my parents, thank you for shaping my scientific curiosity and my deep appreciation for explorative travels since early childhood.

The manuscript has been accepted as Ph.D. thesis at the Faculty of Mathematics and Natural Sciences at the University of Cologne, Germany, on June 29, 2016.

Contents

Abbreviations

BLK Bund-Länder-Kommission für Bildungsplanung und Forschungsförderung (German Federal Commission for Educational Planning and Research Support)
BVIEER Bharati Vidyapeeth Institute of Environment Education and Research
$C^{+/-}$ Strong/Weak Classification (within the pedagogic device)
CBSE Central Board of Secondary Education
CCE Continuous and Comprehensive Examination
CEE Center of Environment Education
CEFR Common European Framework of References for Languages
CSE Center for Science and Environment
DESD UN Decade of Education for Sustainable Development (2005–2014)
DIET District Institute of Education and Training
EE Environment Education
EFA Education for All movement
EMS English-medium School
ESD Education for Sustainable Development
$F^{+/-}$ Strong/Weak Framing (within the pedagogic device)
FGD Focus Group Discussion
GDP Gross Domestic Product
GoI Government of India
I Interview
ICSE Indian Certificate of Secondary Education
ID Instructional Discourse (within the pedagogic discourse)
INR Indian Rupee (currency)
KG Kindergarten
M Teaching Module/Method
MC Municipal Corporation
MDG Millennium Development Goals
MHRD Ministry of Human Resource Development
MoEFCC Ministry of Environment, Forests and Climate Change

MSBSHSE	Maharashtra State Board for Secondary and Higher Secondary Education
MSBTPCR	Maharashtra State Bureau of Textbook Production and Curriculum Research
MSCERT	Maharashtra State Council of Educational Research and Training
NCERT	National Council of Educational Research and Training
NCF	National Curriculum Framework
NCTE	National Council for Teacher Education
NGC	National Green Corps
NGO	Non-governmental Organization
NPE	National Policy on Education
OECD	Organisation for Economic Co-operation and Development
P	Principal at a Secondary School
PD	Pedagogic Device
PIL	Public Interest Litigation
PMC	Pune Municipal Corporation
RD	Regulative Discourse (within the pedagogic discourse)
RTE	Right of Children to Free and Compulsory Education Act
S	Secondary School
SC	Scheduled Caste
SCERT	State Council of Educational Research and Training
SD	Sustainable Development
SDG	Sustainable Development Goals
SSA	Sarva Shiksha Abhiyan (India's Education for All movement)
SSC	Secondary School Certificate
ST	Scheduled Tribe
St	Student at a Secondary School
T	Teacher at a Secondary School
TIMSS	Trends in International Mathematics and Science Study
UN	United Nations
UNCED	United Nations Conference on Environment and Development 1992
UNCSD	United Nations Conference on Sustainable Development 2012
UNDESD	UN Decade for Education for Sustainable Development (2005–2014)
UNDP	United Nations Development Programme
UNEP	United Nations Environmental Programme
UNESCO	United Nations Educational, Scientific and Cultural Organization
UNGA	United Nations General Assembly
WCED	World Commission on Environment and Development
WHO	World Health Organization
WWDR	World Water Development Report
WWAP	World Water Assessment Programme

List of Figures

List of Tables

List of Photographs

Series Editors' Introduction

While the Decade of Education for Sustainable Development (DESD) (2005–2014) has been completed, the status and advocacy of Education for Sustainable Development (ESD) remains prominent. The United Nations (UN) goals of Education for All (EFA) and the Millennium Development Goals (MDGs 2000–2015) were complementary and provided a rationale for the importance of environmental education (EE) and Education for Sustainable Development (ESD). The United Nations Educational, Scientific and Cultural Organization's (UNESCO) Muscat Agreement in 2014 advocated seven global education targets, one of which was to cultivate skills for global citizenship and environmental sustainability. As part of the UN Sustainable Development Goals (SDGs 2015–2030) and as echoed by the Aichi-Nagoya Declaration on Education for Sustainable Development, education is embedded in goals which pertain to biodiversity, sustainable consumption and production, and climate change. Supporting these goals, there is a call for research and development as well as coordinated actions with an emphasis on the principles of human rights, gender equality, democracy, and social justice. There is also a call for attention to the importance and relevance of traditional knowledge and indigenous wisdom in various geographical, socio-cultural, and educational contexts.

With this background, and in light of UNESCO's Education 2030 Agenda (2017), this *Education for Sustainability* book series has been launched. Its purpose is to echo and enhance the global importance of education for a sustainable future as an educational vision. The series provides insights on a broad range of issues related to the intersection of, and interaction between, sustainability and education. The series showcases updated and innovative practice, discusses salient theoretical topics, and uses cases as examples.

The series adopts international, environmental education, and lifelong learning perspectives and explores connections with the agenda of education for sustainability and of Education for Sustainable Development. The intended audience includes university academics in educational studies, environmental education, geographical education, science education, curriculum studies, comparative education, educational leadership, and teacher education; the staff of international agencies with responsibilities for education; and school teachers in primary and secondary schools.

Supported by the expertise of a distinguished and diverse International Advisory Board, this series features authoritative and comprehensive global coverage, as well as diversified local, regional, national, and transnational perspectives. As a complement to the *Schooling for Sustainable Development* book series, it explores issues that go beyond primary and secondary schooling into university, vocational, and community education settings. These educational issues involve multiple stakeholders ranging from international agencies, governmental and non-governmental organizations, educational and business leaders to teachers, students, and parents. The research topics covered include global themes related to environment such as climate change education, disaster prevention and risk reduction, biodiversity education, and ecological education. They also include human ecological issues such as global citizenship, peace education, childhood development, intergenerational equity, gender studies, and human rights education. Further, they include society-oriented issues such as governance, green skills for sustainable development, sustainability leadership, and applied learning.

July 2018

John Chi-Kin Lee
The Education University of Hong Kong
Hong Kong, Hong Kong SAR

Rupert Maclean
Office of Applied Research and Innovation
College of the North Atlantic-Qatar
Doha, Qatar

Peter Blaze Corcoran
Florida Gulf Coast University
Fort Myers, USA

Chapter 1
Introduction: Education for Sustainable Development in India

Abstract The transnational policy *Education for Sustainable Development* (ESD) aims to facilitate the transformation of the pedagogic practice toward critical thinking and environmental action. I argue that the normative objectives of ESD have far-reaching consequences for formal educational systems as didactic approaches require transforming prevalent reproductive pedagogies which emphasize memorization. In this chapter, I outline how I develop a transformative approach to pedagogic practice which focuses on promoting argumentation skills on water resource conflicts. I explore the challenges and opportunities that exist for the translation of ESD principles into pedagogic practice in the newly industrializing country of India. As a highly stratified society with structural inequalities in the form of class, caste, age and gender, the implementation of ESD with a democratizing and Critical Pedagogy perspective requires engaging with cultural values and power relations in pedagogic discourse and practice in India. To achieve a transformative approach, and its democratizing potential, it is necessary to achieve a broader structural transformation of the educational system, with particular attention to the factors such as the structuring role of the textbook, the limited agency of the teacher, and the development of a broader normative vision across scales of what education is expected to achieve.

The sufficient supply of water is a global challenge, which is particularly relevant in developing contexts (UN World Water Assessment Programme 2016; Comprehensive Assessment of Water Management in Agriculture 2007). Although the MDG's goal to halve the population without safe drinking water was met in 2010, the use of improved drinking water in urban areas has not increased as much as population growth (World Health Organization and UNICEF 2010: 18). The increasing pressure on water resources due to urbanization, economic and population growth, land use, and climate change requires sustainable water use practices for water and food security as well as poverty elimination. On a local scale, water conflicts can emerge between competing water users, and socio-economic disparities between water users often determine water access (Comprehensive Assessment of Water Management in Agriculture 2007). As changing lifestyles, increasing population, and new technologies lead to global freshwater use and other "planetary boundaries" (Rockström

S. Leder, *Transformative Pedagogic Practice*, Education for Sustainability,
https://doi.org/10.1007/978-981-13-2369-0_1

et al. 2009) being increasingly exceeded, one important task of formal education is to promote environmental concerns, and individual as well as collective environmental action. To anticipate and react to quickly changing social and environmental conditions, the causes and impacts of insufficient access to and control over water in emerging countries are pressing problems that need to be understood and managed (UNESCO 2005). Education can promote critical environmental awareness and skills, and on this basis, people can understand the environmental and social consequences of their own (local) actions and deliberately make decisions about their lifestyles. Thereby, education is one factor that can contribute to sustainability.

Several national and transnational educational policies aim to strengthen and reform environmental education. The United Nations Decade of "Education for Sustainable Development" (UNDESD, 2005–2014) encourages the rethinking of existing environmental values, attitudes, and behavior. The central premise of Education for Sustainable Development (ESD) is that "everyone [hereby] has the opportunity to benefit from quality education and learn the values, behaviour and lifestyles required for a sustainable future and for positive societal transformation" (UNESCO 2005: 6). With this, 172 UN member states, including Germany and India, agreed to promote ESD for children, youths, and adults. Following the precepts of ESD to interlink ecological, economic, and social factors, people should be enabled to understand and manage human–environment interactions of present and future concern relevant to their everyday lives, such as water supply conflicts. Methodologically, ESD is aimed at promoting systemic and critical thinking and communication skills to solve problems and to encourage responsible participation in decision-making processes in society (UNESCO 2005).

These transnational educational aims could have particularly far-reaching consequences for formal educational systems. Since the proclamation of the UN Decade of ESD, several countries have developed and launched initiatives within their national educational systems to promote awareness, skill development, and action for a sustainable future (UNESCO 2009). Currently, these efforts are continuing in the discussion on the important role of education in the development agenda for the post-2015 Sustainable Development Goals (SDGs). As ESD is primarily a political and not a scientific and educational objective, an unambiguous interpretation of the definition, a grounded theoretical framework, and a consensual methodological approach are not available (Krautz 2013). These are, however, necessary to channel political efforts toward transforming pedagogic practice.

Developing concrete pedagogical and didactic directives for the normative objectives of ESD poses a challenge to strongly contextual pedagogic practice. Multiple reviews of ESD critically assess the hegemonic educational discourse and Western epistemology of ESD, which overshadow the need for regionally relevant and valued educational approaches (Manteaw 2012). To avoid these criticisms, I examine ESD in the present study as a transnational "travelling model," which is illustrative of how "ideas assembled in one site connect with meanings and practices in another" (Behrends et al. 2014: 4). I build on the argument that within each country's cultural systems and historical developments, transnational educational policies are "reworked, reinterpreted, and reenacted contextually" (Mukhopadhyay and Sriprakash

2011: 323) by educational stakeholders. Hence, I argue that ESD needs a "translation" (Merry 2006) to the socio-cultural context in which educational systems are embedded. The aim of this study is to explore how the transnational educational objective ESD relates to pedagogic practice within a socio-cultural context.

To examine the socio-cultural context, I recognize that pedagogic practice involves power relations (Giroux 2004: 33). Central to this understanding is that educational systems reflect notions of political and cultural values (such as "hierarchical" or "democratic principles"), which are transmitted in pedagogic practice. I build on the approach of Bernstein (1975, 1990) which views pedagogic practice as a socio-cultural reproduction of dominant ideologies in society. Micro-level interactions in the classroom can be linked to overlying power structures and control mechanisms at the macro-level. In that sense, teaching techniques meant to encourage critical thinking can be interpreted by educational stakeholders and teachers in such a way as to not challenge the established teacher–student relationship and the culturally and historically founded authority of the teacher. The present in-depth analysis of cultural values and social hierarchies relayed in pedagogic practice will explain how the tenets of ESD are interpreted within classroom and teaching structures in the respective contexts. To make educational policies such as ESD "locally relevant and culturally appropriate" (McKeown 2002: 7), it is thus essential to consider pedagogic practice within the context of larger social processes and to take into account long-standing cultural values and social hierarchies.

In addition to embedding ESD theoretically with a perspective on power relations in pedagogic practice, I conceptually link the pedagogic principles of ESD to the didactic approach of promoting argumentation skills. The ability to argue can contribute to achieving ESD's objectives of critical thinking on sustainable development and decision-making for sustainable lifestyles. With the growth of plural lifestyles and opportunities for decision-making, argumentation skills are a power resource to help enforce people's own economic, political, ecological, or social interests (Budke and Uhlenwinkel 2011; Budke et al. 2010). The slow dissolution of traditional class structures in societies, increasing globalization, transnational structures, and individualization has led to a "risk society" (Beck 1986) in which prevalent life pre-structuring norms are changing. To negotiate sustainable development in this setting, today's democratic societies as well as individuals must be able to make decisions and solve conflicts predominantly through argumentations.

Further to the social imperative for argumentations on sustainable development, argumentation skills promote learner-centered education and help students to develop a deeper understanding of the subject in question's contents, as constructivist learning theories propose (cf. Budke 2012; Reich 2007; Vester 2002). This hypothesis is also supported by neurodidactical and psychological research that promotes brain-based learning in which student-oriented, cooperative, and communicative teaching approaches are central to successful learning (Spitzer 2007). However, the didactic approach of argumentation has not been linked to ESD principles as yet. I argue that the political postulations for ESD principles can be realized in the didactic approach for argumentation skill development. In the study at hand, I develop an ESD approach grounded in state-of-the-art teaching methodology research, apply argumentation to structure, and translate ESD principles into classroom teaching.

The case of India, which is studied in this book, highlights the difficulties of translating this approach of ESD into pedagogic practice. As a highly stratified society with deeply entrenched structural inequalities in the form of class, caste, age, and gender, the implementation of ESD with a Critical Pedagogy perspective implies questioning the political connotations of ESD and, to a greater degree, engaging with social and cultural power relations. Traditional values transmitted in schooling contexts such as obedience, respect, discipline, and frugality (Thapan 2014) shape pedagogic practice in India and intermingle with new pedagogic principles. While ESD postulates critical thinking on controversial topics through argumentative and participatory teaching approaches, pedagogic practice in India is strongly influenced by studying for examinations and a "textbook culture" (Kumar 1988). Therefore, I recognize that ESD requires a multi-level approach that involves rethinking pedagogic principles and content transmitted in teacher education and training, as well as syllabi and textbook design.

While ESD aims to improve the quality of teaching and learning (UNESCO 2005), the Indian educational system is struggling with increasing student enrollment, limited resources, and complex socio-cultural conditions. The recent *Right to Education Act* focuses on providing access to elementary education for millions of children and shapes the educational landscape of India through several policies. With a national gross enrollment rate of 73.6% in classes 9 and 10 in the academic year 2013–2014, the enrollment rate has drastically increased since 2005–2006, when it was 52.2% (Government of India 2015). However, the respective conditions and success of education in the form of acquired knowledge and skills are highly context-dependent. Although each child should receive compulsory schooling from the age of 6–14 through the access to free elementary education (classes 1–8), there is a great discrepancy between urban and rural education concerning the provision and quality of schooling. Moreover, within the cities of India, profound disparities endure on a small spatial scale. Two-tier educational systems (government/private, local vernacular schools/English-medium) contribute their own idiosyncrasies to the country's teaching and learning processes. Private schools provide high-level education, while government schools struggle with limited financial and human resources such as overcrowded classrooms, insufficient teaching material, and inadequate pre- and in-service teacher training. This produces a great educational heterogeneity within India which I take into account while investigating the pedagogic practice.

The empirical research was conducted in the emerging megacity Pune in the state of Maharashtra. Pune mirrors the country's stark socio-economic disparities and the complex urban locality in which children live. Because of its numerous universities, Pune is known as the educational capital of India. A general trend relating to primary and secondary education which is similar to other Indian cities can be noticed: Intra-urban disparities in access to schooling and quality of education are closely connected to their socio-economic background (Govinda 2002; National University of Educational Planning and Administration 2008). In this context, sustainable development, for example, through education, requires a comprehensive problem-oriented, process-oriented, and people-oriented approach in order to address the environmental and social challenges of global change (cf. Kraas 2007). A critical educational approach to human–environment relations is particularly needed in the emerging megacities of India such as Pune, which are globalizing market-driven urban settings.

Every subject qualifies for ESD coverage, but specifically geography includes human–environment relations and resource utilization—particularly with regard to the topic of "water." Geography education is especially relevant as a subject for ESD because of its immanent interdisciplinary topics, the focus of content on resources, and societal processes, as well as its constant space and time reference. The *Lucerne Declaration on Geographical Education for Sustainable Development* (2007) for the International Geographical Union argues that "the paradigm of sustainable development should be integrated into the teaching of geography at all levels and in all regions of the world" (Haubrich 2007). In this book, I examine the wider socio-cultural implications of this claim for a paradigmatic shift in geography education in India.

1.1 Research Questions and Objectives of This Study

The implementation of Education for Sustainable Development as a transnational educational objective in local teaching contexts has hitherto been explored by only a small number of academic studies under context-specific conditions and with alternate objectives (Mulà and Tilbury 2009). Several empirical studies have evaluated ESD programs and opportunities for classroom teaching, but didactic concepts and models, which are theoretically and empirically grounded and transferable, are not sufficiently derived for ESD teaching (Fien and Maclean 2000: 37). As stated in the expert review by Tilbury (2011: 9) for UNESCO, ESD's objectives for critical reflective thinking on the economic, political, and social implications of sustainability and social change through changing consumer choices remain "poorly researched and weakly evidenced" (ibid.). Similarly, ESD research in Germany barely influences disciplinary and international discourses and is able to match established methodological standards and theoretical grounding in educational research (Gräsel et al. 2012). While most research on pedagogic practice and ESD is conducted in industrialized countries, the implementation of ESD principles in the Global South has barely been systematically analyzed (Manteaw 2012). However, several in-depth studies on the implementation of other learner-centered educational policies than ESD are available and point out significant gaps between educational policy discourse and its interpretation and implementation in different socio-cultural contexts (Clarke 2003; Sriprakash 2012; Schweisfurth 2011; Berndt 2010).

To fill this research gap, I will explore the challenges that exist for the translation of ESD principles into pedagogic practice in the newly industrializing country of India. The objective of this book is to identify and analyze possible linkages between the transnational educational policy ESD, academic frameworks, and current pedagogic

Research Questions

Guiding Research Question
Which challenges exist for the translation of the transnational educational policy *Education for Sustainable Development* (ESD) in pedagogic practice in geography teaching at English-medium secondary schools in Pune, India?

Subordinate Research Questions
1) How do institutional regulations (curricula, syllabi, textbooks) on the topic of water in geography education in India relate to ESD principles?
2) How do power relations and cultural values structure pedagogic practice on water in Indian geography education, and how are these linked to ESD principles?
3) How can ESD principles be interpreted through argumentation on water conflicts, and how can this approach be applied to pedagogic practice in Indian geography education?

Fig. 1.1 Research questions for this study

practice in India's formal educational system. I argue that geographical education research needs to be embedded in its socio-cultural context by considering how power relations shape pedagogic practice. Therefore, I analyze how the democratizing teaching approach of ESD relates to the power relations and cultural values reproduced in pedagogic practice. I examine how values of teaching and learning structure teacher–student interaction, and how norms and beliefs of respect toward and authority of the teacher, recitation as well as the governing role of the textbook shape the translation of ESD principles into pedagogic practice. The book aims to examine the transformative potential of ESD for pedagogic practice in Indian geography education. The outcome is a theoretically and empirically grounded concept for the introduction of ESD principles through argumentation in the Indian context.

The guiding research question (Fig. 1.1) examines how the transnational educational policy of ESD challenges pedagogic practice in geography education in English-medium secondary schools (classes 7–12) in Pune.

The detailed research questions aim to analyze how ESD principles link to pedagogic practice in geography education within the educational system of India. Adapting a multi-level analytical approach, the first two subordinate research questions investigate how ESD principles link to regulations at the institutional level and to pedagogic practice at the classroom level. Building on this knowledge, the third subordinate research question identifies institutional, structural, and socio-cultural opportunities and challenges to translate ESD through the didactic approach of argumentation into Indian geography education.

The first subordinate research question examines how academic frameworks on the topic of water in geography education in India relate to ESD principles. For this purpose, teaching contents and methods promoted in educational policies, curricula, syllabi, and geography textbooks are analyzed applying a conceptual framework for ESD developed for this study. The second subordinate research question investigates how power relations and cultural values of teaching and learning structure pedagogic practice on water in Indian geography education. For this research question, the status quo of teaching content and methods in geography teaching in English-medium secondary schools in Pune on the topic of water is examined. Furthermore, I analyze how pedagogic practice links to the ESD principles of network thinking on social, economic, and ecological sustainability as well as argumentation and student orientation. Additionally, the perceptions and attitudes of educational stakeholders, teachers, and students are examined in regard to ESD in geography education. The third detailed research question examines how ESD principles could be interpreted through the didactic approach of argumentation and how this approach translates into geography education. To examine opportunities for implementing ESD principles in geography education, an intervention study was conducted for which three ESD teaching modules were developed. These ESD teaching modules aim to promote argumentation skills relating to urban water conflicts. They were introduced at a teacher workshop and implemented by teachers in their geography lessons. I examine this implementation process and identify challenges for the transformative potential of pedagogic practice in geography education in India through the introduction of ESD principles.

As a theoretical framework for the analysis of pedagogic practice and transnational educational policies in developing contexts is not available, I develop a theoretical and methodological concept to examine the translation of ESD as a competence-oriented transnational educational policy in performance-oriented pedagogic practice in Indian geography education. In order to identify underlying challenges for transforming pedagogic practice in India, I draw on a Sociological Theory of Education, merging the code theory of Bernstein (1975, 1990) with the approach of Critical Pedagogy developed by Freire (1996). I further build on the perspective that power relations and political discourses within the educational system strongly shape pedagogic practice, as suggested by Archer (1995). As the didactic realization of ESD, I use the approach of argumentation skill development put forward by Budke (2010, 2012). Bringing together these theoretical components, I develop a new conceptual approach to examine power relations and socio-cultural structures in pedagogic practice and the transformative potential of educational policies in developing contexts. This theoretical grounding could be transferred to or modified for other teaching contexts to explore how educational policies relate to pedagogic practice.

1.2 Structure of the Book

This book is divided into ten chapters (cf. Fig. 1.2). This chapter, the introduction, focuses on the wider relevance of this study, my research questions, and the structure of the book.

Chapter 2 theoretically consolidates my empirical geographical education research on pedagogic practice and transnational educational policies with insights from sociological theory and a Critical Pedagogy perspective. I view pedagogic practice from different theoretical perspectives of reproduction, transformation, and recontextualization. Primarily, I conceptualize the potential of education for social transformation as well as power relations inherent in pedagogic practice through the lens of Bernstein (1975, 1990), Freire (1996), Archer (1995). Furthermore, I theoretically explore the translation of transnational educational policies in pedagogic practice in local contexts.

In Chap. 3, the transnational educational policy of ESD is examined. The objectives, origin and development, and more specifically, different principles, competences, and indicators of ESD are reviewed and their implications for formal educational systems are outlined. For a geography didactic grounding of ESD, the argumentation approach put forward by Budke (2012) provides concrete principles as to how this normative policy can translate into pedagogic practice in geography education.

In Chap. 4, institutional linkages between the development and structure of the Indian educational system and ESD are explored. Special focus is given to the pressing problem of quality education. The status of geography teaching in India is highlighted, as this is the subject of interest. Furthermore, the importance of ESD-oriented water education in schools is depicted. The growing water demands as well as intra-urban disparities of water access and controversies on causes of insufficient water supply are outlined to demonstrate the relevance of the topic "water conflicts" for geography education.

Chapter 5 introduces a methodological approach to research the translation of the transnational educational policy of ESD in pedagogic practice through fieldwork, document analyses, and an intervention study. The applied research methods for data documentation, preparation, and analysis are discussed and critically reflected in their application to geographical education research in the cultural context.

In Chap. 6, academic frameworks on water-related topics in geography education in India are presented and analyzed in regard to ESD principles. Indian educational policies, curricula, syllabi, and geography textbooks are examined with respect to addressing social, environmental, and economic sustainability, as well as teaching approaches focusing on argumentation, network thinking, and student orientation. Particularly, the content and didactic analyses of national and state geography textbooks give great insights into the structuring principles for classroom interaction in geography education.

In Chap. 7, pedagogic practice observed in geography lessons at English-medium schools in Pune is analyzed with a Bernsteinian lens on the use of classroom space

Research Process	Structure of the Book
Identification of research objective and relevant scope	**1) Introduction** • Motivation, research questions and research objectives • Overview of the structure of the thesis
Interpretation of relevant theories and their interlinkages	**2) Theoretical considerations on transformative pedagogic practice** • Definitions • Discussion of existing approaches • Integration of research in theoretical framework
Interpretation of transnational educational policy and didactic approach	**3) Education for Sustainable Development and argumentation skills** • Outlining, discussing and linking existing research on ESD to argumentation skills
Identification of relevance of research topic in a different socio-cultural context	**4) Research context: education and water conflicts in Pune** • ESD in India • Water conflicts in Pune
Derivation of methodological framework	**5) Methodology and Methods** • Document analysis • Field research • Intervention Study
Empirical results on institutional regulations for education	**6) Institutional regulations for formal school education and ESD principles** • Thematic and methodological analysis of educational policies, curricula, syllabi, and geography textbooks
Empirical results on pedagogic practice	**7) ESD principles and pedagogic practice** • Analysis of observed pedagogic practice in geography lessons • Analysis of student, teacher and educational stakeholder perspectives
Empirical results of the intervention study	**8) Opportunities of Education for Sustainable Development** • Development of ESD teaching modules • Teacher workshop and classroom implementation • Analysis of changing and sustaining principles of pedagogic practice
Reflection of the theoretical approach and re-interpretation of the empirical results in regard to the research questions	**9) Discussion** • Reflection and synthesis of empirical results • Revisiting the research questions • Revision of theoretical concept • Reflection of the methodological approach
Concluding remarks and prospects	**10) Conclusion** • Synthesis • Research prospects • Strategies for implementing ESD in geography education

Fig. 1.2 Structure of the book in relation to the research process

and resources, and the selection, sequence, pacing, and evaluation of knowledge and skills. Students' perceptions, attitudes, and knowledge relating to the subject of geography and the topic of water, as well as teachers and educational stakeholders' perspectives on the opportunities and barriers of ESD and potential curriculum change, are analyzed.

In Chap. 8, opportunities for ESD are examined through the analysis of the development and implementation of the three ESD teaching modules "Visual Network," (cf. Leder 2014) "Position Bar," (cf. Mayenfels & Lücke 2012) and "Rainbow Discussion" (cf. Kreuzberger 2012) on water conflicts in Pune. The implementation process of the ESD teaching modules exemplifies changes in the use of space and resources in the classrooms, whereas the underlying principles of pedagogic practice sustain during the implementation.

Chapter 9 introduces the concept of transformative pedagogic practice. The practicability of the didactic framework and the transferability of the theoretical and methodological approaches to other contexts are reflected upon. Furthermore, strategies and perspectives for implementing ESD in Indian geography teaching are discussed.

Chapter 10 synthesizes the key empirical findings and outlines the theoretical and policy implications as well as future directions for research. The book ends with a conclusion on the transformative potential of ESD for pedagogic practice in Indian geography education.

References

Archer, M. (1995). The neglect of the educational system by Bernstein. In A. R. Sadovnik (Ed.), *Knowledge and pedagogy: The sociology of Bernstein* (pp. 211–235). Norwood, NJ: Ablex Publishing Corporation. http://books.google.de/books?id=3tgXQ_ISJHYC&printsec=frontcover&hl=de&source=gbs_ge_summary_r&cad=0-v=onepage&q&f=false.

Beck, U. (1986). *Risikogesellschaft. Auf dem Weg in eine andere Moderne*. Frankfurt am Main: Suhrkamp.

Behrends, A., Park, S.-J., & Rottenburg, R. (2014). Travelling models: Introducing an analytical concept to globalisation studies. In A. Behrends, S.-J. Park, & R. Rottenburg (Eds.), *Travelling models in African conflict management: Translating technologies of social ordering* (pp. 1–40). Leiden: Brill.

Bernstein, B. (1990). *Class, codes and control: the structuring of pedagogic discourse* (Vol.IV). London: Routledge.

Berndt, C. (2010). *Elementarbildung in Indien im Spannungsverhältnis von Macht und Kultur. Eine Mikrostudie in Andhra Pradesh und West Bengalen*. Berlin: Logos Verlag.

Bernstein, B. (1975). *Class and pedagogies: Visible and invisible*. Paris: OECD.

Budke, A., Schiefele, U., & Uhlenwinkel, A. (2010). Entwicklung eines Argumentationskompetenzmodells für den Geographieunterricht. *Geographie und ihre Didaktik, 3,* 180–190.

Budke, A. (2012). Argumentationen im Geographieunterricht. *Geographie und ihre Didaktik, 1,* 23–34.

Budke, A., & Uhlenwinkel, A. (2011). Argumentieren im Geographieunterricht - Theoretische Grundlagen und unterrichtspraktische Umsetzungen. In C. Meyer, R. Henry, & G. Stöber (Eds.), *Geographische Bildung* (pp. 114–129). Braunschweig: Westermann.

Budke, A., Schiefele, U., & Uhlenwinkel, A. (2010). Entwicklung eines Argumentationskompetenzmodells für den Geographieunterricht. *Geographie und ihre Didaktik, 3,* 180–190.

Clarke, P. (2003). Culture and classroom reform: The case of the district primary education project, India. *Comparative Education, 39*(1), 27–44. https://doi.org/10.1080/0305006032000044922.

Comprehensive Assessment of Water Management in Agriculture. (2007). *Water for food, water for life: A comprehensive assessment of water management in agriculture.* London, Colombo: Earthscan, International Water Management Institute.

Fien, J., & Maclean, R. (2000). Teacher education for sustainability II. Two teacher education projects from Asia and the Pacific. *Journal of Science Education and Technology, 9*(1), 37–48. http://www.jstor.org/stable/40188539.

Freire, P. (1996). *Pedagogy of the oppressed.* London: Penguin Books Ltd.

Giroux, H. A. (2004). Critical pedagogy and the postmodern/modern divide: Towards a pedagogy of democratization. *Teacher Education Quarterly, 31*(1), 31–47.

Government of India. (2015). *'India in Figures 2015' Ministry of Statistics and Program Implementation.* New Delhi: Central Statistics Office.

Govinda, R. (2002). *India education report.* New Delhi: Oxford University Press.

Gräsel, C., Bormann, I., Schütte, K., Trempler, K., Fischbach, R., & Asseburg, R. (2012). Perspektiven der forschung im Bereich Bildung für nachhaltige Entwicklung. *Bildungsforschung. Bundesministerium für Bildung und Forschung (BMBF), 39,* 7–24.

Haubrich, H. (2007). Geography education for sustainable development. In S. Reinfried, Y. Schleicher, & A. Rempfler (Eds.), *Geographical views on education for sustainable development* (pp. 27–38). Geographiedidaktische Forschungen.

Kraas, F. (2007). Megacities and global change: Key priorities. *Geographical Journal, 173*(1), 79–82. https://doi.org/10.1111/J.1475-4959.2007.232_2.X.

Krautz, J. (2013). Stellungnahme zur Entwurfsfassung des Kernlehrplans Kunst (Sek II) NRW vom 09.04.2013.

Kreuzberger, C. (2012). Regenbogen-Vierer - Diskussion mit Redekarten In A. Budke (Ed.), *Kommunikation und Argumentation* Braunschweig: Westermann.

Kumar, K. (1988). Origins of India's "Textbook Culture". *Comparative Education Review, 32*(4), 452–464. http://www.jstor.org/stable/1188251.

Leder, S. (2014). Barrieren und Möglichkeiten einer Bildung für nachhaltige Entwicklung in Indien. In I. Hemmer, M. Müller, & M. Trappe (Eds.), *Nachhaltigkeit neu denken. Rio+X. Impulse für Bildung und Wissenschaft* München: Oekom-Verlag.

Manteaw, O. O. (2012). Education for sustainable development in Africa: The search for pedagogical logic. *International Journal of Educational Development, 32*(3), 376–383.<Go to ISI>://000301698300003 http://ac.els-cdn.com/S0738059311001301/1-s2.0-S0738059311001301-main.pdf?_tid=e247c694-1df6-11e2-a864-00000aab0f6b&acdnat=1351095877_200e77592eca28a76d7de4643351e023.

Mayenfels, J., & Lücke, C. (2012). Einen Standpunkt "verorten" - der Meinungsstrahl als Argumentationshilfe. In A. Budke (Ed.), *Kommunikation und Argumentation* (pp. 64–68). Braunschweig: Westermann.

McKeown, R. (2002). Education for sustainable development toolkit. http://www.esdtoolkit.org/. Accessed March 10, 2014.

Merry, S. E. (2006). Transnational human rights and local activism: Mapping the middle. *American Anthropologist, 108*(1), 38–51.

Mukhopadhyay, R., & Sriprakash, A. (2011). Global frameworks, local contingencies: Policy translations and education development in India. *Compare—A Journal of Comparative and International Education, 41*(3), 311–326. https://doi.org/10.1080/03057925.2010.534668.

Mulà, I., & Tilbury, D. (2009). A United Nations decade of education for sustainable development (2005–14). *Journal of Education for Sustainable Development, 3*(1), 87–97. https://doi.org/10.1177/097340820900300116.

National University of Educational Planning and Administration. (2008). *Status of education in India national report*. New Delhi: Ministry of Human Resource Development, Department of Higher Education.
Reich, K. (2007). Interactive constructivism in education. *Education and Culture, 23*(1), 7–26.
Rockström, J., Steffen, W., Noone, K., Persson, Å., Chapin, F. S. I., Lambin, E., et al. (2009). Planetary boundaries: Exploring the safe operating space for humanity. *Ecology and Society, 14*(2), 32.
Schweisfurth, M. (2011). Learner-centred education in developing country contexts: From solution to problem? *International Journal of Educational Development, 31*(5), 425–432. https://doi.org/10.1016/j.ijedudev.2011.03.005.
Spitzer, M. (2007). *Lernen. Gehirnforschung und die Schule des Lebens*. Berlin: Spektrum Akademischer Verlag.
Sriprakash, A. (2012). *Pedagogies for development: The politics and practice of child-centred education in India*. New York: Springer.
Thapan, M. (2014). *Ethnographies of Schooling in Contemporary India*. Delhi: Sage.
Tilbury, D. (2011). *Education for sustainable development. An expert review of processes and learning*. Paris: UNESCO.
UN World Water Assessment Programme. (2016). *The United Nations world water development report 2016: Water and jobs*. Paris: UNESCO.
UNESCO. (2005). United Nations decade of education for sustainable development (2005–2014): International Implementation Scheme. Paris.
UNESCO (2009). UNESCO World conference on ESD: Bonn declaration.
Vester, F. (2002). *Unsere Welt - ein vernetztes System*. München: Deutscher Taschenbuchverlag.
World Health Organization, & UNICEF. (2010). *Progress on sanitation and drinking water*. Geneva: WHO Press.

Chapter 2
Theoretical Considerations on Pedagogic Practice and Its Transformative Potential

Abstract This chapter develops a theoretical framework for *transformative pedagogic practice* to analyze the transformative potential of Education for Sustainable Development. The framework draws on insights of Basil Bernstein's Sociological Theory of Education (Bernstein 1975a, 1975b, 1990), Paolo Freire's Critical Pedagogy (Freire 1996), and the didactic approach of argumentation skill development. The combination of these approaches allows an integrative multi-level analysis of possibilities for a paradigm change toward transformative pedagogic practice. First, I review the tension in educational policy discourses between human capital and human rights approaches by discussing how several dominant traditions of social theory, Critical Pedagogy, and post-colonial perspectives view the role of education. Then, I conceptualize the tension between reproductive and transformative pedagogic practice by focusing on weakening power relations inherent in teaching. I offer a novel synthesis that allows us to understand the translation of standardized educational policies such as Education for Sustainable Development in local contexts by reflecting on the role of cultural values for the reinterpretation of dominant "Western" ideas in education. I use this framework as a basis to analyze the empirical findings on academic frameworks (Chap. 6), pedagogic practice (Chap. 7), and ESD teaching modules (Chap. 8).

The postulations of ESD in formal educational settings pose a great challenge as they reach beyond the scope of current classroom routine: ESD's transformative objectives interfere with power relations and cultural values in pedagogic practice. ESD plays a key role in questioning existing pedagogic practices related to the environment. By promoting a change of human–environment interactions through education, ESD involves the transformation of social relations.[1] More than through the contents of what is to be learned, this change works through the communication

[1]For an overview of the link between human–environment relations and socio-ecological research, see Becker and Jahn (2006).

S. Leder, *Transformative Pedagogic Practice*, Education for Sustainability,
https://doi.org/10.1007/978-981-13-2369-0_2

skills which students develop. When, for example, critical-thinking-based learning is to be introduced (UNESCO 2012b), this can help challenge prevailing power relations and requires different forms of classroom communication. Student-centered teaching methods shift the focus from the teacher to the students, who become more involved in selecting and structuring learning contents. Students are encouraged to voice their opinion in the classroom, which is designed to prepare them for social participation. This may challenge the steering role and authority of the teacher, particularly in textbook-centered learning contexts. Hence, ESD is a political means to support transforming classroom routines through altering cultural constructions of teaching and learning. This conveys how far-reaching and challenging ESD's transformative aspirations are for pedagogic practice, and for the cultural context in which they take place. Thus, ESD aims to trigger a holistic and lengthy transformative social process. This chapter will outline the interwovenness of pedagogic practice and desired social transformation.

Since ESD as a primarily political objective lacks a solid theoretical grounding, I merge concepts of Basil Bernstein's *Sociological Theory of Education* (1971–1990) and Paulo Freire's *Critical Pedagogy* (1996) to build a theoretical framework for a *transformative pedagogic practice*, which I use as conceptual foundation in this book. I build on conceptualizations of power relations and cultural values in pedagogic practice, which are central in Bernstein and Freire's theories. Combining these two theories can advance a comprehensive understanding between the poles of reproductive and transformative pedagogic practice and examine how ESD can strive toward "societal transformation" (UNESCO 2005: 6). Freire (1996) states that a *critical consciousness* through education plays a central role for *social transformation*. I transfer this perspective to evaluate ESD's postulation for critical thinking on sustainable development in pedagogic practice. Bernstein (1975a, b, 1990) provides the concepts of classification and framing which are beneficial for the description and analysis of power relations and cultural values in pedagogic practice. Hence, the Sociological Theory of Education as a descriptive–analytical approach and Critical Pedagogy as a normative approach are suitable to ground this study theoretically. The combination of Bernstein and Freire's approaches provides a conceptual framework for the analysis of social reproduction and transformation through pedagogic practice on sustainable development.

However, this theoretical approach contains some gaps concerning the influence of globalization, socio-cultural change, and institutional structures on pedagogic practice. Hence, I integrate further concepts of other theories: The concept of *translation* (Merry 2006) explains the recontextualization process of educational policies in local contexts and interprets ESD as a "traveling model" as put forward by Behrends et al. (2014). Moreover, the neglected role of the educational system in Bernstein's approach is countered, as suggested by Archer (1995), by including the educational system and its central institutions and stakeholders as influential factors for pedagogic practice in the theoretical framework. The combination of these theories can later be linked to the didactic approach of argumentation (cf. Chap. 3) in order to enhance this framework and to examine the research question empirically.

This theory chapter is structured as follows: The *first* subchapter links the role of education to social transformation. To embed the approaches of Bernstein (1975a, b, 1990) and Freire (1996) theoretically, the role of education is outlined within larger social theories and within a post-colonial context. This subchapter also describes the role of education for social justice and in capability approaches. To promote critical thinking as promoted by ESD, education includes not only skill development for employability, as the discourse around the human capital approach emphasizes, but especially skill development for questioning social and moral values, according to the human rights discourse.

In the *second* subchapter, the relevance of power relations and cultural values in pedagogic practice is depicted. Following Bernstein (1975a, b, 1990), I view pedagogic practice against the backdrop of power relations which structure communicative interaction between the teacher and the students. Bernstein's abstract concepts of classification and framing are introduced as they provide a language to analyze teaching in its particular context and the fundamental socio-cultural and political challenge the educational policy ESD poses. Further, pedagogic practice is viewed from the perspective of its transformative potential. Building on the Critical Pedagogy developed by Freire (1996), concepts of critical consciousness, problem-posing education, and the role of dialogue are introduced as one approach to interpreting ESD objectives for critical thinking on sustainable development.

In the *third* subchapter, the concept of *translation* (Behrends et al. 2014; Merry 2006) of transnational policies in local developing contexts is introduced to identify the need and challenges of contextualizing educational policies such as ESD theoretically. Translation refers to the relocation and reinterpretation of dominant ideas in a cultural context to make them relevant for educational stakeholders, teachers, and students. This subchapter also explains how ESD can only function as a template which needs to be reconceptualized at a local level. The chapter concludes with pointing out the relevance of these theoretical approaches for this study on pedagogic practice in the Indian context.

2.1 The Role of Education for Social Transformation

The role of education for individuals and in society has been widely discussed from pedagogical, didactic, psychological, sociological, and political perspectives. Micro-level studies focus on the learner's individual development and how the relationship between teacher and learner can support this. Macro-level approaches comprise the function of education for society. This study interrelates both micro-level and macro-level as it is concerned with contents and methods of pedagogic practice in the context of social transformation. In the following sections, the societal role of education is discussed drawing concepts from wider social theory. This is relevant for the overarching research question as it places transnational educational policies such as ESD within a greater school of thought and helps analyze underlying concepts and tensions of education relevant for pedagogic practice and discourse.

The first section outlines the tension in educational policy discourses between human capital and human rights approaches. Particularly in developing contexts, the debate on the functionality of education for economic development is put forward by international development agencies such as the World Bank (Schultz 1989). From this human capital perspective, education has primarily political and economic objectives, namely to develop skills that enhance a person's productivity. In contrast, a human rights perspective views the task of education to enhance skills for a more holistically understood individual development and thus relates to social and moral dimensions of education. These approaches indicate basic directions in educational policy discourses, within which ESD as transnational educational policy can be embedded.

The second section theoretically examines how education contributes to societal development by reviewing how larger social theories view the role of education. I summarize the central arguments of functionalist, conflict, and interactionist theories as these coined relevant concepts, which are integrated into Bernstein's Sociological Theory of Education (1975a, b, 1990).

The third section reviews perspectives on the role of education in post-colonial contexts. The conflicting priorities of modernization and dependency theory influence the discourse on the role of education and its policies in developing contexts in the past and offer insights for perspectives on translating transnational educational policies in the Indian context. This understanding elucidates the roots of the critique of transnational educational policies, and ESD in particular, as "Western" concepts, and highlights the need to deliberately translate transnational educational policies. An overview of these theoretical approaches relevant for educational research in India is illustrated in Fig. 2.1.

2.1.1 Educational Policy Discourses Between Human Capital and Human Rights Approaches

For the study of transnational educational policies, two dominant approaches can be distinguished in educational policy discourses: the human capital and the human rights approach. The human capital approach is based on the assumption that education contributes to economic growth, and thus, students should acquire skills in the formal educational system needed in professions for the economic market (Schultz 1989; Fagerlind and Saha 1989; Ilon 1996). This human capital approach promoting skills based on economic needs is especially followed by international agencies such as the World Bank. This approach also found strong support from the Nobel Prize winner of economics and schooling, Schultz (1989), who investigated the World Bank's discourse on investments in education to raise human capital. He argues that investments, especially in elementary schooling, are cheap and efficient from an economic viewpoint, as human capital enhances the productivity of labor and physical capital. The educational programs of major developing agencies (e.g., the World

Theoretical Approach	Authors	Central Concepts Relevant for Own Research Approach
Education within larger social theories		
Functionalist theories	DURKHEIM (1951)	education as medium for social cohesion
Conflict theories	MARX (1867), WEBER (1922)	imbalanced power relations in education
Education in the postcolonial context		
Modernization theory	INKELES & SMITH (1974)	"Becoming modern" through a resocialization process
Dependency theory	CARNOY (1974)	"Education as cultural imperialism"
Education for micro-level contexts		
Human capital approaches	SCHULTZ (1989)	skill development for economic development
Human rights approaches	PIGOZZI (2008)	knowledge, skills and values for individual development and societal participation
Education for micro-level contexts		
Critical Pedagogy	FREIRE (1970)	education for social transformation
Social justice and capabilities approach	SEN (1999, 2009), FRASER (2008), TIKLY & BARRETT (2011)	3 dimensions of quality education: inclusion, relevance, democracy; measurable through capabilities as functioning individuals have reason to value
Interactionist theories	GOFFMAN (1956)	communicative and symbolic interaction
Synthesis of macro and micro-level theories		
Sociological theory of education	BERNSTEIN (1976, 1990)	social reproduction and cultural transmission through education

Fig. 2.1 Theoretical approaches to research education in developing context

Bank, UNDP, and UNESCO) focus on supporting measurable learning outcomes, primarily for literacy and arithmetic. After the post-war period and until the late 1970s, educational programs emphasized technical skill development to strengthen the industrialization process in "low-income countries" (Ilon 1996). As the focus was to increase the GNP and to foster "modernization" (Fagerlind and Saha 1989), social, cultural, and political contexts were largely excluded. In the context of post-independence, the emancipatory potential of economic growth is one driving force for this approach. Critics such as Tikly (2004: 189) pointed out that "human capital theory contributed to the de-politicization of development discourse [...] through removing reference to the role of education in relation to reproducing social inequality." This indicates that education is meant to prepare students to become active citizens, but conceptions of active citizenship and the role of education in citizenship vary according to context, stakeholder, and discourse (Nagda et al. 2003: 165).

ESD, the transnational educational policy examined in this study, can be broadly classified as a human rights approach. In contrast to the human capital approach with the objective of accrediting students with skills for employability, the human rights approach emphasizes the normative principle of education as an inherent right for humanity leading to the "full development of the human personality" (United Nations General Assembly 1948, Article 26). Initially in the 1980s, and up to today, the educational policy discourse has increasingly included "welfare" (Ilon 1996) and thus considered social rates of return. The human rights approach views education as a fundamental individual right to foster societal participation "through learner-centered and democratic school structures" (Tikly and Barrett 2011: 5). Students shall acquire skills and values to understand and not only react to economic, but also social, environmental, and political challenges (cf. Pigozzi 2008). This also includes critical thinking to strengthen personal independence and societal participation. Through socialization in educational institutions, students are prepared for public democracy, civic leadership, and public service (Nagda et al. 2003: 165). ESD is associated with this human rights approach, as it presents a democratizing educational approach, which is aimed at critical thinking and societal participation. Tikly (2004) notes that this social capital perspective still implies a "Western notion." He critically observes that "discourses around education and development have the effect of rendering populations economically useful and politically docile in relation to dominant global interests" (Tikly 2004: 174). This critique can be linked to similar ESD criticism, which is reviewed in detail in Sect. 3.1.4.

To detach educational approaches from hegemonial perspectives, Tikly and Barrett (2011) provide a valuable approach to rethinking perspectives on education quality. They draw on the work of Fraser (2008) on social justice and Sen's (1999, 2009) capability approach. Their approach interrelates three dimensions: *inclusion, relevance,* and *democracy*. Inclusion is concerned with overcoming barriers to enhance access to education for different societal groups. Relevance means that the learning process is meaningful and valued by societies, communities, and individuals. The democratic dimension is concerned with opportunities for participation in educational debates. In this approach, quality education promotes "capabilities that learners, parents, communities and governments have reason to value" (Tikly and Barrett

2011: 9). The terms "reason to value" based on Sen (1999: 10) is not prescriptive but focuses on cultural norms and values of different societies and groups. Public debate on the means and measures of successful education is encouraged on local, national, and global levels. Thus, this approach stresses the importance of context and provides a normative ground to relate education to development.

Similarly, Bernstein (1996) formulates three democratic pedagogic rights: *enhancement, inclusion,* and *participation*. Enhancement is the right to achieve a "critical understanding and a sense of possibility" and the condition for developing the confidence to act at an individual level. The right to be included whether socially, intellectually, culturally, or personally as an individual and a member of a group operates at the social level. The right to participate in the procedures involved in the construction, maintenance, and transformation of order is the condition for civic discourse and practice and operates at the level of politics. This social justice perspective fostering capabilities as well as democratic principles indicates how strongly political and value-laden education is.

In this book, I argue that ESD can have a transformative potential if ESD objectives are translated into pedagogic practice through a participatory, meaningful, and critical approach. To escape ESD's "Western notion" (Tikly 2004), I attribute ESD to Bernstein's democratic pedagogic rights and Tikly and Barrett's social justice perspective on education. Hence, Bernstein and Tikly and Barrett's approach provide the basis for the understanding of the role of education for this study. Bernstein's work, which will be fundamental for this study, integrates concepts of functionalist, conflict, and interactionist theories on education. Before outlining Bernstein's Sociological Theory of Education (1975a, b, 1990) in Sect. 2.2.1, these sociological theories are briefly described in the following subchapter to understand the roots of his thought.

2.1.2 Linking Education and Societal Development Through Functionalist, Conflict, and Interactionist Theories

To understand how the role of education in society is viewed in larger social theories, this subchapter will introduce functionalist, conflict, and interactionist theories. As the editors Michael Apple, Stephen Ball, and Luis Gandin state in The *Routledge International Handbook of the Sociology of Education* (Apple et al. 2010: 1), the sociology of education is a "diverse, messy, dynamic, somewhat elusive and invariable disputatious field of work." Outlining three major theories will help to understand how Bernstein (1975a, b, 1990) fused ideas from these theories into his *Sociological Theory of Education* to conceptualize the relationship between education and societal development.

Education can be seen as contributing to social cohesion and order, as described in *functionalist theories* based on Émile Durkheim. From a functionalist view, education contributes to the stabilization of "modern, democratic societies through the production of moral values shared by the majority" (Sadovnik 2011b: 3). The purposes of education are democratically agreed intellectual, political, social, and economic objectives (Sadovnik 2011a). These are interpreted from various perspectives and given different importance by political, economic, social, and educational stakeholders. The balance of political, economic, intellectual, and social objectives fosters social order regulated by legitimated moral values agreed upon in a democratic process. Due to industrialization and urbanization, Durkheim anticipated an "anomie" (Durkheim 1951), the breakdown of social cohesion and the increase of individualism. This leads to a mismatch between the practices and norms of individuals and societal standards, and thus the disintegration and instability of societies. An imbalance of these values is a sign of societal disorder. While one task of the educational system is to transmit moral values, the other task is to promote skills and select students on the basis of their abilities to maintain a stable social order. Hence, functionalist theories see the role of education in retaining social cohesion and order.[2]

Functionalist theories are criticized for failing "to recognize divergent interests, ideologies, and conflicting group values" (Ballantine and Hammack 2012: 13) in educational spheres. Furthermore, change is considered as slow and deliberate, which is often not true for the rapid and complex changes in societies. They are also not suitable for analyzing classroom interaction, as they do not engage with the consequences for pedagogic practice.

By contrast, *conflict theories* assume tensions and social conflicts due to the unequal distribution of power which becomes relayed in schools. Conflict theorists believe that "society is [not] held together by shared values and collective agreements alone, rather than the ability of dominant groups to impose their will on subordinate groups through force, cooptation, and manipulation" (Sadovnik 2011b: 6). Conflict theory is based on the thought of Karl Marx (1818–1883), who believed that with the rise of industrialization and urbanization, workers ("proletariat") become separated from the benefits of their own work which leads to alienation. The differential access to the means of production leads to a class struggle. The anticipation of a class system in society is reflected in the organization of schools. Focusing on the social conditions of exploited workers in the class system, Marx' approach has a strong economic emphasis and excludes cultural and political factors shaping institutions (e.g., educational systems) and actors. Max Weber (1864–1920) examined not only the class rank but also status competition in schools, determined by consumption patterns and socializing habits. Thus, Weber's approach opened Marx' purely economic perspective to cultural and social aspects in schooling.

According to conflict theory, schools are depicted as "social battlefields" (Sadovnik 2011b) in which struggles between antagonists (students, teachers, administrators, etc.) exist. Thus, power relations in schools mirror power relations in society. In the Indian context, authority and respect is closely tied to social divides such as caste, class, gender, and age. These stratify Indian society, and these hierarchies are reflected in schooling experiences within the educational system (Thapan 2014). For example, the school type attended with the heterogeneous and highly competitive

[2]For a further overview of the functionalist theory, read Morais (2002).

educational system is closely tied to class, and caste, particularly ST and SC, plays a role in terms of school access through quotas, as well as social networks, which are often limited to the same caste. The role of power relations is substantiated by Bernstein (1975a, b, 1990), as he considers power relations relayed in classroom communication.

While functionalist theory and conflict theory focus on the role of education for society on a macro-level, interactionist theory focuses on social interaction on a micro-level. As prominent exponent of this theory, Goffman (1959) investigated how daily social interaction maintains social order. In his book "The Presentation of Self in Everyday Life" (Goffman 1959), he describes social interaction through a dramaturgical approach in which people take social roles according to existing role expectations. He contrasts "in front of house/onstage," where actors present themselves to leave a desired impression through their behavior, dress, and language in front of an audience, conforming with formal expectations, versus "backstage" in which individuals are in their private sphere and withdraw from their official role. Thus, the interaction in schools as public institutions is seen as an onstage theatrical presentation in which students and teachers take specific roles they believe to be desired by the audience, representing institutional expectations. In the classroom, teachers perform in front of their spectators, the students. Backstage, certain tensions of interaction may become obvious.[3]

Basil Bernstein synthesizes Durkheim's functionalist theory and Marx and Weber's conflict theory, as well as elements of interactionist theory. He combines the assumption of conflict theory that tensions due to unequal power relations exist in society and that these are relayed in classroom interaction. Despite or rather because of that, Bernstein ascribes schools the role of maintaining social order, which is compliant to the functionalist approach. In his *Sociological Theory of Education*, he views pedagogic practice as social and cultural reproduction and offers opportunities to examine *how* reproduction and transformation through pedagogy takes place. Bernstein developed concepts to explore social interaction in classrooms, and building on Goffman, teachers are seen as taking on specific roles conforming to institutional expectations. Hence, Bernstein relates teacher–student interaction at a micro-level, namely the classroom, with societal power relations at a macro-level which makes it a valuable approach for emphasizing the socio-cultural context within which this empirical educational study takes place. I adopt Bernstein's theoretical perspective of combing these theories as it provides a framework to analyze pedagogic practice in India's highly stratified society. Before his approach is further elaborated in Sect. 2.2, I will outline the role attributed to education in post-colonial contexts, as these are influential for researching transnational influences on pedagogic practice in India.

[3]In a similar approach, Mead (1934) was one of the first authors to examine how social constructions of reality are produced through symbolic interactions.

2.1.3 *Perspectives on the Role of Education in Post-colonial Contexts*

In order to examine the role of the transnational educational policy Education for Sustainable Development (ESD) in post-colonial contexts, a review of modernization and dependency theories demonstrates educational perspectives which have been historically relevant but developed toward more differentiated post-colonial theories. These theories are relevant to consider as they convey the need to differentiate perspectives on the role of education in developing contexts from the role of education in the "West" (Ball 1981). The perspectives of functionalist and conflict theories outlined earlier influence post-colonial debates on education. While functionalist perspectives shape modernization theory, in which education shall contribute to social order, conflict theory influences the view of dependency theory, which criticizes education being used as a tool for economic and political control and interests. These two contradicting theories influence the discourse on the role of education and its policies in developing contexts and offer insights for perspectives on translating transnational educational policies in the Indian context.[4]

Modernization theory includes a functionalist perspective on education to create and maintain social order. It follows the human capital approach and views education as necessary investment in human capital to enhance economic growth. Modernization theory is critiqued for its strong focus on manpower planning, promoting the selection of elites and downplaying education for the masses, and reinforcing "tribal, social-class and regional disparities existing within the system" (Ball 1981: 303).

Dependency theory developed as a reaction to modernization theory. Dependency theorists argue that development is "conditioned by dependence and interaction with the metropolitan society [or societies]" (Ball 1981: 303). Education is viewed "not as neutral process of value transformation and the dissemination of Western functional rationality, but as a form of economic and political control. […] it is no longer taken to be a liberating process but a basis for, and a medium of, *cultural imperialism*" (Ball 1981: 304). This perspective is rooted in conflict theory's assumption of social tensions due to the unequal distribution of power.[5] This view, especially that "schools are *colonialistic* in that they attempt to impose economic and political relationships in the society" (Carnoy 1974: 10), is critiqued by Ball (1981: 309) for being too deterministic and "concentrating almost entirely upon the structure of provision, ignoring the struggle and "relative autonomy of the school in colonial society ."

[4]Post-colonial theories contribute to interpreting the roles assigned to education in developing contexts. The critical approach of post-colonial theory offers a range of views on the cultural legacy of colonial rule and identity construction in schooling beyond the binary relationship of colonized and colonist. However, these are not further considered for this study although they provide another interesting theoretical lens on influences on contemporary pedagogic practice.

[5]From a dependency theory perspective, personal characteristics and values of the "modern man," as depicted by Inkeles and Smith (1974, are only achieved through a fundamental resocialization process. This view is based on the social psychology of structural functionalism by Parsons (1951, 1991).

Thus, dependency theorists fail to "extend the arguments presented into the period of post-independence" (Ball 1981: 308). Furthermore, differences of localities are ignored.

Considering the heterogeneous development in educational contexts, modernization and dependency theories are too generalizing and single-sided for viewing the role of education as well as for critically analyzing the implementation of ESD as a policy of "Western" origin. Yet, they present interesting aspects for the translation of a transnational educational policy to India. On the one hand, the Indian educational system needs to be understood in its historical context including colonial influences on its formation (Kumar 2005). On the other hand, international influence through educational policies (such as the Education For All movement and ESD) needs to be critically examined against historically shaped definitions of the learners' needs and national discourses on education (Kumar 2004). In the following subchapter, I review how Bernstein's *Sociological Theory of Education* and Freire's Critical Pedagogy approach conceptualize power relations and cultural values in pedagogic practice. I argue that combined they are a worthwhile theoretical framework to explore the tensions between and within pedagogic practice and educational discourse.

2.2 Power Relations and Cultural Values in Pedagogic Practice

To understand the transformative potential of education in the contexts of social inequality, pedagogic practice has to be viewed against the background of its specific cultural setting. For this purpose, I will theoretically examine how power relations are relayed in classroom interaction, and how cultural values of teaching and learning shape pedagogic practice. This subchapter will approach these questions by reviewing two influential thinkers who theorized power relations and cultural values in pedagogic practice.

The first subchapter depicts how the *Sociological Theory of Education* developed by the British structuralist Basil Bernstein views pedagogic practice as the cultural reproduction of social hierarchies (Sect. 2.2.1). I outline Bernstein's *Code Theory* to analyze power and control in the classroom as I use his codes for the analysis of the empirical data in this study. He offers *codes* to analyze *classroom communication* in regard to *relayed power structures* and thus embeds pedagogic practice in a socio-cultural context. Bernstein offers a holistic perspective for understanding the socio-cultural tensions ESD provokes when learner-centered teaching methods and contents are introduced in classrooms. Fusing functionalist, conflict, and interactionist theories, Bernstein anticipates that communication transmits culturally shaped dominant and dominated codes regulating the relationships within and between social groups. His theory is critically reviewed, before the application in my study is outlined.

This approach is followed by a second subchapter depicting the lines of thought in critical educational theory and the *Critical Pedagogy* of the Brazilian educationist Paulo Freire (Sect. 2.2.2). His normative perspective on education can be linked to the transformative objectives of ESD. The Critical Pedagogy of Freire is summarized and critically examined in terms of its shortcomings and potential for analyzing pedagogic practice in India. Bernstein's and Freire's approaches theoretically frame my study on the role of power relations and cultural values in pedagogic practice which is relevant in considering the implementation of the transnational educational policy ESD in the Indian context.

2.2.1 Sociological Theory of Education by Basil Bernstein

In order to analyze pedagogic practice and the transformative potential of educational policies, I will examine in this book how new teaching approaches encouraged by educational policies relate to and aim to change teaching contents and methods, as well as interaction patterns between teachers and students. Basil Bernstein developed codes to reveal how power relations and control mechanisms are relayed through pedagogic practice. In his four-volume book series "Class, Codes and Control" (1971–1990), he developed a code theory[6] based on observations of the English educational system during the 1970s and 1980s. Through codes, he analyzes underlying messages between *transmitter* and *acquirer* to uncover social and cultural power relations in pedagogic practice.

In Bernsteinian analyses, the educational system as an institution "plays a key role in transmitting dominant ideologies of the society" (Clark 2005: 32) within which it is located. Bernstein's *Theory of Pedagogic Discourse* and his *Theory of Cultural Transmission* argue that pedagogic discourses and pedagogic practices are a "relay for power relations external to it" (Bernstein 1990: 168) as they procreate social structure. Bernstein (1975a: 85) states "how a society selects, classifies, distributes, transmits and evaluates the educational knowledge it considers to be public, reflects both the distribution of power and the principles of social control." Pedagogic discourse has often been understood by cultural theorists such as Michel Foucault (1969) as "a medium for other voices: class, gender, race" (Bernstein 1990: 165), but Bernstein argues that it is not considered *how* the medium of reproduction functions. Therefore, Bernstein distinguishes the *message* from the *carrier of the message*. In his theory, he provides principles of description for the carrier and analyzes *how* pedagogic discourse is socially constructed: Through language, notions of social class are structured and reproduced within society, and thus, discourses sustain existing inequalities. I extend this perception beyond notions of social class, as in the Indian

[6]Bernstein developed in his book series "Class, codes and control" several intertwined theories, comprising code theory, e.g., Sociological Theory of Education (1971), Theory of educational transmission (1975, Vol. III), and Theory of pedagogic discourse (1990, 2000, Vol. IV). For further reading, I especially recommend Volumes III and IV.

context, gender, caste, age, and other social differentiations are transmitted in pedagogic practice. The content of a subject, for example, depends on who controls the curriculum and *how* contents and competences are transmitted and evaluated. With the anticipation of pedagogic practice as cultural relay, I examine the asymmetrical pedagogic relation of *transmitters* and *acquirers* as put forward by Bernstein to understand how cultural reproduction occurs.

2.2.1.1 The Role of Language in Acquisition Contexts of Pedagogic Practice

In his earlier approaches, Bernstein examined how language in educational contexts reproduces social inequality. He derived from his observations of the British school system that educational failure was closely tied to social class and the specific type of language associated with it. Following Bernstein's theory, educational failure is related to linguistic competence and performance: Children cannot recognize and produce the expected legitimate text. This is highly relevant for examining in pedagogic practice in India, as multilingualism in society is closely bound to other social divides such as ethnicity, religion, and social identity. In urban areas, for example, children who are exposed to language and discussions on politics and science will have advantages in geography education, as the subject covers topics close to everyday life. Furthermore, when communication skills are meant to be promoted, the acquisition contexts in pedagogic practice have to be understood.

Children have two different sets of communicative rules for acquisition: *recognition and realization rules*. *Recognition rules* create the means of recognizing a context by distinguishing between contexts. *Realization rules* are concerned with the production of specialized relationships by expressing the language specific to a context. Bernstein (1990) quotes a study by Holland (1981) to prove his hypothesis that distinguishes on an empirical basis context-related, *restricted* language and context-independent, *elaborated* language. When children of different social classes were asked to tacitly group food items, one group of students used a language with direct relations to a specific material base (e.g., "It's what Mum makes"—mainly used by working class children) and the other group used a language which was more abstract and only indirectly related to a specific material base (e.g., "These come from the ground"). In schooling contexts, context-independent language is usually considered legitimate. Several studies have proved that linguistic codes differ by social class and are therefore important especially when considering the educational success of students with a lower socio-economic background. For example, students with a lower socio-economic background or marginalized students tend to have problems with the legitimate forms of language (Lubienski 2004). Many working class children have not acquired context-independent, elaborated language and thus cannot recognize and realize the language legitimate in this context. Thus, the students have different sets of recognizing and realization rules causing educational disadvantages. Muller (2004: 8) states based on his observations that some children use more context-dependent, implicit language, as they are used to listening to orders and obeying them. Students need the appropriate set of recognition rules to distinguish between contexts. Further, they also need realization rules to produce specialized communication within contexts to provide the educational system with what they have understood.

Morais (2002) and Morais and Neves (2001, 2006, 2011) developed a sociological research methodology adapted from Bernstein (1996) to analyze how specific power and control relations lead to differential access to recognition and realization rules (Neves and Morais 2001: 185). They examined how these rules regulate learning contexts that characterize teacher training interaction and pedagogic practices in Portuguese schools. Furthermore, they analyzed children and teachers' recognition and realization rules as mediating sociological factors in the relation between family and school discourses and practices. For each context, a model was developed with indicators for evaluation criteria such as "students ask questions." These indicators were classified in a 4-point scale of framing: F++/F+ for very strong/strong framing and F^-/F^- for weak/very weak framing. With this scale, they operationalized Bernstein's internal language of description, the concepts of framing and classification, into empirical data analysis, and thus provided an external language of description for a specific context. Through this, they considered empirical evidence and theoretical principles in a dialectical relationship, making evident "the diagnostic, predictive, descriptive, explanatory, and transferability potential of the theory" (Neves and Morais 2001: 212). As Bernstein (1990) had already suggested, their study demonstrates how children's success in scientific and social learning in school depends on the children's acquisition of recognition and realization rules in distinctive micro-contexts. If weak framing of pacing was given, it permits teachers to explicate the expected strong framing that helps students to acquire recognition and realization rules (Neves and Morais 2001: 214). In another study, Neves and Morais (2001) applied the model of pedagogic discourse by Bernstein (1990) to the analysis of natural science curriculum reforms, syllabi, and textbooks. Through an analysis in a 3-point scale (F++, F+, F−), they demonstrate how dominant principles of society are recontextualized in sociological messages transmitted by reformed educational syllabuses, and how changes in syllabus guidelines are recontextualized in textbook guidelines. These findings are relevant for this study, as they highlight and measure recontextualization processes in educational settings.

Another study exemplified how scientific knowledge and competences are acquired by students of different social backgrounds in elementary schools (Morais et al. 2004). The theoretical framework combines Bernstein's *Theory of Pedagogic Discourse* (Bernstein 1990) with the social constructivism of Vygotsky (1978) which considers the child as an active learner and the teacher as a creator of social contexts that enhance learning. The analysis revealed the influence of characteristics of pedagogic practice and the socio-economic backgrounds of the children's families on scientific achievements. Major factors were the scientific competence of the teachers, explicit evaluation criteria and weakly classified teacher–child space and open and intense communication between children.

Based on the results of their studies on learning in families and schools, on teacher education, and on the construction of syllabi and textbooks, Morais and Neves (2011)

Table 2.1 Characteristics of mixed pedagogic practice based on Morais and Neves (2011: 192)

Sociological characteristic	Modality of pedagogic practice
Teacher–student relation (clear distinction of status)	C++
Selection and sequencing of knowledge, competences, and classroom activities (teacher control)	F+
Pacing (student control over time of acquisition)	F−−
Evaluation criteria (explicit by teacher)	F++
Hierarchical rules between communication relations of teachers and students and students between themselves	F−−
Intra-disciplinary relations	C−−
Boundaries between teacher–student and student–student spaces	C−−

developed a model of a mixed pedagogic practice for overcoming the effect of diverse social backgrounds which proved successful for students' scientific learning. Provided that teachers have a high level of scientific knowledge and competences, the following modalities of sociological characteristics on pedagogic practice are fundamental for students' scientific learning (Table 2.1).

To link academic theory and educational policy, Clark (2005) applies Bernstein's theory of pedagogic discourse (1990, 1996) to the national curriculum in English in England in the late 1980s and the National Literacy Strategy in the 1990s. Her analysis shows that the teaching of English grammar reproduces notions of national identity and social class and transmits dominant ideologies of national and cultural identity conflicting with language and culture at home (Clark 2005: 32). The principles of distribution, recontextualization, and evaluation governing pedagogic discourse are used for the empirical description of how cultural reproduction works and social order is established through the teaching of Standard English.

The role of language in pedagogic practice is relevant for this study, as in the Indian educational system the word-by-word reproduction of definitions is central to achieving high marks in examinations. If these definitions also include technical terms, this may lead to the exclusion of those students who are less familiar with a particular vocabulary. Furthermore, the increasing number of English-medium schools in India may additionally challenge those students who are not as frequently exposed to the English language in their parental home. This calls for structural support for learning "elaborate" language. The insights of these Bernsteinian studies demonstrate the explanatory value and in-depth understanding his concepts can provide to inform educational policies.

2.2.1.2 Reproduction and Transformation of Social Structures Through Pedagogic Practice

Apart from a linguistic approach, a sociological perspective on education is particularly relevant here to understand the mechanisms of transformative processes in pedagogic practice in a highly stratified India. The later Bernstein argues that social order is reproduced in the classroom and structures pedagogic culture. Yet at the same time, educational systems can also "create the possibility of change" (Bernstein 1990: 13). Clark (2005) applies Bernstein's *Theory of Pedagogic Discourse* (1990, 1996) to analyzing educational policy and practice in England in the 1980s and 1990s and notes:

> Education is thus never neutral, but a site of cultural reproduction as much as any other. Nevertheless, systems of education also contain within them the seeds of their own transformation. […] today's acquirers become, in turn, tomorrow's reproducers and producers of knowledge. Schools, then, and the curriculum which they teach, are at one and the same time sites of cultural reproduction and of potential future transformations. (Clark 2005: 44)

A transformation of pedagogic practice can only take place, as Bernstein argues, if the internal logic of the *pedagogic device* (PD) is analyzed. He criticizes theories of cultural reproduction, resistance, or transformation as "relatively weak on analysis of 'relations within'," and, thus not providing "strong rules for the description of the agencies/processes of its concerns" (Bernstein 1990: 178). He further argues that in an educational context, it is not only *what* is transmitted which is relevant, but also *how* the pedagogic device is constituted. Through the understanding of principles and codes in message systems, one can develop prospects for the transformative potential of education. To analyze Indian geography education on cultural reproduction and transformation, I will elucidate Bernstein's pedagogic device (cf. Fig. 2.2).

In order to examine potential transformative processes in Indian pedagogic practice, it is necessary to understand three rules which constitute the internal order of the pedagogic device. Distributive (1), recontextualizing (2), and evaluative (3) rules are hierarchical to each other. *Distributive rules* (1) relate to the power of social groups and thus legitimize the production of a specific educational knowledge, and hence a pedagogic culture. For example, analysis of the subject geography in the German Democratic Republic (GDR) by Budke (2010) displays how geopolitical and ideological interests of the state are communicated and disseminated through geography education, and hence, how pedagogic practice can politically manipulate students to become "new socialist citizens." This example demonstrates an extreme case of how educational knowledge and dissemination can become politicized.

Recontextualizing rules (2) are concerned with the transmission of pedagogical knowledge through pedagogic discourse. Pedagogic discourse underlies rules for embedding and relating two discourses: the instructional discourse (ID) and the regulative discourse (RD). The question of "*what is transmitted?*" refers to the instructional discourse (ID) and is transmitting desired scientific knowledge and cognitive competences. The question "*how is knowledge transmitted?*" refers to the regulative discourse (RD) and translates the dominant values of society. The regulative

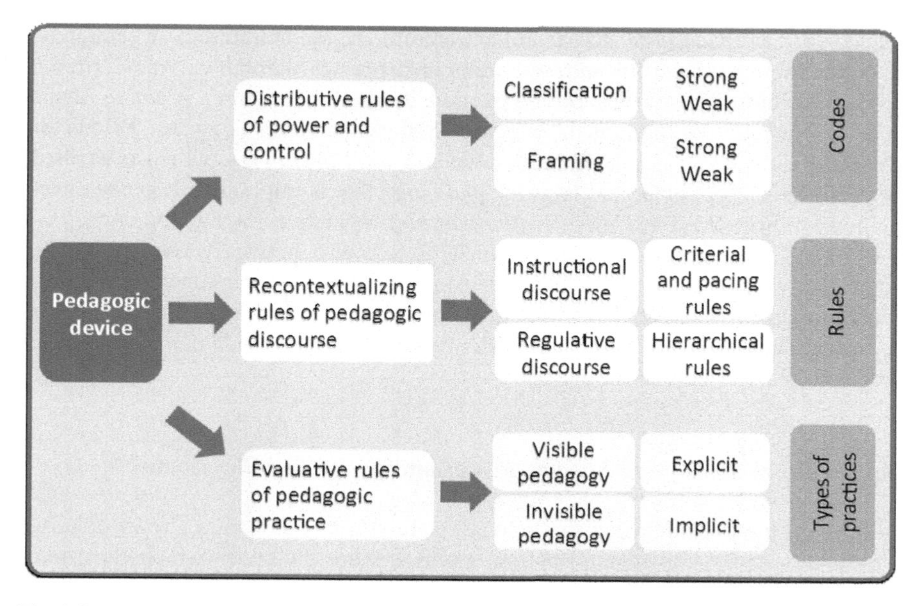

Fig. 2.2 Bernsteinian (1990) concepts of the pedagogic device relevant to this study (own draft)

discourse includes "attitudes and values, rules of conduct and principles of social order" (Morais 2002: 562). The instructional discourse is embedded in the regulative discourse; thus, the regulative discourse is dominant over the instructional discourse. If the instructional discourse is weak, the regulative discourse is also weak. Bernstein (1996: 28) demonstrates this in the following way:

$$\text{Framing} = \frac{\text{Instructional Discource (ID)}}{\text{Regulative Discource (RD)}}$$

The regulative discourse comprises *hierarchical rules*, in which the transmitters and acquirers recognize their social position and function, and includes their expectations about conduct, character, and manner. The instructional discourse underlies two types of rules: *criterial rules* comprising the aforementioned acquirer's recognition and realization rules of legitimate communication and social relations, and *pacing rules* which determine the "rate of expected acquisition of *sequencing rules*, that is, how much you have to learn in a given amount of time" (Bernstein 1990: 66). For example, the current performance-oriented regulative discourse in the Indian educational system assigns particular roles to the teacher and the student; thus, the former is the transmitter of the knowledge in textbooks, and the latter the reproducer of this knowledge. The teacher in the position of correcting students is hence attributed with defining the legitimate answers, the instructional discourse, and hence has a clear hierarchically distinct position above the student.

Pedagogic practice is the actual realization of pedagogic culture. Any pedagogic relation between a transmitter and an acquirer incorporates hierarchical rules, criterial and pacing rules. These rules can be explicit, in that the acquirer is aware of the expected behavior, or implicit, in which only the transmitter is aware. This leads to two different types of pedagogic practices: *visible pedagogy*, which is explicit and performance-oriented, and *invisible pedagogy* that is implicit and competence-oriented (cf. Sect. 2.2.1.4). These *evaluative rules* regulate pedagogic practice.

The analysis of this study provides insights into distributive, recontextualizing, and evaluative rules of the pedagogic device in Indian geography education. Aimed at changing in existing pedagogic practice, transformation of the distributive rules is necessary. Thus, the complexity and difficulty of aiming for fundamental change in pedagogic practice becomes obvious.

2.2.1.3 Bernstein's Code Theory: Power and Control in Classrooms

Building on the consideration that educational systems are both a site of cultural reproduction and social transformation, I will explore the codes which Bernstein (1975a: 85) suggests to analyze *messages* and its meaning and assumptions of social order. He provides codes to analyze the underlying structure of three message systems: *curriculum* (what counts as valid knowledge?), *pedagogy* (what counts as valid transmission of knowledge?), and *evaluation* (what counts as valid realization of this knowledge?). Bernstein's codes examine the structure and changes of the organization, transmission, and evaluation of educational knowledge. A *code* is a "regulative principle, tacitly acquired, which selects and integrates" (Bernstein 1990: 14 f.) relevant meanings, forms of their realization, and evoking contexts. Thus, codes regulate the relationship *between* and *within* contexts. With the codes of *classification* and *framing*, I will analyze relayed power relations and control mechanisms in pedagogic practice. The concept of classification defines the role of power structures, and framing represents interactional control mechanisms.

Classification (cf. Fig. 2.3) refers to the "degree of insulation between categories of discourse, agents, practices, context and provides recognition rules for both transmitters and acquirers for the degree of specialization of their texts" (Bernstein 1990: 214). Thus, classification refers not to the definition of a category or content, as the term is semantically commonly used, but to the relations *between* contents. In secondary school, for example, separate subjects, such as physics, geography, history, have their own identity, voice, and specialized rules of internal relations. When subjects or curricular contents are clearly distinguished, or textbooks contain a high number of separate chapters, Bernstein calls this "strong classification," as power relations between subjects are strong. Strong classification is particularly found in an authoritative environment. When contents are taught in an interdisciplinary or phenomenological manner, they have a "weak classification."

To investigate the structure of the message system of the *curriculum*, Bernstein distinguishes two types of curricula, the *collection* and the *integrated* types. Bernstein defines the *curriculum* as the "principle by which units of time and their contents are

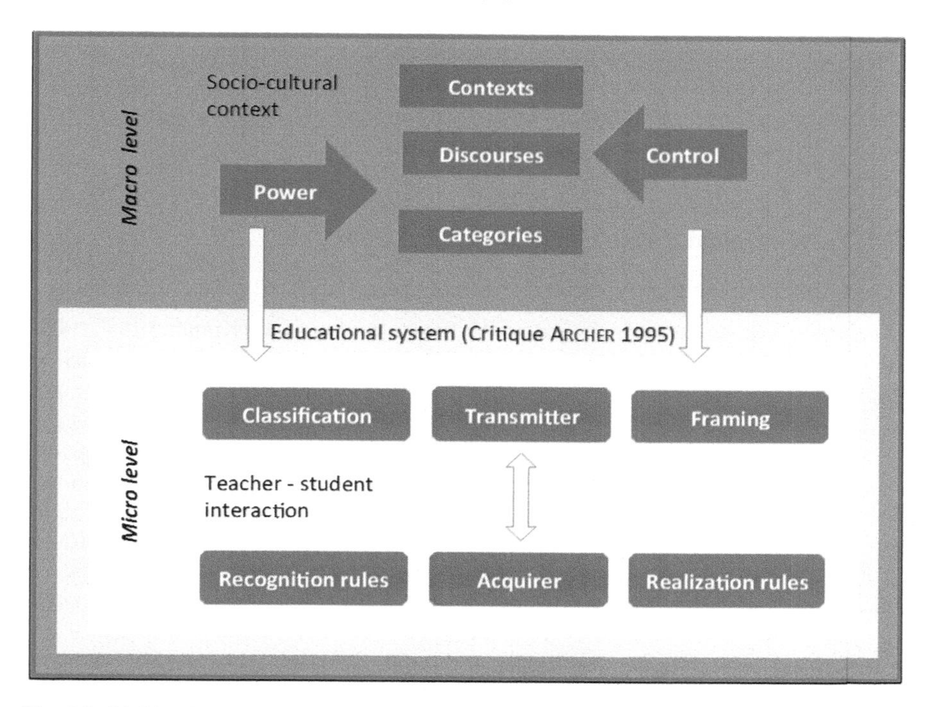

Fig. 2.3 Linking Bernstein's concepts of classification and framing to power and control [altered from Sertl and Leufer (2012: 17)]

brought into a special relationship with each other" (Bernstein 1975a: 86). A highly differentiated *collection curriculum* relates to strong classification (C+) shown in separated subjects and contents being explicitly insulated from each other. Weak classification (C^-) relates to an *integrated curriculum* in which content "boundaries between subjects are fragile" (Sadovnik 1991: 52). If classification is weak, discourses, identities, and voices are less specialized and power relations are weaker. The space or insulation between categories has an external function as well as an internal function. The external function is to create social order, whereas the internal function is a "system of psychic defense to maintain the integrity of a category" (Bernstein 1996: 21). The subject geography, for example, struggles to keep its identity as a separate subject in India, as it is at danger of becoming subordinated to the subject of social or environmental sciences.

Pedagogic practice is controlled through different forms of communication. Thus, *framing* regulates the communication between the transmitter and the acquirer (Bernstein 1996: 26). When framing is strong (F+), the transmitter (teacher) controls the selection, organization, and pacing criteria of transmitted and acquired educational knowledge. The position, posture, and dress of the communicants, as well as the arrangement of the physical location, indicate whether framing is strong or weak. If framing is weak (F^-), the acquirer (student) has more control over selection, organization, and pacing criteria of pedagogic communication. In a strong framing, the teacher has the highest amount of speaking, while in a weak framing, students also have control and can participate in steering the classroom communication. Hence, framing defines the relationship of teacher and student.

2.2.1.4 Visible and Invisible Pedagogy

To distinguish pedagogic principles promoted by ESD and those prevalent in pedagogic practice in India as described in the literature reviewed (Kumar 1988, 2003, 2005, 2007; Berndt 2010), I will use the two ideal types of visible and invisible pedagogy (Bernstein 1975a, b). In visible pedagogy, framing and classification are strong. Hierarchical rules are explicit, and sequencing and criterial rules, thus the selection, organization, and pacing of knowledge, are prescribed. Student performance is important and measured by the external product of the child. Due to the strong emphasis of the students' graded production with "an evaluation orientation that focuses on absences (of content, skill, etc.)" (Sriprakash 2010: 298), Bernstein calls this model the *performance model* of pedagogic practice. Visible pedagogy focuses on transmission and is related to conservative, behaviorist models of pedagogy.

In *invisible* pedagogy, framing and classification are weak. Hierarchical, sequencing, and criterial rules are implicit. Invisible pedagogies, which have a rather progressive connotation, focus more on internal cognitive, linguistic, affective, and motivational procedures (Bernstein 1990: 71). Invisible pedagogy indicates that boundaries of social relationships and learning are implicit. The student is given space to explore, select, structure, regulate, and rearrange, while the teacher becomes a facilitator of these processes. In contrast to visible pedagogy, hierarchical rules are implicit in that power relations are less obvious. Teaching material is "less likely to be pre-packaged (for instance, in textbooks) as the degree of teacher autonomy over the interaction is expanded" (Bernstein 1975b: 9). Invisible pedagogy is a *competence model* of pedagogic practice, with a strong focus on acquisition and relating to progressive and radical pedagogies such as Critical Pedagogy. When classification and framing are strong in pedagogic practice, they will also be strong at the pedagogic discourse level (curriculum). If power relations are strong, pedagogic practice is also strong.[7] Thus, one can draw inferences from the generic type of pedagogy (visible or invisible) about the distributive and recontextualizing rules.

Bernstein assumed that visible pedagogy and invisible pedagogy carry social class assumptions which he derived from the English educational system. Concepts of *space, time,* and *control* in classrooms can be operationalized to examine which model of pedagogy is practiced in classrooms. For example, if the space per child is small and the possibility of movement is low, this would relate to visible pedagogy. The symbolic value can be interpreted as that there are explicit rules "regulating the movement of objects, practices, communication from one space to another" (Bern-

[7] Archer (1995) criticized this approach for transposing cause and effect (cf. Sect. 2.2.1).

stein 1990: 80). Concerning time, a strictly clocked timetable with little flexible time also represents a visible pedagogy. In contrast, invisible pedagogy is related to concepts of space, time, and control with weaker boundaries. Since the economic costs of space and time are high, invisible pedagogy is related to more expensive pedagogy; thus, the pedagogy of the English middle class.[8] Visible pedagogy can be used to weaken social hierarchies. Mixing visible and invisible pedagogic elements could provide marginalized students with access to privileged codes and curricular content. This is not only applicable to schools; homes can also function as a second site of acquisition of codes. Bernstein (1996: 170) argues that strongly framed pedagogic discourses should be embedded in informal local discourses.

I will investigate empirically the principles of classification and framing in curriculum and pedagogic practice and also analyze recognition and realization rules of the acquirers. From the analysis of classification and framing, I can describe and analyze pedagogic principles of ESD and in the Indian geography education. This helps to understand power relations and control mechanisms, and thus, the cultural reproduction of the distribution of power. From this understanding, I can derive opportunities for transformation in pedagogic practice through identifying triggers that change existing power relations and control mechanisms. Bernstein's code theory supports the approach of Critical Pedagogy holistically by investigating underlying relays of power structures and using codes to embed them into empirical research. Through this analysis, potentials for transformation can be identified. I will apply the mentioned concepts and codes of strong framing and classification to predominant pedagogic practice in India, as cited in the literature (Kumar 1988, 2003, 2005, 2007; Berndt 2010) and weak framing and classification to the postulations of ESD.

2.2.1.5 Critical Review of Bernstein's Code Theory

As pedagogic practice is strongly structured by the Indian educational system through its textbooks and curricula, for example, it is of the utmost importance to examine these influences on the teacher and student interaction in classrooms. However, Bernstein's code theory neglects the structure of the educational system in which pedagogic practice takes place. He views education as a permeable social institution, as "the power relationships created outside the school penetrate the organization, distribution and evaluation of knowledge" (Bernstein 1990: 200). Conversely, the direct translation of classification and framing into pedagogic practice misses the mediating social structure in between, namely the educational system. Influential for pedagogic practice are educational politics, such as struggles of individual teachers and educational authorities, belonging to different interest groups. Archer (1995: 214) criticizes Bernstein's theory for homogenizing educational systems and omitting educational politics. She argues that he "attaches primacy to cultural universals which override

[8]Time and space concepts and their interpretation of visible and invisible pedagogies may vary in the Indian context, as in Himalayan villages, for example, space and time are implicitly dealt with (Leder 2012).

comparative differences in educational structures" (Archer 1995: 213). For a macro-sociological theory, Archer suggests including "the conditions under which different social groups can influence the prevailing definition of instruction" (Archer 1995: 222). I take this criticism into account as in the heterogeneous country of India, it is particularly important to analyze the curriculum design and textbook development processes to understand how the dominant instructional discourse is shaped. Young (1971) states:

> It is or should be the central task of the sociology of education to relate the principles of selection and organisation that underline curricula to their institutional and interactional settings in schools and classrooms and to the wider social structure. (Young 1971: 66)

Agreeing that Bernstein's *Sociological Theory of Education* leaves out the specific historical and contemporary development of educational systems, I include Archer's remark on the structure of the educational system into my approach. Thus, I incorporate processes and power struggles of educational authorities and do not limit my analysis to the classroom level. However, I contradict Archer's assumption that Bernstein's attempt at a universal theory overarching cross-cultural variations fails due to "insidious ethnocentrism" (Archer 1995: 223). Although Bernstein focuses on class relations that are relayed in pedagogic practice, he also notes that "culture cannot be wholly identified with class relations" (Bernstein 1990: 168). Several studies (Barrett 2007; Hoadley 2008; Neves and Morais 2001; Mukhopadhyay and Sriprakash 2013) show that Bernstein's concepts can be applied to different cultural contexts reinterpreting culture-specific reproductions through pedagogic practice. These studies examine relayed power relations and social control through pedagogic practice in contexts differing fundamentally in their respective educational systems, for example, Tanzania, South Africa, Portugal and India. Thus, Bernstein's theory proves useful for other cultural contexts although the neglect of the educational system admittedly "reduces its explanatory power" (Archer 1995: 212). With Archer's approach, teachers, and educational stakeholders such as curriculum designers, textbook authors, and teacher trainers can be seen as social agents through whose interactions emergent properties of pedagogic practice develop.[9] Thus, I extend and complement Bernstein's concepts with an analysis of stakeholder perceptions. To include a historical dimension to the analysis, as Archer (1995) suggests, I will include a structural and developmental perspective on the Indian educational system in my empirical research approach.

---

[9] As Giddens' *Theory of Structuration* (1984) suggests that agency and social structures coevolve, agents are influenced, but not determined by social structures, since they can change structures through their agency. In my approach, teachers in their position as multiplicators of social structures are seen as the drivers of reproduction and transformation.

2.2.2 *Critical Pedagogy by Paolo Freire*

Similar to Bernstein's perspective that power relations are present in pedagogic practice, Critical Pedagogy makes power relations in educational settings, a subject of discussion. In addition to Bernstein's code theory, which primarily analyzes reproduction, Critical Pedagogy adds a normative perspective to transformation. In the following, I introduce Critical Pedagogy and its radical founder Paulo Freire—the central concept of an empowering *critical consciousness* in his pedagogy will be used as an analytical lens for this book.

Critical Pedagogy originated from the critical theory of the Frankfurt School,[10] the post-structuralist and postmodern works of Derrida (1978), Foucault (1969), Baudrillard (1968), and Lyotard (1984). Critical educational theory has evolved from "an interdisciplinary mixture of social theory, sociology, and philosophy [... and] has been profoundly affected by postmodernist thought" (Sadovnik 2011b: 16). Combining different strands of theories and disciplines, critical theory offers a corpus of ideas and concepts relevant to educational research. One central point of critical theory is that:

> Critical theorists from Herbert Marcuse to Theodor Adorno have always recognized that the most important forms of domination are not simply economic but also cultural and that the pedagogical force of the culture with its emphasis on belief and persuasion is a crucial element of how we both think about politics and enact forms of resistance and social transformation. (Giroux 2004: 32)

The Brazilian educationist Freire (1996) laid the foundations for Critical Pedagogy with the publication of *Pedagogy of the Oppressed* (1996). Through his pedagogy, he aims to empower "the oppressed," broadly considered to be "the poor," by resolving the relationship between them and their "oppressors," those "in power." To become freed from "oppression," Freire (1996) states the need to develop a *critical consciousness* of individual experiences and oppressive social contexts as a basis to promote skills to overcome oppressive conditions. Freire describes the existing educational system in Brazil as an instrument to further oppress marginalized groups. The central objective of Critical Pedagogy is individual liberation through promoting capabilities for individual development as well as the collective struggle for social justice (Freire 1996). Freire (1996) presents two radical types of cultural action operating in and upon social structure, and creating the dialectical relation of *permanence* and *change*: antidialogical action, characterized by conquest, divide, rule, manipulation, and cultural invasion; and dialogical action for the oppressed, marked by cooperation, unity, organization, and cultural synthesis (Freire 1996: 162). Thus, Critical Pedagogy is closely linked to anti-imperialist social movements strongly opposing all forms of social, economic, and cultural oppression such as racism, terrorism ,

---

[10]The Frankfurt School is the original source of critical theory, and in the second generation, it was especially influenced by Jürgen Habermas, who defines communicative interactions as the basis of society, in which rational grounds of validity are raised and approved. For further information: http://www.iep.utm.edu/frankfur/.

and the class system. Apart from his radical political binary notions, his educational approach is worthwhile studying, particularly the concept of an empowering *critical consciousness*.

Through a pedagogy oriented toward social justice, "the oppressed" are meant to become transformative democratic citizens and thus contribute to national development. Instead of following prescriptions, "pedagogy makes oppression and its causes objects of reflection by the oppressed and from that reflection will come their necessary engagement in the struggle for their liberation" (Freire 1996: 30). He suggests countering reproduction through reflection and dialogue for an "educational awakening." Freire developed teaching methods concerned with the actual experiences of rural peasants in Brazil and available material from their cultural background. This approach led to an influential social and educational movement to create empowerment through *critical consciousness*, "la conscientização" (Freire 1996: 17).

The objectives of Critical Pedagogy are that students learn to recognize and question power relations. Freire distinguishes between two opposing concepts of pedagogy: *banking* and *problem-posing* education. The metaphor of banking depicts the unknowing student as a recipient of the teacher's knowledge; knowledge is deposited and collected like money in a bank. In contrast to this, problem-posing education takes place in a dialogue and includes the knowledge, perspectives, and experiences of both teacher and student (Nagda et al. 2003: 168). Through dialogue, education becomes a democratic and emancipatory process, in which the oppressed are freed from their *culture of silence* (Freire 1996: 12): "There is no true word that is not at the same time a praxis. Thus, to speak a true word is to transform the world" (Freire 1996: 68). The essence of dialogue is the authentic word, which is constituted of the two elements *action* and *reflection*. If action is sacrificed, "the word is changed into idle chatter, into *verbalism*"; if reflection is sacrificed, the word converts to *activism*, which "negates the true practice and makes dialogue impossible" (Freire 1996: 68 f.):

$$\text{Action} + \text{Reflection} = \text{True Word} = \text{Praxis} = \text{Transformation}$$

Following Freire's thoughts, the teaching methods and contents of a school subject are not "something unique or logical, but defined by what those who regulate and control the curriculum believe to be the most useful and desirable to benefit society. They are social, not logical facts" (Clark 2005: 36). With such an analytical lens, I will examine academic frameworks for education in India, namely curricula, syllabi, and textbook contents.

Many critical educationists argue in a normative manner and view pedagogy as a tool to counterbalance economic hegemony and to overcome the divide between postmodern cultural politics and modernist material politics. Henri Giroux is one prominent critical educationist building his arguments on Freire's ideas. He describes Critical Pedagogy as a project of intervention in which teachers as public intellectuals are responsible for rethinking what is perceived as legitimate (Giroux 2004: 35). Curricula are seen as the result of negotiations between conflicting societal groups interested in influencing the adolescent generation (Meyer 1997: 359). Involved curricula designers provide directives oriented toward scientific and pedagogic theories and reflect national interests. Curricula should justify their suggested objectives, contents, and methods pedagogically, didactically, and scientifically. However, rather often, political and economic interests interfere; "what counts as legitimate knowledge is the result of complex power relations and struggles among identifiable class, race, gender, and religious groups" (Apple 2000: 44).

These reviews of Critical Pedagogy demonstrate how far-reaching, political, and, in an extreme case, radical ESD's claim for critical thinking and social transformation could be interpreted. I incorporate moderate elements of Critical Pedagogy in my interpretation of ESD and argue that it is the teachers' task in schools to promote critical consciousness, problem-posing education, and language skills as a prerequisite for social agency (cf. Giroux 2004: 40).

In order to investigate the transformative potential through ESD, it is necessary to understand how one can integrate not only social, but also ecology and place aspects into a Critical Pedagogy, something that has been attempted by Gruenewald (2003). On the basis of Freire and Giroux, he has developed a *critical pedagogy of place*, which takes spatial aspects of social experiences into account. He blends the discourse of "Critical Pedagogy" and "place-based education" as mutually supportive educational traditions and thus emphasizes the inclusion of social and ecological places in Critical Pedagogy. As I examine geography education, I will particularly focus on the constructions of place depicted in textbooks and transmitted by teachers. Do places (e.g., rivers or cities) have fixed presentations, or do diverse social groups construct places differently? For adapting a Critical Pedagogy approach for environmental education, Kyburz-Graber (1999) develops a critical environmental education which, among others, focuses on interaction, construction of contextual knowledge, and critical understanding of human action. I will draw on these concepts of social construction, similar to the attempt of environmentalists in India to use ESD as an opportunity to reform the existing subject of environmental education. A similar "paradigm shift" (Fien 1995: 23) for environmental education has been initiated to move from science-oriented nature education to the inclusion of development education with a focus on changes in social paradigms and worldviews. This transition can only take place if a critical consciousness is promoted in geography education.

2.2.2.1 Critical Review of Critical Pedagogy

Freire's approach represents a radical pedagogy perspective operating with the dialectic relation of oppressors and oppressed. This dichotomy simplifies social structures and neglects complex social relations. Freire's educational approach is primarily politically motivated as it considers education as a tool to transform oppressive social structures. However, his approach is not sufficiently supported with empirical data and the concrete measures that teachers and educators can use to reach the state of *critical consciousness* are focused on rural peasants in Brazil. Similarly, Sadovnik (2011b) censures Critical Pedagogy for its missing link between its theory and research and practice. Apple (2004) further makes the criticism that "rhetorical flourishes […] need to come to grips with […] changing material and ideological conditions." Bernstein (1990) argues Critical Pedagogy is only concerned with the *message* of dominant groups, but does not describe the structure that makes the message possible. He argues that critical theories analyze pedagogic practice "in *relation to* class, gender, and race, but are less concerned with analyzing *relations within*" (Bernstein 1990: 178). Bernstein himself provides a theory not only concerned with *what* is reproduced, but also *how* reproduction takes place. Davies (2001) concurs with Bernstein and describes him as "an analyst of power rather than a prescriber of policy," in contrast to critical pedagogists.

Despite this criticism, Freire's concept of critical consciousness fills the gap between Bernstein and the critical postulations of ESD. The normative approach of Critical Pedagogy envisions a desirable future, encourages a critical consciousness, and thus parallels the transformative objectives of ESD. Hence, it provides a necessary element to guide research on the implementation of ESD, while Bernstein provides the means for in-depth analyses.

2.3 The Translation of Transnational Educational Policies in Local Contexts

Power relations and cultural values of teaching and learning in pedagogic practice can contrast the objectives of democratizing educational reforms. Hence, transnational educational objectives can represent a challenge to local contexts, especially in contexts with deeply entrenched social stratifications. This holds particularly true in countries with limited financial and human resources, as well as with ingrained resistance or indifference to changes being made in existing hierarchical power structures and institutional frameworks. The demand for learner-centered teaching methods and broader quality of education appears subordinate if basic equipment and sufficient teacher training are lacking. That said, the realization of policies, such as ESD, is, however, a problem not only of resources, but also of structures within the state which enable educational debates and possibilities of implementing educational reforms.

Countries and regions have their own structural challenges, living conditions, capacities, and willingness to address these, as "environmental problems and sustainable solutions are historically, geographically and culturally local" (Bonnett 1999: 319). Hence, every society has "to produce its own solutions, for no other society has precisely its capacities, faces its problems, nor has the same possibilities for 'internal' insight into them" (Bonnett 1999: 319). Taking over principles and formal procedures to solve environmental or social problems of one culture into another, particularly from highly industrialized countries to non-industrial cultures, is not only difficult but also frequently inappropriate; therefore, a country's own cultural resources are of primary importance.

Already in 1981, Guthrie (1986) summarized the main findings of curriculum reform research in formerly colonized countries from 1976 to 1981 and found that curriculum change policies were limited and irregular with mixed success because innovations were inappropriate to the context. He points out the "uncritical adoption of models and change strategies from other countries, particularly those in the developed world" (Guthrie 1986: 82). Based on a meta-study of 72 studies within a period of 30 years (1981–2010), Schweisfurth (2011) identified as limitations of reforms and implementation of learner-centered education in developing countries influential factors such as materialistic and human resources, as well as culturally shaped power and agency structures. Hence, the availability of teacher training and infrastructure alone does not limit and challenge learner-centered principles, but also the role of the teacher, examinations, donors, and learners. Nevertheless, the "template transfer" (Manteaw 2012) of programs originating from "Western" industrialized cultures to "developing" and newly industrialized countries is still common practice and can be expected to become even more relevant due to the growing importance of transnational educational organizations as a product of globalization. In pursuance of prescriptive standards, local conditions such as a lack of infrastructure, robust educational institutions and ownership as well as methodological knowledge and skill priorities, however, can make it difficult to sustain the implementation or application of standardized program frameworks in the long run (Mukhopadhyay and Sriprakash 2011; Leder and Bharucha 2015; Clarke 2003).

As a consequence of being an idea originating out of multilateral conferences, I argue that ESD undergoes a process of "translation" (Merry 2006) through different stakeholders operating in different local contexts around the world. As a transnational "traveling model," ESD is illustrative of how "ideas assembled in one site connect with meanings and practices in another" (Behrends et al. 2014: 4). On different levels, policies are "reworked, reinterpreted, and reenacted contextually" (Mukhopadhyay and Sriprakash 2011: 323) by a variety of national, state, regional, and local stakeholders. Stakeholders from politics and science influence the public education discourse, which becomes implemented in schools by political decision-makers and administrative regulations. Thompson (2013) notes that this translation is taking place rather through small-scale institutional relationships than national governments.

Rather than "borrowing" policies from one context to another, educational policies "undergo a process of 'translation' involving the contextualization and inevitable transformation of policies" (Mukhopadhyay and Sriprakash 2011: 311). Through this, transnational educational objectives are interpreted and transformed in the context of each country's cultural systems and historical developments, both having a significant influence on the educational system. For example, British colonialism had major structural, contextual, and methodological effects on India's educational system—ones that continue to influence its institutions, processes, and outputs (syllabi, textbooks, and examinations) to this day (Kumar 1988; Rothermund 2008).

Educational systems not only reflect principles that are regional or national in origin, but also intermingle with the latest international political trends. These principles can contain notions of political values (such as "hierarchic" or "democratic principles") that are subsequently reinterpreted in the respective educational systems. As outlined in the prior chapters, it is fundamentally important to understand that "pedagogy is a moral and political practice that is always implicated in power relations" (Giroux 2004). This is particularly relevant concerning the implementation of new pedagogic principles and curriculum change that can be subsumed under policies which focus on *learner-centered education* (LCE). This highlights that the cultural context within pedagogic practice needs to be studied more closely to better link educational objectives to social realities.

Learner-centered education is not necessarily a "Western" product per se and is growing in relevance in developing contexts if adapted to the specific cultural settings through "cultural translation" (Thompson 2013). Social constructivism, the educational theory grounding learner-centered education, can, for example, translate in developing contexts into what Vavrus (2009) calls *contingent constructivism.* Based on a year-long ethnographical study of a Tanzanian teacher training college, she examined the cultural politics of pedagogy, comprising cultural, economic, and political forces that ignored the regional variability of the principles of quality teaching. Her concept of *contingent constructivism* aims to maintain "an appropriate degree of student-teacher interaction for the contexts in which they were teaching" (Vavrus 2009: 310). One central suggestion is to integrate problem-based peer learning activities and critical thinking in authoritative and formalistic methods such as inquiry–response cycles, instead of replacing them. This exemplifies how educational ideas do "not just migrate; in speaking to different cultural histories and conditions they also change" (Alexander 2001: 546).

To elaborate on pedagogic practice beyond the oversimplified dichotomous categorization of teacher-centered and learner-centered pedagogy, Barrett (2007) used Bernstein's performance and competence modes in Tanzanian primary schools. Her analysis of classroom observations provides a "nuanced understanding of primary school teacher's [sic] practice" (Barrett 2007: 273) and stresses the limitations of "Western" analytical frameworks for low-income countries. Conditions of scarcity should not lead to "underestimating the pedagogic traditions and debate that do exist within Tanzania and other low-income countries" (Barrett 2007: 292).

This book aims to highlight how the translation of the transnational educational policy ESD in Indian school education is shaped by cultural values.

2.3.1 The Role of Cultural Values in Pedagogic Practice

To target conditions for targeted educational policies and curriculum change, the understanding of cultural values in pedagogic practice is relevant for the implementation of learner-centered education in developing contexts (Sriprakash 2010; Schweisfurth 2011). Several studies have examined pedagogic practice and the challenges and opportunities of learner-centered education in classrooms in Africa, particularly in Tanzania and Nigeria (cf. Alexander 2001; Barrett 2007; Manteaw 2012; Schweisfurth 2011; Thompson 2013; Vavrus 2009). A few studies provide in-depth analyses of pedagogic practice in Indian classrooms or wider schooling principles (Berndt 2010; Clarke 2003; Kumar 1988, 2005; Sriprakash 2012; Thapan 1984, 2014). Disciplines of these studies cover educational sciences, comparative educational research, and sociologies of education (cf. Table 2.2). Most studies used qualitative methodology with observations, interviews, and ethnographic methods. Interestingly, subject-specific in-depth research, which analyzes the teaching methodology of a particular subject, specifically geography didactics, has hardly been conducted. Hence, subject-specific didactic consequences for learner-centered education have not been derived for specific socio-cultural contexts.

As India is the study's focus, an understanding of the role of cultural values, values, and hierarchies in teaching and learning is relevant for successful educational reforms. In the following, an overview of principles and factors identified by subject-independent studies on learner-centered pedagogic practice in India and African countries, as well as the learning of applying Bernstein's concepts and theories, will be summarized to review possible challenges to consider in translating the transnational educational policy of ESD.

One case study concluded that learner-centered reforms could not change the essential nature of traditional pedagogic practice (Clarke 2003). Although instructional aids and activities were integrated in classroom teaching in the context of the District Primary Education Project (DPEP) in Karnataka, teaching characteristics such as rote learning and repetition remained unchanged. Clarke (2003) identified four cultural constructs which frame teacher thinking and thus, teaching style: a holistic worldview, hierarchical structure, collective knowledge, and instruction as duty. Similarly, Alexander (2001: 546) points out that "it is not yet clear how far an individualistic, inquiry-based ideology is compatible with either the deeply rooted collective orientation of Indian primary teaching or the unassailable fact of very large classes." Whereas openness to regulation and the conception of the teacher's task as duty enables teachers to be receptive for change processes, "certain other cultural constructs, namely hierarchy as a social framework and the collective construction of knowledge, limit teachers' appropriation of the new pedagogy" (Clarke 2003: 37). Barrett (2008) identified four teacher identities based on their practice in relation to gender, length of service, and geographical setting. She named these identities relaters, self-improvers, gazers, and storytellers. These identities differently respond to educational innovations. Thus, child-oriented policies and ideas should be introduced "with a sense of deference that simultaneously respects how teachers construct their identity" (Barrett 2008: 506). The role of diverse experiences and life histories influence teachers' performances (Neves et al. 2004).

The sociological, ethnographic study of Sriprakash (2012) examined the complexities of change in pedagogic practice through learner-centered education in rural, under-resourced Indian primary schools in Karnataka by applying Bernstein's codings to classroom teaching. She analyzed existing cultural constructs of teaching and the importance of contexts by investigating how new forms of instruction were interpreted through teachers. She demonstrated that learning is understood as knowledge assimilation instead of knowledge construction, as educational reforms envision. She observed "tensions" experienced by teachers when introducing child-centered pedagogic approaches to the classroom because of the changing social order in the classroom. Her study exemplified that learner-centered pedagogic models rather "reinforce social messages of control and hierarchy relayed to children" (Sriprakash

Table 2.2 Studies on learner-centered education and ESD in India and Africa

Region	Authors (year)	Discipline	Thematic focus, central terms, and concepts	Methodological approach	Countries
Comparative studies	Alexander (2001)	Comparative Education	Multi-level comparative study on the culture of primary education systems, schools, and pedagogy	Ethnographic study of five countries	India, Russia, England, USA, and France
	Schweisfurth (2011)	Comparative Education	Practical, material, and cultural barriers to learner-centered education in developing contexts	Meta-study of 72 studies of 30 years	Multiple
Studies in India	Thapan (1984)	Sociology of Education	School as socio-cultural system; education and ideology in a Krishnamurthi school	Ethnographic study	India (Andhra Pradesh)
	Kumar (1988, 2003, 2007)	Sociology of Education	Colonial and nationalist influences on education content and methods in India	Textbook analyses	India
	Clarke (2003)	Cultural Psychology/Social Anthropology	Cultural construction of teacher thinking on teaching and learning and educational reform impact	Qualitative and quantitative methodologies	India (Kerala)
	Berndt (2010)	Educational Sciences	Power and cultural barriers to quality primary Education for All (EFA)	Interviews, questionnaires, observations	India (Andhra Pradesh, West Bengal)

(continued)

Table 2.2 (continued)

Region	Authors (year)	Discipline	Thematic focus, central terms, and concepts	Methodological approach	Countries
	Sriprakash (2010, 2011, 2012)	Sociology of Education	Politics and practice of child-centered pedagogies for development with Bernstein's concepts of framing and classification	Ethnographic study	India (Kerala)
Studies in Africa	Barrett (2007, 2008)	Comparative Education	Models of classroom practice and teacher identities through Bernstein's concepts	Interviews and observations	Tanzania
	Vavrus (2009)	Comparative Education	Cultural politics of constructivist learning theories and contingent pedagogy	Ethnographic study	Tanzania
	Manteaw (2012)	Environment Education	Education and sustainable development in Africa	Discourse analysis	African countries
	Thompson (2013)	Educational Sciences	Dialectical model of cultural translation of learner-centered education in developing countries	Interviews and questionnaires	Nigeria

2010: 304) instead of democratizing the classroom. For example, closed questions and repetition do not allow individual answers; instead, students answered collectively with single-word responses. *Nali Kali*, a District Primary Education Project (DPEP), which focused on bringing student-centered approaches into the classroom, failed in that students were given time to work independently, but the task given was strongly framed by fixed outcomes as they had to reproduce pre-given sentences. She further derived educational implications of educational reforms. Thus, the integration of weakly framed tasks in terms of space and explicit evaluation criteria subverted existing strongly framed teaching styles.

Berndt (2010) analyzed the socio-cultural and historical embeddedness of education in India and exemplified how social stratification through caste, class, and gender contrasts with the reform approaches of *Education For All* (EFA). She points out the vertical and elite structure of education marked by traditional social hierarchies defined by class, caste, and gender in contrast to the demands of international and national educational reforms and programs. These frame the fundamental social and political conditions under which teaching realities are influenced and strongly socially stratified—most dominantly through the two-tier educational system in India which reproduces inequity (Berndt 2010: 288). Steiner (2011: 74) points out the importance of learning conditions, institutional frameworks, and interactional partners in India's educational system. Building on Giddens' *Theory of Structuration*, she argues that these social structures in interdependency with individuals will only slowly change.

The sociological, ethnographic study conducted by Thapan (1984) is one of the first well-known in-depth studies in India analyzing the interaction of teachers and students, and the structural conditions of the internal and external orders of the school as a social institution. She applies an interactionist approach to reveal the complex interlinkages of the holistic ideas of the philosopher Krishnamurthi and the Rishi Valley School in Andhra Pradesh. In her recent publication *Ethnographies of Schooling in Contemporary India*, Thapan (2014) uncovers schooling ideals and student culture against the backdrop of a society highly stratified by caste, class, and gender (cf. Leder 2015b). Rich ethnographies in eight Indian schools highlight how citizenship is translated into schooling processes, and how values of obedience, equality, discipline, spirituality, and frugality are partly adopted, but also questioned and subverted by students under the influence of peer cultures, media, and marketing. While students are submissive to the school's declared values, they also challenge and reject these ideals, being involved in a parallel construction of identity which may not necessarily match the ideals laid down by the official tenets of their school (Leder 2015b). One study of a private school in rural Andhra Pradesh demonstrates how the school's "bubble" created by the school's rejection of "urban" or "Western" values of consumerism regarding food, clothes, and entertainment is resisted by students by favoring and sharing these products. This demonstrates how identities of being urban or rural, frugal or consumerist, are intermingled despite the schools' proclaimed ideals. This example conveys the contrasts of publicity pronounced and hidden values and identities within a schooling environment, and a lack of dialogue between the principal, teachers, and the students. This brings into focus the role and duty of teachers, and the ways in which they negotiate with expectations from the school and the pupils. The ethnographies open the debate whether the values, principles, and rules transmitted in schools truly meet the skills needed by children and youths from diverse socio-economic backgrounds in contemporary urban India. Memorization and rote learning, as practiced in Indian schools, do not promote intellectual curiosity, criticism, and communicative skills as required in a rapidly changing society, in which decision-making is becoming increasingly important.

These studies highlight the importance of adjusting the transnational educational policy ESD to the cultural context through learner-centered educational principles.

This includes studying cultural norms and institutions in teaching, as well as taking the conditions of the educational infrastructure as well as available human resources into account. When educational reforms are linked to these cultural constructs, and teacher training is appropriately designed, new forms of instructions can be integrated and transform teaching and learning in the classroom. Instructional aids and activities need to be understood only as a "means to an end" to be integrated into the cultural patterning of teaching. Understanding teachers' cultural values of teaching and learning could facilitate a process of change in pedagogic practice.

2.4 Theoretical Framework for Transformative Pedagogic Practice

Since the transnational educational policy of Education for Sustainable Development promotes education and critical thinking for social transformation, subchapter two theoretically consolidates the link between pedagogic practice and larger social processes. The role of education both for social transformation and within a post-colonial context is reviewed within larger social theories and educational policy discourses. These approaches are often coined by binary notions, e.g., human rights versus human capital, modernization versus dependency, functionalist versus conflict approaches. This book aims to overcome dualistic concepts by developing an approach to bridging the conflicting priorities of reproductive and transformative roles assigned to pedagogic practice.

For the theoretical framework for this study, I integrate several previously discussed approaches in order to frame the concept of *transformative pedagogic practice* (Fig. 2.4). The concept of transformative pedagogic practice locates the research question of this study between the notions of reproduction and transformation and illustrates how the transnational objective of ESD poses both a challenge and an opportunity to rethink the structuring elements of pedagogic practice.

I theoretically ground the investigation of pedagogic practice in geography in the socio-cultural context of India by combining the descriptive–analytical concepts of Bernstein's *Sociological Theory of Education* (1975–1990) with the Critical Pedagogy perspective of Freire (1996). Similar to the postulations of ESD, Freire (1996) demands a critical consciousness for social transformation and provides a normative analytical lens on empowering for dialogue which is particularly interesting within a society as highly stratified as that of India.

To analyze how principles in pedagogic practice in Indian geography education relate to the principles of ESD, I will use Bernstein's concepts such as classification and framing. These concepts help examine relayed power relations in pedagogic practice and link the micro-level of classroom teaching to the macro-level of the socio-cultural context in which pedagogic practice takes place. Bernstein's educational modes of visible and invisible pedagogy help to understand the challenges of learner-centered approaches in hierarchically structured teaching contexts in detail.

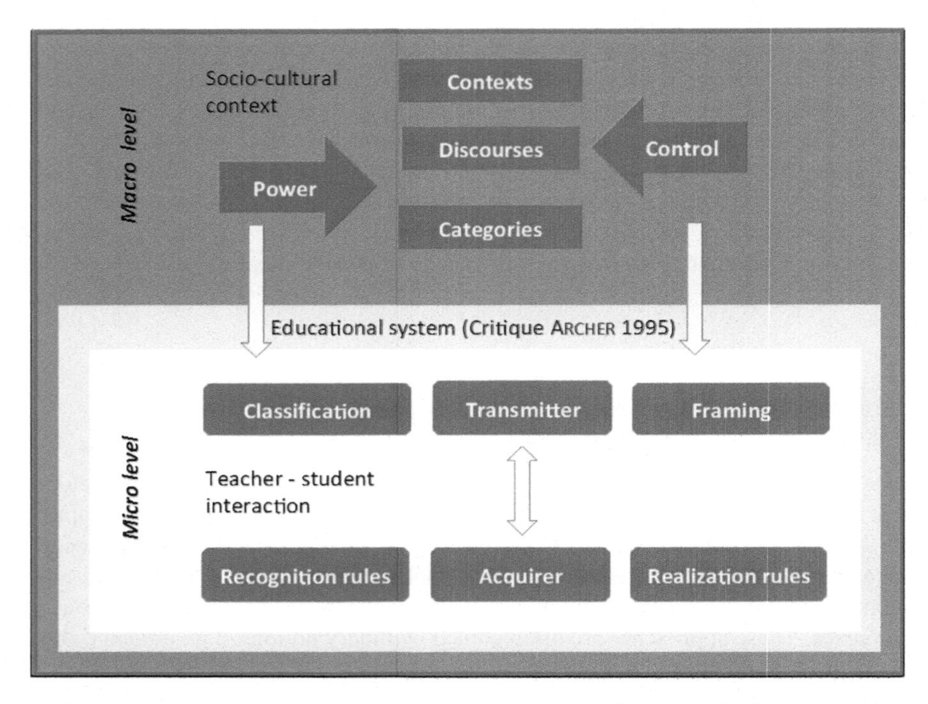

Fig. 2.4 Theoretical framework to study transformative pedagogic practice for this study

With abstract terms, simple dichotomies, such as authoritarian and learner-centered, participative teaching styles, can be viewed in a more differentiated way. The application of these concepts facilitates the in-depth analysis of monologic and dialogic texts such as textbooks, syllabi, classroom practices, and teacher training (cf. Neves and Morais 2001: 187). Hence, Bernstein's theory offers a language for description and explanation, diagnosis, prediction, and, most importantly, transferability by viewing the empirical and the theoretical dialectically (cf. Morais 2002: 564).

In addition to these two approaches, I integrate the perspective of Archer (1995), who suggests that the institutional level, the structure and politics of the educational system, and educational authorities add explanatory value to the analysis of pedagogic practice. Furthermore, an analysis of socio-cultural constructs in classroom teaching (Sriprakash 2012; cf. Berndt 2010; Schweisfurth 2011) can help to understand how new forms of instruction are interpreted by teachers. The setting is especially important concerning the complexities of pedagogic practice in under-resourced contexts.

Educational reforms, such as the transnational educational objective of ESD, intervene on several levels when aimed at influencing pedagogic practice. To understand the process of how the transnational educational policy ESD is transformed into pedagogic practices in geography education, I use the concept of *translation* (Merry 2006; Behrends et al. 2014). Building on the idea of ESD as a traveling model, educational stakeholders on different levels construct, interpret, and translate principles of ESD differently into practice.

The integration of these approaches into the concept of transformative pedagogic practice helps to understand the multi-level challenges of implementing ESD in pedagogic practice. I merge these theoretical approaches to examine how ESD relates to and challenges power relations in pedagogic practice, and to identify institutional, structural, and socio-cultural barriers and opportunities for translating ESD in local geographical pedagogic practice in India. To translate ESD objectives into pedagogic practice, the didactic approach of argumentation (Budke 2012a) promotes critical thinking and helps to guide teaching methodology in classrooms. In the following chapter, ESD and the didactic approach of argumentation will be outlined and discussed.

References

Alexander, R. (2001). *Culture and pedagogy. International comparisons in primary education*. Singapore: Blackwell Publishing.

Allen, M., & Preiss, R. W. (1997). Comparing the persuasiveness of narrative and statistical evidence using meta-analysis. *Communication Research Reports, 14*(2), 125–131.

Andrews, R. (2009). *The importance of argument in education*. London: Institute of Education.

Apple, M. (2000). Cultural politics and the text. In *Official knowledge* (pp. 42–60). London: Routledge.

Apple, M., Ball, S., & Gandin, L. (2010). *The Routledge handbook of sociology of education*. London: Routledge.

Apple, M. W. (2004). Creating difference: Neo-liberalism, neo-conservatism and the politics of educational reform. *Educational Policy, 18*(1), 12–44. https://doi.org/10.1177/0895904803260022.

Archer, M. (1995). The neglect of the educational system by Bernstein. In A. R. Sadovnik (Ed.), *Knowledge and pedagogy: The sociology of Bernstein* (pp. 211–235). Norwood, NJ: Ablex Publishing Corporation. http://books.google.de/books?id=3tgXQ_ISJHYC&printsec=frontcover&hl=de&source=gbs_ge_summary_r&cad=0#v=onepage&q&f=false.

Aufschnaiter, C., Erduran, S., Osborne, J., & Simon, S. (2008). Arguing to learn and learning to argue: Case studies of how students' argumentation relates to their scientific knowledge. *Journal of Research in Science Teaching, 45*(1), 101–131.

Baccini, P. (2007). Cultural evolution and the concept of sustainable development: From global to local scale and back. In S. Reinfried, Y. Schleicher, & A. Rempfler (Eds.), *Geographical Views on Education for Sustainable Development* (pp. 11–26). Geographiedidaktische Forschungen.

Bagoly-Simó, P. (2014a). Implementierung von Bildung für nachhaltige Entwicklung in den Fachunterricht im internationalen Vergleich [Implementation of education for sustainable development into formal education: A comparative analysis of German, Mexican and Romanian curricula]. In M. M. Müller, I. Hemmer, & M. Trappe (Eds.), *Nachhaltigkeit neu denken. Rio+X: Impulse für Bildung und Wissenschaft*. München: Oekom verlag.

Bagoly-Simó, P. (2014b). Tracing sustainability: An international comparison of ESD implementation into lower secondary education. *Journal of Education for Sustainable Development, 7*(1), 95–112.

Baker, S. (2006). *Sustainable development*. London: Routledge.

Ball, S. (1981). The sociology of education in developing countries. *British Journal of Sociology of Education, 2*(3), 301–313.

Ballantine, J. H., & Hammack, F. M. (2012). *The sociology of education. A systematic analysis*. Boston: Pearson.

Barrett, A. M. (2007). Beyond the polarization of pedagogy: Models of classroom practice in Tanzanian primary schools. *Comparative Education, 43*(2), 273–294. https://doi.org/10.1080/03050060701362623.

Barrett, A. M. (2008). Capturing the différance: Primary school teacher identity in Tanzania. *International Journal of Educational Development, 28*(5), 496–507. https://doi.org/10.1016/j.ijedudev.2007.09.005.

Barth, M. (2011). How to deal constructively with the challenges of our times—Education for sustainable development as an educational objective. *Sws-Rundschau, 51*(3), 275–291. <Go to ISI>://000295428700003.

Baudrillard, J. (1968). *Das System der Dinge. Über unser Verhältnis zu den alltäglichen Gegenständen*. Frankfurt: Campus.

Beck, U. (1986). *Risikogesellschaft. Auf dem Weg in eine andere Moderne*. Frankfurt am Main: Suhrkamp.

Beck, U., Giddens, A., & Lash, S. (1995). *Reflexive modernization. Politics, traditions and aesthetics in the modern social order*. Cambridge: Polity Press.

Becker, E., & Jahn, T. (2006). *Soziale Ökologie. Grundzüge einer Wissenschaft von den gesellschatlichen Naturverhältnissen*. Frankfurt: Campus Verlag.

Behrends, A., Park, S.-J., & Rottenburg, R. (2014). Travelling models: Introducing an analytical concept to globalisation studies. In A. Behrends, S.-J. Park, & R. Rottenburg (Eds.), *Travelling models in African conflict management: Translating technologies of social ordering* (pp. 1–40). Leiden: Brill.

Benedict, F. (1999). A systemic approach to sustainable environmental education. *Cambridge Journal of Education, 29*(3), 433.

Berndt, C. (2010). *Elementarbildung in Indien im Spannungsverhältnis von Macht und Kultur. Eine Mikrostudie in Andhra Pradesh und West Bengalen*. Berlin: Logos Verlag.

Bernstein, B. (1975a). *Class and pedagogies: Visible and invisible*. Paris: OECD.

Bernstein, B. (1975b). *Class, codes and control. Towards a theory of educational transmission*. London: Routledge.

Bernstein, B. (1990). *Class, codes and control. The structuring of pedagogic discourse*. London: Routledge.

Bernstein, B. (1996). *Pedagogy, symbolic control and identity: Theory, research, critique*. Boston: Rowman & Littlefield Publishers.

Bonnett, M. (1999). Education for sustainable development: A coherent philosophy for environmental education? *Cambridge Journal of Education, 29*(3), 313–324.

Brunotte, E. (2005). Nachhaltigkeit. In C. Martin, E. Brunotte, H. Gebhardt, M. Meurer, P. Meusburger, & J. Nipper (Eds.), *Lexikon der Geographie*. Wiesbaden: Spektrum Akademischer Verlag.

Budke, A. (2010). *Und der Zukunft abgewandt - Ideologische Erziehung im Geographieunterricht der DDR*. Göttingen: V&R unipress.

Budke, A. (2012a). Argumentationen im Geographieunterricht. *Geographie und ihre Didaktik, 1*, 23–34.

Budke, A. (2012b). Ich argumentiere, also verstehe ich. - Über die Bedeutung von Kommunikation und Argumentation für den Geographieunterricht. In A. Budke (Ed.), *Kommunkation und Argumentation* (pp. 5–18). Braunschweig: Westermann Verlag. Geo Di 14 (KGF), Alexandras einführungsartikel ausgdruckt.

Budke, A., & Meyer, M. (2015). Fachlich argumentieren lernen - Die Bedeutung der Argumentation in den unterschiedlichen Schulfächern. In A. Budke, M. Kuckuck, M. Meyer, F. Schäbitz, K. Schlüter, & G. Weiss (Eds.), *Fachlich argumentieren lernen*. Münster: Waxmann.

Budke, A., Schiefele, U., & Uhlenwinkel, A. (2010). Entwicklung eines Argumentationskompetenzmodells für den Geographieunterricht. *Geographie und ihre Didaktik, 3*, 180–190.

Budke, A., & Uhlenwinkel, A. (2011). Argumentieren im Geographieunterricht - Theoretische Grundlagen und unterrichtspraktische Umsetzungen. In C. Meyer, R. Henry, & G. Stöber (Eds.), *Geographische Bildung* (pp. 114–129). Braunschweig: Westermann.

Carnoy, M. (1974). *Education as cultural imperialism*. New York: McKay.

Clark, U. (2005). Bernstein's theory of pedagogic discourse. *English Teaching: Practice and Critique, 4*(3), 32–47.

Clarke, P. (2003). Culture and classroom reform: The case of the district primary education project, India. *Comparative Education, 39*(1), 27–44. https://doi.org/10.1080/0305006032000044922.

Combs, S. C. (2004). The useless-/usefulness of argumentation: The DAO of disputation. *Argumentation and Advocacy, 41*(2), 58.

Cotton, D. R. R., & Winter, J. (2010). It's not just bits of paper and light bulbs: A review of sustainability pedagogies and their potential for use in higher education. In P. Jones, D. Selb, & S. Sterling (Eds.), *Sustainability education: Perspectives and practice across higher education*. London: Earthscan.

Davies, B. (2001). Introduction. In A. M. Morais, I. Neves, B. Davies, & H. Daniels (Eds.), *Towards a sociology of pedagogy. The contribution of Basil Bernstein to research* (pp. 1–14). New York: Peter Lang Publishing.

De Haan, G. (2006). Bildung für nachhaltige Entwicklung– ein neues Lern- und Handlungsfeld. *UNESCO heute, 1*, 4–8.

de Haan, G., Bormann, I., & Leicht, A. (2010). Introduction: The midway point of the UN decade of education for sustainable development: Current research and practice in ESD. *International Review of Education, 56*(2), 199–206. https://doi.org/10.1007/s11159-010-9162-z.

de Haan, G., & Harenberg, D. (1999). Expertise Förderprogramm Bildung für eine nachhaltige Entwicklung: Gutachten zum Programm. *Bund-Länder-Kommission für Bildungsplanung und Forschungsförderung*.

Derrida, J. (1978). *Writing and difference*. London: Routledge.

Dewey, J. (1916). *Democracy and education*. New York: The Macmillan Company.

Durkheim, E. (1951). *Suicide: A study in sociology*. New York: Free Press.

Ekins, P. (2000). *Economic growth and environmnetal sustainability: The prospects of green growth*. London: Routledge.

Fagerlind, I., & Saha, L. (1989). *Education and national development: A comparative perspective*. Oxford: Pergamon.

Fien, J. (1995). Teaching for a sustainable world: The environmental and development education project for teacher education. *Environmental Education Research, 1*(1), 21–33. https://doi.org/10.1080/1350462950010102.

Fien, J., & Maclean, R. (2000). Teacher education for sustainability II. Two teacher education projects from Asia and the Pacific. *Journal of Science Education and Technology, 9*(1), 37–48. http://www.jstor.org/stable/40188539.

Foucault, M. (1969). *Archeology of knowledge*. London: Routledge.

Fraser, N. (2008). *Scales of justice: Reimagining political space in a globalizing world*. Cambridge: Polity Press.

Freire, P. (1996). *Pedagogy of the oppressed*. London: Penguin Books Ltd.

Giddens, A. (1984). *The constitution of society: Outline of the theory of structuration*. Cambridge: Polity Press.

Giroux, H. A. (2004). Critical pedagogy and the postmodern/modern divide: Towards a pedagogy of democratization. *Teacher Education Quarterly, 31*(1), 31–47.

Goffman, E. (1959). *The presentation of self in everyday life*. New York: Doubleday.

Graupe, S., & Krautz, J. (2014). Die Macht der Messung. Wie die OECD mit PISA ein neues Bildungskonzept durchsetzt. *COINCIDENTIA – Zeitschrift für europäische Geistesgeschichte, 4*, 139–146.

Gruenewald, D. A. (2003). The best of both worlds: A critical pedagogy of place. *Educational Researcher, 32*(4), 3–12.

Gruenewald, D. A. (2004). A foucauldian analysis of environmental education: Toward the socioecological challenge of the Earth Charter. *Curriculum Inquiry, 34*(1), 71–107. https://doi.org/10.1111/j.1467-873X.2004.00281.x.

Guthrie, G. (1986). Current research in developing countries: The impact of curriculum reform on teaching. *Teaching & Teacher Education, 2*(1), 81–89.

Habermas, J. (1984). *The theory of communicative action*. Boston, MA: Beacon Press.
Hasse, J. (2010). Globales Lernen. Zum ideologischen Gehalt einer Leer-Programmatik. In G. Schrüfer & I. Schwarz (Eds.), *Globales Lernen. Ein geographischer Diskursbeitrag*. Münster: Waxmann.
Haubrich, H. (1992). *International charter on geographical education for sustainable development*. Nürnberg: Hochschulverband für Geographie und seine Didaktik.
Haubrich, H. (2007). Geography education for sustainable development. In S. Reinfried, Y. Schleicher, & A. Rempfler (Eds.), *Geographical views on education for sustainable development* (pp. 27–38). Geographiedidaktische Forschungen.
Haubrich, H., Reinfried, S., & Schleicher, Y. (2007). *Lucerne declaration of geographical education for sustainable development*. Switzerland: Geographiedidaktische Forschungen.
Hawkes, J. (2001). *The fourth pillar of sustainability: Culture's essential role in public planning*. Victoria, Australia: Cultural Development Network.
Hellberg-Rode, G., Schrüfer, G., & Hemmer, M. (2014). Brauchen Lehrkräfte für die Umsetzung von Bildung für nachhaltige Entwicklung (BNE) spezifische professionelle Handlungskompetenzen? *Zeitschrift für Geographiedidaktik, 4*, 257–281.
Hoadley, U. (2008). Social class and pedagogy: A model for the investigation of pedagogic variation. *British Journal of Sociology of Education, 29*(1), 63–78.
Hoffmann, K. W., Dickel, M., Gryl, I., & Hemmer, M. (2012). Bildung und Unterricht im Fokus der Kompetenzorientierung. *Geographie und Schule, 195*(34), 4–14.
Hoffmann, T., & Bharucha, E. (2013). Education for Sustainable Development (ESD) as modern education—A glimpse on India. Accessed August 8, 2014.
Holland, J. (1981). Social class and changes in orientations to meanings. *Sociology, 15*(1), 1–18.
Hoogen, A. (2016). *Didaktische Rekonstruktion des Themas illegale Migration. Argumentationsanalytische Untersuchung von Schüler*innenvorstellungen im Fach Geographie*. Münster: MV-Verlag.
Hopkins, C., & McKeown, R. (2002). Education for sustainable development: An international perspective. In D. Tilbury, R. B. Stevenson, J. Fien, & D. Schreuder (Eds.), *Education and sustainability. Responding to the global challenge* (pp. 13–24). Cambridge: IUCN.
Hornikx, J., & de Best, J. (2011). Persuasive evidence in India: An investigation of the impact of evidence types and evidence quality. *Argumentation and Advocacy, 47*, 246–257.
Huckle, J. (1993). Environmental education and sustainability: A view from critical theory. In J. Fien (Ed.), *Environmental education: A pathway to sustainability* (pp. 43–68). Geelong, VIC: Deakin University.
Ilon, L. (1996). The changing role of the World Bank: Education policy as global welfare. *Policy and Politics, 24*(4), 413–424.
Inkeles, A., & Smith, D. (1974). *Becoming modern*. London: Heinemann.
Jickling, B. (1994). Why I don't want my children to be educated for sustainable development: Sustainable belief. *Trumpeter, 11*(3). http://trumpeter.athabascau.ca/index.php/trumpet/article/viewArticle/325/497.
Kienpointer, M. (1983). *Argumentationsanalyse*. Innsbruck: Verlag des Instituts für Sprachwissenschaft.
Klafki, W. (2013). *Kategoriale Bildung: Konzeption und Praxis reformpädagogischer Schularbeit zwischen 1948 und 1952*. Bad Heilbrunn: Klinkhardt.
Kopperschmidt, J. (1995). Grundfragen einer allgemeinen Argumentationstheorie unter besonderer Berücksichtigung formaler Argumentationsmuster. In H. Wohlrapp (Ed.), *Wege der Argumentationsforschung* (pp. 50–73). Stuttgart: Frommann-Holzboog.
Kopperschmidt, J. (2000). *Argumentationstheorie. Zur Einführung*. Hamburg: Junius.
Kuckuck, M. (2014). *Konflikte im Raum - Verständnis von gesellschaftlichen Diskursen durch Argumentation im Geographieunterricht*. Münster: MV-Verlag.
Kuhn, D. (1992). Thinking as argument. *Harvard Educational Review, 62*(2), 155–178.
Kumar, K. (1988). Origins of India's "textbook culture". *Comparative Education Review, 32*(4), 452–464. http://www.jstor.org/stable/1188251.

Kumar, K. (2003). *Quality of education at the beginning of the 21st century*. Lessons from India: UNESCO.

Kumar, K. (2004). Educational quality and new economic regime. In A. Vaugier-Chatterjee (Ed.), *Education and democracy in India*. New Delhi: Manohar.

Kumar, K. (2005). *Political agenda of education. A study of colonialist and nationalist ideas*. New Delhi: Sage Publications.

Kumar, K., & Oesterheld, J. (2007). *Education and social change in South Asia*: Orient Longman.

Kyburz-Graber, R. (1999). Environmental education as critical education: How teachers and students handle the challenge (Article). *Cambridge Journal of Education, 29*(3), 415. http://search.ebscohost.com/login.aspx?direct=true&db=a9h&AN=2666088&site=ehost-live.

Leder, S. (2012) Educational issues in Himalayan Villages: Observations in the Kumaon region, Uttarakhand.

Leder, S. (2015a). Bildung für nachhaltige Entwicklung durch Argumentation im Geographieunterricht. In A. Budke, M. Kuckuck, M. Meyer, F. Schäbitz, K. Schlüter, & G. Weiss (Eds.), *Fachlich argumentieren lernen*. Münster: Waxmann.

Leder, S. (2015b). Uncovering schooling ideals and student culture: The case of India. In M Thapan (Ed.), Ethnographies of schooling in contemporary India. http://www.booksandideas.net/Schooling-Ideals-and-Student-Culture-the-Case-of-India.html. Accessed May 25, 2015.

Leder, S., & Bharucha, E. (2015). Changing the educational landscape in India by transnational policies: New perspectives promoted through Education for Sustainable Development (ESD). *ASIEN, 134,* 167–192.

Leng, M. (2009). *Bildung für nachhaltige Entwicklung in europäischen Großschutzgebieten. Möglichkeiten und Grenzen von Bildungskonzepten*. Hamburg: Verlag Dr. Kovac.

Lubienski, S. T. (2004). Decoding mathematics instruction: A critical examination of an invisible pedagogy. In J. Muller, B. Davies, & A. M. Morais (Eds.), *Reading Bernstein, researching Bernstein* (pp. 108–122). London: Routlege Falmer.

Lyotard, J.-F. (1984). *The postmodern condition. A report on knowledge*. Minneapolis: Minnesota Press.

Malatesta, S., & Camuffo, M. (2007). Geography, education for sustainable development and primary school curricula: A complex triangle. In S. Reinfried, Y. Schleicher, & A. Rempfler (Eds.), *Geographical views on education for sustainable development* (pp. 58–65). Geographiedidaktische Forschunge.

Mandelbaum, D. G. (1975). *Society in India*. Noida: Popular Prakashan.

Mandl, H., & Gerstenmaier, J. (2000). *Die Kluft zwischen Wissen und Handeln - Empirische und theoretische Lösungsansätze*. Göttingen: Hogrefe-Verlag.

Manteaw, O. O. (2012). Education for sustainable development in Africa: The search for pedagogical logic. *International Journal of Educational Development, 32*(3), 376–383. <Go to ISI>://000301698300003 http://ac.els-cdn.com/S0738059311001301/1-s2.0-S0738059311001301-main.pdf?_tid=e247c694-1df6-11e2-a864-00000aab0f6b&acdnat=1351095877_200e77592eca28a76d7de4643351e023.

Marrow, J. (2008). *Psychiatry, modernity and family values: Clenched teeth illness in North India*. Chicago: ProQuest.

Marx, K. (1867). Das Kapital. Kritik der politischen Oekonomie.

McKeown-ice, R. (1994). Environmental education: A geographical perspective. *Journal of Geography, 93*(1), 40–42. https://doi.org/10.1080/00221349408979684.

Mead, G. H. (1934). *Mind, self and society from the standpoint of a social behaviorist*. Chicago: University of Chicago Press.

Meadows, D. H., Meadows, D. L., Randers, J., & Behrens, W. W., III. (1972). *The limits to growth*. New York: Universe Books.

Merry, S. E. (2006). Transnational human rights and local activism: Mapping the middle. *American Anthropologist, 108*(1), 38–51.

Meyer, H. (1997). *Schulpädagogik. Band I: Für Anfänger*. Berlin: Cornelsen.

Moore, R. (2013). *Basil Bernstein: The thinker and the field*. London: Routeledge.

Morais, A. M. (2002). Basil Bernstein at the micro level of the classroom. *British Journal of Sociology of Education, 23*(4), 559–569. https://doi.org/10.2307/1393312.

Morais, A. M., & Neves, I. (2001). Pedagogic social contexts: studies for a sociology of learning. In A. M. Morais, I. Neves, B. Davies, & H. Daniels (Eds.), *Towards a sociology of pedagogy. The contribution of Basil Bernstein to research* (pp. 185–221). New York: Peter Lang.

Morais, A. M., & Neves, I. (2006). Teachers as creators of social contexts for scientific learning: New approaches for teacher education. In R. Moore, M. Arnot, J. Beck, & H. Daniels (Eds.), *Knowledge, power and educational reform. Applying the sociology of Basil Bernstein*. Abingdon: Routledge.

Morais, A. M., & Neves, I. (2011). Educational texts and contexts that work discussing the optimization of a model of pedagogic practice. In D. Frandji & P. Vitale (Eds.), *Knowledge, pedagogy and society. International perspectives on Basil Bernstein's sociology of education* (pp. 191–207). Abingdon: Routledge.

Morais, A. M., Neves, I., & Pires, D. (2004). The what and the how of teaching and learning. Going deeper into sociological analysis and intervention. In J. Muller, B. Davies, & A. M. Morais (Eds.), *Reading Bernstein, Researching Bernstein* (pp. 75–90). London: Routledge.

Mukhopadhyay, R., & Sriprakash, A. (2011). Global frameworks, local contingencies: Policy translations and education development in India. *Compare—A Journal of Comparative and International Education, 41*(3), 311–326. https://doi.org/10.1080/03057925.2010.534668.

Mukhopadhyay, R., & Sriprakash, A. (2013). Target-driven reforms: Education for all and the translations of equity and inclusion in India. *Journal of Education Policy, 28*(3), 306–321. https://doi.org/10.1080/02680939.2012.718362.

Mulà, I., & Tilbury, D. (2009). A United Nations decade of education for sustainable development (2005–14). *Journal of Education for Sustainable Development, 3*(1), 87–97. https://doi.org/10.1177/097340820900300116.

Muller, J. (2004). The possibilities of Basil Bernstein. In J. Muller, B. Davies, & A. M. Morais (Eds.), *Reading Bernstein, researching Bernstein* (pp. 1–12). London: RoutledgeFalmer.

Nagda, B. A., Gurin, P., & Lopez, G. E. (2003). Transformative pedagogy for democracy and social justice. *Race Ethnicity and Edcuation, 6*(2), 165–191.

Neumayer, E. (2003). *Weak versus strong sustainability: Exploring the limits of two opposing paradigms*. Bodmin: MPG Books Ltd.

Neves, I., & Morais, A. M. (2001). Texts and contexts in educational systems: Studies of recontextualising spaces. In A. M. Morais, I. Neves, B. Davies, & H. Daniels (Eds.), *Towards a sociology of pedagogy. The contribution of Basil Bernstein to research* (pp. 223–249). New York: Peter Lang.

Neves, I., Morais, A. M., & Afonso, M. (2004). Teacher training contexts. Study of specific sociological characteristics. In J. Muller, B. Davies, & A. M. Morais (Eds.), *Reading Bernstein, researching Bernstein* (pp. 168–186). London: Routledge.

Nisbett, R. E., Peng, K., Choi, I., & Norenzayan, A. (2001). Culture and systems of thought: Holistic versus analytic cognition. *Psychological Review, 108*(2), 291–310.

OECD. (2005). *The definition and selection of key competencies*. Paris: OECD.

Oetke, C. (1996). Ancient Indian logic as a theory of non-monotonic reasoning. *Journal of Indian Philosophy, 24*(5), 447–539.

Parsons, T. (1951, 1991). *The social system*. London: Routledge.

Piaget, J. (1962). *The language and thought of the child*. London: Routledge & Kegan Paul.

Pigozzi, M. J. (2008). Towards an index of quality education. *Paper prepared for the International Working Group on Education (IWGE)*. http://www.iiep.unesco.org/fileadmin/user_upload/CapDev_Networking/pdf/2008/pigozzi_IWGE_GlenCoveJune2008.pdf.

Raschke, N. (2015). *Umweltbildung in China. Explorative Studien an Grünen Schulen*. Münster: Geographiedidaktische Forschungen.

Reich, K. (2007). Interactive constructivism in education. *Education and Culture, 23*(1), 7–26.

Reich, K. (2010). *Systemisch-konstruktivistische Pädagogik - Einführung in die Grundlagen einer interaktionistisch-konstruktivistischen Pädagogik*. Weinheim: Beltz Verlag.

Reuber, P. (2012). *Politische Geographie*. Münster: Schöningh UTB.
Rhode-Jüchtern, T. (1995). *Raum als Text*. Wien: Perspektiven einer konstruktiven Erdkunde.
Rothermund, D. (2008). *Indien. Aufstieg einer asiatischen Weltmacht*. Bonn: C.H. Beck.
Rousseau, J.-J. (1979). *Emil or on education*. New York: Basic Books.
Rychen, D. S. (2008). OECD Referenzrahmen für Schlüsselkompetenzen — ein Überblick. Kompetenzen der Bildung für nachhaltige Entwicklung. In I. Bormann & G. Haan (Eds.), *Kompetenzen der Bildung für nachhaltige Entwicklung. Operationalisierung, Messung, Rahmenbedingungen, Befunde* (pp. 15–22). Wiesbaden: VS Verlag für Sozialwissenschaften. https://doi.org/10.1007/978-3-531-90832-8_3.
Sadovnik, A. (2011a). *Sociology of education. A critical reader*. New York: Routeledge.
Sadovnik, A. (2011b). Theory and research in the sociology of education. In A. R. Sadovnik (Ed.), *Sociology of education. A critical reader*. Routeledge: New York.
Sadovnik, A. R. (1991). Basil Bernstein's theory of pedagogic practice: A structuralist approach. *Sociology of Education, 64*(1, Special Issue on Sociology of the Curriculum), 48–63.
Schockemöhle, J. (2009). *Außerschulisches regionales Lernen als Bildungsstrategie für eine nachhaltige Entwicklung. Entwicklung und Evaluierung des Konzeptes "Regionales Lernen 21+"*. Münster: Geographiedidaktische Forschungen.
Schrüfer, G. (2010). Förderung interkultureller Kompetenz im Geographieunterricht - Ein Beitrag zum Globalen Lernen. In G. Schrüfer & I. Schwarz (Eds.), *Globales Lernen* (pp. 101–110). Münster: Waxmann.
Schultz, T. W. (1989). Investing in people: Schooling in low income countries. *Economics of Education Review, 8*(3), 219–223. https://doi.org/10.1016/0272-7757(82)90001-2.
Schweisfurth, M. (2011). Learner-centred education in developing country contexts: From solution to problem? *International Journal of Educational Development, 31*(5), 425–432. https://doi.org/10.1016/j.ijedudev.2011.03.005.
Sen, A. (1999). *Development as freedom*. Oxford: Oxford University Press.
Sen, A. (2005). *The argumentative Indian*. Noida: Penguin.
Sen, A. (2009). *The idea of justice*. London: Penguin.
Sertl, M., & Leufer, N. (2012). Bernsteins Theorie der pädagogischen Codes und des pädagogischen Diskurses: eine Zusammenschau (Bernstein's theory of pedagogic codes and pedagogic discourse: an overview). In M. Sertl & N. Leufer (Eds.), *Zur Soziologie des Unterrichts: Arbeiten mit Basil Bernsteins Theorie des pädagogischen Diskurses (Sociology of teaching: working with Basil Bernstein's theory of pedagogic discourse)* (pp. 15–62). Weinheim: Belt.
Spitzer, M. (2007). *Lernen. Gehirnforschung und die Schule des Lebens*. Berlin: Spektrum Akademischer Verlag.
Spivak, G. (2008). *Can the subaltern speak?*. Wien: Turia + Kan.
Sriprakash, A. (2010). Child-centered education and the promise of democratic learning: Pedagogic messages in rural Indian primary schools. *International Journal of Educational Development, 30*(3), 297–304.
Sriprakash, A. (2011). The contributions of Bernstein's sociology to education development research. *British Journal of Sociology of Education, 32*(4), 521–539.
Sriprakash, A. (2012). *Pedagogies for development: The politics and practice of child-centred education in India*. New York: Springer.
Steiner, R. (2011). *Kompetenzorientierte Lehrer/innenbildung für Bildung für Nachhaltige Entwicklung. Kompetenzmodell, Fallstudien und Empfehlungen*. Münster: MV-Verlag.
Thapan, M. (1984). *Life at school: An ethnographic study*. New Delhi: Oxford University Press.
Thapan, M. (2014). *Ethnographies of schooling in contemporary India*. Delhi: Sage.
Thompson, P. (2013). Learner-centred education and 'cultural translation'. *International Journal of Educational Development, 33*(1), 48–58. https://doi.org/10.1016/j.ijedudev.2012.02.009.
Tikly, L. (2004). Education and the new imperialism. *Comparative Education, 40*(2), 173–198.
Tikly, L., & Barrett, A. M. (2011). Social justice, capabilities and the quality of education in low income countries. *International Journal of Educational Development, 31*(1), 3–14. https://doi.org/10.1016/j.ijedudev.2010.06.001.

Tilbury, D. (2007). Asia-Pacific contributions to the UN decade of education for sustainable development. *Journal for Education for Sustainable Development, 1*, 133–141.
Tilbury, D. (2011). *Education for sustainable development. An expert review of processes and learning*. Paris: UNESCO.
Toulmin, S. E. (1996). *The uses of argument*. New York: Cambridge University Press.
UNESCO. (2002). *Education for sustainability. From Rio to Johannesburg: Lessons learnt from a decade of commitment*. Paris: UNESCO.
UNESCO. (2005). *United Nations Decade of education for sustainable development (2005–2014): International implementation scheme*. Paris: UNESCO.
UNESCO. (2006). *Water. A shared responsibility. The United Nations World Water Development Report 2*. Barcelona: UNESCO, Berghahn Books.
UNESCO. (2009a). *Review of contexts and structures for education for sustainable development 2009*. Paris: UNESCO.
UNESCO. (2009b). *UNESCO World Conference on ESD: Bonn Declaration.*
UNESCO. (2011). *Education for sustainable development. An expert review of processes and learning*. Paris: UNESCO.
UNESCO. (2012a). *EFA Global Monitoring Report. Youth and skills—Putting education to work.*
UNESCO. (2012b). *Shaping the education of tomorrow. 2012 report on the UN decade of education for sustainable development*. Paris: UNESCO.
UNESCO, & UNEP. (1977). Tbilisi Declaration.
United Nations Conference on Sustainable Development. (2012). The future we want. http://daccess-dds-ny.un.org/doc/UNDOC/GEN/N11/476/10/PDF/N1147610.pdf?OpenElement. Accessed March 10, 2013.
United Nations General Assembly. (1948). The universal declaration of human rights. Article 26.
Vavrus, F. (2009). The cultural politics of constructivist pedagogies: Teacher education reform in the United Republic of Tanzania. *International Journal of Educational Development, 29*(3), 303–311. https://doi.org/10.1016/j.ijedudev.2008.05.002.
Vester, F. (2002). *Unsere Welt - ein vernetztes System*. München: Deutscher Taschenbuchverlag.
Voß, J.-P., Bauknecht, D., & Kemp, R. (2006). *Reflexive governance for sustainable development*. Edward Elgar Publishing.
Vygotsky, L. S. (1962). *Thought and language*. Cambridge, Mass.: MIT Press.
Vygotsky, L. S. (1978). *Mind in society. The development of higher psychological processes*. Cambridge, MA: Harvard University Press.
Wals, A. E. J. (2009). A mid-DESD review. *Journal of Education for Sustainable Development, 3*(2), 195–204. https://doi.org/10.1177/097340820900300216.
Weber, M. (2002). *Wirtschaft und Gesellschaft. Grundriss der verstehenden Soziologie*. Tübingen: J.C.B. Mohr (Paul Siebeck).
Werlen, B. (1999). *Zur Ontologie von Gesellschaft und Raum. Sozialgeographie alltäglicher Regionalisierungen* (Bd. 1). [*On the ontology of society and space. Social geography of everyday regionalizations* (Vol. 1)]. Stuttgart: Franz Steiner Verlag.
Wohlrapp, H. (2006). Was heißt und zu welchem Ende sollte Argumentationsforschung betrieben werden? In E. Grundler & R. Vogt (Eds.), *Argumentieren in der Schule und Hochschule. Interdisziplinäre Studien* (pp. 29–40). Tübingen: Stauffenburg Verlag Brigitte Narrr.
World Commission on Environment and Development. (1987). Report of the World Commission on Environment and Development: Our common future. http://www.un-documents.net/our-common-future.pdf. Accessed March 03, 2014.
Wuttke, E. (2005). *Unterrichtskommunikation und Wissenserwerb. Zum Einfluss von Kummunikation auf den Prozess der Wissensgenerierung*. Frankfurt a. M: Peter Lang.
Young, M. F. D. (1971). *Knowledge and control: New directions for the sociology of education*. London: Collier. Macmillan.

Chapter 3
Education for Sustainable Development and Argumentation

Abstract This chapter elaborates on my approach of transformative pedagogic practice for Education for Sustainable Development (ESD). I explore how ESD principles can contribute to overcome memorization and the fragmentation of knowledge by postulating network and critical thinking, student orientation and environmental action. I argue that the development of critical consciousness (Freire 1996) through the promotion of argumentation skills is one important way to work toward reflective decision-making on sustainable development. In this chapter, I develop a didactic framework for the implementation of ESD in geography education by linking the principles of ESD with an approach that emphasizes argumentation (Toulmin 2010). Firstly, I critically review the political and conceptual development of sustainable development and ESD. Secondly, I discuss the societal relevance of argumentation and its role in the Indian subcontinent. Furthermore, I examine the didactic relevance of argumentation skills and their promotion in classrooms before merging this approach to the concept of ESD for a didactic framework. Finally, I develop a refined definition of ESD as "pedagogic practice that promotes critical consciousness through argumentation to empower decision-making on sustainable natural resource use by facilitating learner-centered, problem-posing, and network-thinking teaching approaches."

Education for Sustainable Development (ESD) is a transnational educational policy that has a normative and comprehensive approach, in which principles and purpose need to be critically reflected for implementation in pedagogic practice. ESD promotes critical awareness on access to and use of natural resources; ecological and economic processes are considered to be integrated with a social dimension and under the norm of sustainability (UNESCO 2011). As ESD postulates network and critical thinking as well as student orientation and participation, I argue that the linkage between the principles of ESD and the approach of argumentation (Budke et al. 2010) can form a didactic framework for the implementation of ESD in geography education. This study is built upon the argument that the development of *critical consciousness* (Freire 1996) through the promotion of argumentation skills in geography education is one important way to work toward reflective decision-making on sus-

S. Leder, *Transformative Pedagogic Practice*, Education for Sustainability,
https://doi.org/10.1007/978-981-13-2369-0_3

tainable development. ESD's objectives of environmental awareness and action focus on skill development and thus represent a competence-oriented model of instruction (Bernstein 1975). Through argumentation skills, different interests of and multiple conflicting perspectives on resource access and use, especially water, can be critically questioned and understood (cf. Budke and Meyer 2015). On this basis, one's own environmental positions can be determined, which can potentially lead to environmental action (Mandl and Gerstenmaier 2000). Education, in particular geography education, plays a significant role not only in raising awareness on environmental, social, and economic implications of resource depletion, but also in developing skills, especially argumentation skills, to reason for environmental action and lifestyles for oneself and toward others in society.

The following subchapter introduces the political and conceptual development of sustainable development and ESD and links it to the didactic approach of promoting argumentation skills. This link can be used to display opportunities for translating the transformative objectives of ESD in pedagogic practices in geography classes. At first, the definition, objectives, development, and implementation studies of ESD are discussed and critically reviewed. Secondly, the societal relevance of argumentation and its role in the Indian subcontinent are discussed. Furthermore, the didactic relevance of argumentation skills and their promotion in classrooms is examined, before merging this approach to the concept of ESD as didactic framework for this study.

3.1 Education for Sustainable Development

In the era of globalization, education is not only marked by current societal, political, and economic contexts within one specific country (Hoffmann et al. 2012), but also by global processes. A prominent outcome that is politically driven on a global stage is the UN Decade *Education for Sustainable Development* (ESD). This new paradigm appeared in the 1990s at international conferences in which political leaders agreed upon sustainable development through education as a new evolving path. UN member states agreed to implement this educational policy, and, for example, to revise educational curricula to comply with it. Hence, this transnational policy exercises public and communicative power on educational processes. This study is aimed at identifying these effects on pedagogic practice in India, specifically in the region of Pune.

The implementation of the transnational policy of ESD in local educational contexts underlies great challenges, which only a few studies have investigated (Mulà and Tilbury 2009). Several empirical studies examine classroom teaching but concrete didactic consequences, theoretically and empirically grounded, are not sufficiently derived for ESD teaching (Fien and Maclean 2000: 37). At the international level, there is no agreement on content and methods for ESD, especially not in classroom teaching (Wals 2009: 197). Studies on ESD in the Asian context focus on extracurricular learning and the institutional level rather than formal classroom education (Tilbury 2007). While classroom teaching has been intensively researched in the

German context subject-wise (Leng 2009: 11), studies on ESD have not been recognized internationally. Educational studies on conceptual frameworks for classroom research in ESD, teacher education, and the reorientation of curricula are rare and focus on best practice examples. Furthermore, research for the identification of needs, challenges, and strategies for ESD has yet to be carried out. A lack of a commonly agreed theoretical framework also hinders researchers to refer to each other, as the theoretical basis is either lacking or divergent for different approaches. Therefore, the implementation of ESD in classrooms needs to be investigated empirically, as this study will do in geography classes in India.

To proceed from case studies to a lasting and widespread change in education, Benedict (1999) suggests a systemic approach to spread environmental education to a large number of schools and change classroom practice. A systemic approach embraces a commitment of educational authorities, curriculum revision, a focus on competences and cooperation. Such a systemic approach needs to be scientifically examined, monitored, and evaluated in depth, which this study aims to do for the Indian context. In order to identify its developmental and educational premise, the political development of ESD is reviewed in this subchapter.

3.1.1 The Sustainable Development Paradigm

Resource access and management, particularly concerning water, are widely considered as social problems, rather than as problems of water scarcity (UNESCO 2006). Meadows et al. (1972) already depicted more than four decades ago *The Limits to Growth* of economy and population through system analysis and inflamed a public debate on global environmental problems, which was carried forward by the *Club of Rome.*[1] The realization that human beings cannot solve all problems technically and that resources are limited was an international turning point in the postwar era. The growing environmental consciousness sought solutions to address this alarming prognosis. In 1987, the term sustainable development[2] evolved and was defined by the World Commission on Environment and Development (WCED) in the Brundtland Report *Our Common Future* as a societal paradigm shift, and a *global agenda for change*: "Sustainable development is development that meets the needs

[1] The Club of Rome is an affiliation of experts of different disciplines and countries concerned about sustainable development and the protection of ecosystems. Their study "The limits to growth" (1972) received great publicity and initiated debates on the future development of the world.

[2] The term sustainability originated in forestry in the eighteenth century, when a prince from the region of Baden in Germany ordered to log only as many trees as could grow back within a specific timeframe. This idea served to sustain the existence of his folk, which mainly lived on wood export Brunotte (2005). In ecological terms, the "management rules of sustainability" by Haubrich et al. (2007) states that the use and consumption rate should be lower than the regeneration rate, and the emission rate should not be higher than the assimilation rate. Over a period of two centuries, the term has gained in scale, linking local to global levels, and now includes a social and cultural dimension to the ecological and economic equilibrium. Yet, the original idea of keeping the balance of resource use and maintenance is still valid and central today.

of the present without compromising the ability of future generations to meet their own needs." (World Commission on Environment and Development 1987, Chap. 2, IV. 1). The claim is that the present resource use and consumption should be distributed and dealt with responsibly by the present generation, especially by those in power, to enable present societies and future generations to have their share of resources. Hence, sustainable development has a temporal and a spatial dimension: Intra-generational equity refers to social and spatial differences within a generation, and intergenerational equity calls for justice between the present and future generations (World Commission on Environment and Development 1987). The Brundtland Report demonstrates the interdependence of environmental, economic, and social development by relating north–south disparities and poverty to environmental problems.

Declarations for implementation that were oriented along principles of sustainable development claim to be not only economic, but also target "education and social commitment" strategies (Haubrich et al. 2007). These so-called strategies for sustainable development present different approaches to ESD. "Education and social commitment" is meant to encourage discussing complex causal relationships and values, thus raising environmental awareness and action that are changing lifestyles toward environment-friendly consumer behavior. Thus, "education and social commitment" can be seen as a bridging function or as the medium, through which principles of efficiency, consistency, permanency, and sufficiency strategies can be discussed and ultimately achieved (Fig. 3.1).

Until today, theoretical frameworks as well as empirical approaches and methods for the investigation and evaluation of sustainable development have been fragmented and controversial. This has led to studies from different disciplines in diverse cultural contexts under various objectives and conditions. Sustainable development's vague framework makes it difficult to operationalize. Therefore, the term is often used as rhetoric without action, or, even worse, misused as a label for actions counterproductive to sustainability ("greenwashing").

Sustainability can broadly be classified into two economic paradigms, *weak* and *strong sustainability*. Weak sustainability suggests that resources are substitutable and man-made capital can counteract the loss of resources. This interpretation can easily allow wider "greenwashing" practices. In strong sustainability, resources are seen as non-substitutable, which means that nature should be protected (Neumayer 2003: 1). These two paradigms can also be classified as *biocentric* and *anthropocentric* (Haubrich 2007: 29). There also are conflicting concepts of sustainability with respect to the dimensions of sustainability. These are shortly introduced in the following section before explaining which concept is used for the application to geography teaching in this study.

The three pillars of sustainability (Ekins 2000) are comprised of ecological balance, social justice, and economic prosperity. In later concepts, these three dimensions were put into reciprocal dependency visualized in the sustainability triangle. Ecological (or here used synonymously: environmental) sustainability refers to the maintenance of natural systems and processes, for example, the water cycle, the climate system, or specific ecosystems (Baker 2006: 26). Economic sustainability

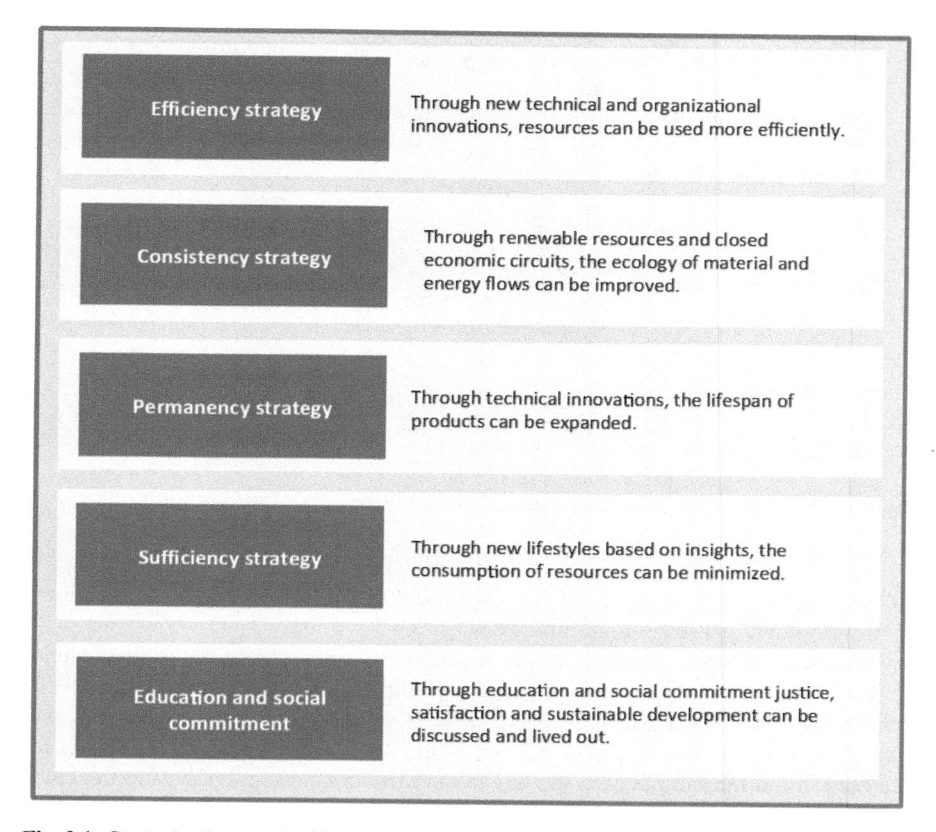

Fig. 3.1 Strategies for sustainable development (Haubrich 2007: 245)

refers to the equilibrium of economic growth through the use of resources and environment preservation (Ekins 2000). Social sustainability includes values, institutions (political, religious, scientific, etc), people and their culture. To strive toward sustainable development, these three dimensions are interrelated and need to be balanced (Fig. 3.2).

Besides striving for ecological balance, social justice, and economic prosperity, in later concepts, Hawkes (2001) includes *culture* as a fourth pillar of sustainable development. He argues that through culture we "make sense of our existence and the environment we inhabit; find common expressions of our values and needs; meet the challenges presented by our continued stewardship of the planet." The idea is to bring the concept closer to communities and their identities, meanings, and beliefs, as well as their way of life—that is customs, faiths, and conventions. Culture encompasses "the processes and mediums through which we develop, receive and transmit […] values and aspirations; the tangible and intangible manifestations of these values and aspirations in the real world" (Hawkes 2001: 13). The role of

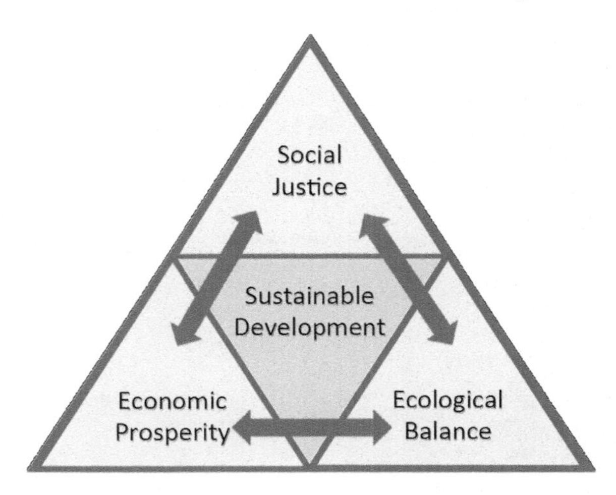

Fig. 3.2 Triangle of sustainable development (UNESCO 2002)

culture in sustainable development highlights cultural perceptions and explanations for development pathways and impacts.

Other approaches include *politics* as an additional concept relevant to sustainable development. Voß et al. (2006: 4) have a procedural perspective on sustainable development and focus on *reflexive governance* as a "societal problem treatment." They state that the value of sustainable development is the problem of framing interconnectedness at different time and spatial scales. Criticizing current policy approaches and planning, they develop a concept questioning the "concepts, practices and institutions by which societal development is governed" (Voß et al. 2006: 4) with reflexive governance.

Baccini (2007) criticizes that the "'magic triangle' […] is not more than a didactically well designed reminder of the framing of SD. It does neither supply the tools to tackle the problems nor does it help to make sound decision with regard to projects in the direction of SD." He further argues that ethical standards need to be pervaded hermeneutically to become scientifically validated. Difficulties with the term sustainable development are different interpretations and concepts, such as strong and weak sustainability, the inflationary use of the term for own and possibly not at all sustainable interests (cf. Baccini 2007). Baker (2006) states a lack of clarity and consistency and therefore the ambivalent use of the term. However, this can also be positive in helping groups with contradictory interests achieve general agreement with which concrete policies can develop, for example, during high-level policy negotiations at the UNCED (Baker 2006: 26, 27).

Nevertheless, the sustainability triangle is didactically useful for this study to focus on the reciprocal dependency of economic, social, and environmental aspects of sustainability. To make the concept of sustainability approachable to students for teaching in this study, the three dimensions of sustainability are used in this analysis and intervention study. Hence, to simplify the dimensions into three, the cultural and political pillar is subsumed, and not excluded, within the social dimension of sustainability. This also links to the interdisciplinary nature of the subject of geography as it is divided into human and physical geography. This thematic framing of the complex concept sustainable development will be used for structuring the empirical investigations on pedagogic practice in this study and is further elaborated in Sect. 3.3. In the following, the political development of Education for Sustainable Development is reviewed.

3.1.2 Political Development of Education for Sustainable Development

ESD came about as the reaction to the paradigm change toward sustainable development within international politics (cf. Table 3.1). In the 1970s, growing international political concern over environmental degradation and the depletion of natural resources initiated international conferences, commissions, and reports on the role of education in respect to the environment. The First Intergovernmental Conference on Environmental Education by the UNESCO and UNEP took place with 66 delegates from member states in Tbilisi, Georgia, in 1977. The Tbilisi Declaration subsumed the important role of environment education "to develop a sense of responsibility and solidarity among countries and regions as the foundation for a new international order which will guarantee the conservation and improvement of the environment" (UNESCO and UNEP 1977). One excerpt of The Tbilisi Declaration describes the objective of environmental education as:

> Environmental education should provide the necessary knowledge for interpretation of the complex phenomena that shape the environment, encourage those ethical, economic, and aesthetic values which, constituting the basis of self-discipline, will further the development of conduct compatible with the preservation and improvement of the environment. It should also provide a wide range of practical skills required in the devising and application of effective solutions to environmental problems. (UNESCO and UNEP 1977)

At the UN Conference on Environment and Development (UNCED) held in Rio de Janeiro in 1992, 172 countries—India and Germany included—agreed on education as an important pillar in the new guiding principle of "sustainable development." In the key action program "Agenda 21" that promotes education, public awareness, and training, it is seen as a "means of implementation" (UNCED 1992, Agenda 21, §36.3) of the now required strengthening of sustainable lifestyles through responsible resource usage. Governments worldwide are being hereby encouraged to promote public participation in decision-making by establishing knowledge, values, and skills on sustainability as integral parts of their respective national school curricula:

> A thorough review of curricula should be undertaken to ensure a multidisciplinary approach, with environment and development issues and their socio-cultural and demographic aspects and linkages [being taken into consideration]. (UNCED 1992, Agenda 21, Paragraph 36.5b)

Table 3.1 Key events in the evolution of ESD in global politics [altered from Leder and Bharucha (2015)]

Year	Conference/Report/Policy	Focus	Details
1972	UN Conference on the Human Environment in Stockholm	1st UN Conference on environment, beginning of international environment politics, resolution to found the UNEP; Inclusion of Environmental Education as one of the 26 principles of the Stockholm Declaration	Principle 19: Education in environmental matters, for the younger generation as well as adults, giving due consideration to the underprivileged, is essential in order to broaden the basis for an enlightened opinion and responsible conduct by individuals, enterprises and communities in protecting and improving the environment in its full human dimension. It is also essential that mass media of communications avoid contributing to the deterioration of the environment, but, on the contrary, disseminates information of an educational nature on the need to project and improve the environment in order to enable man to develop in every respect
1977	UNESCO/UNEP: 1st Intergovernmental Conference on "Environmental Education" in Tiflis	Substantiation of objectives of Environmental Education	Tbilisi Declaration defining the role, objectives and principles of interdisciplinary Environment Education; responsibility and solidarity for the survival and well-being of all people
1987	UN World Commission on Environment and Development (WCED): "Brundtland Report Our Common Future"	"Sustainable Development" is the new guiding principle of global environment and development policy	Definition of Sustainable Development as development that meets the needs of the present without compromising the ability of future generations to meet their own needs." Taken as the basis for the 1992 UNCED conference in Rio de Janeiro and for the adoption of Agenda 21

(continued)

Table 3.1 (continued)

Year	Conference/Report/Policy	Focus	Details
1990	Jomtien Conference	"Education for All"	Targets to be reached by 2000: – Universal access to learning – A focus on equity – Emphasis on learning outcomes – Broadening the means and scope of basic education – Enhancing the environment for learning – Strengthening partnerships
1992	UN Conference on Environment and Development (UNCED) in Rio de Janeiro	Agenda 21 is passed and the new guiding principle is Education for Sustainable Development	Initiation of global environmental politics, reorientation of education toward Sustainable Development
2000	UN Millennium Development Goals (MDGs)	Achieving universal primary education and ensuring environmental sustainability	Eight targets focusing on reducing the gender gap in literacy, biodiversity loss and environmental resources depletion through country policy programs integrating principles of Sustainable Development
2002	World Summit on Sustainable Development (WSSD) in Johannesburg	Decision for the UN Decade of "Education for Sustainable Development" (UNDESD) from 2005–2014	International Implementation Scheme for the UNDESD; Each year a topic, e.g., 2008: Water, 2009: Energy, 2011: City
2012	UN Conference on Sustainable Development (UNCSD) in Rio de Janeiro	Document: "The Future We Want"	Reaffirmation of the right to education and universal access to primary education
2015	UN Post-2015 Agenda	Sustainable Development Goal (SDG) 4 on quality education	Development of a global development framework beyond the MDGs' target date of 2015; Goal 4: Ensure inclusive and equitable quality education and promote lifelong learning opportunities for all

At the UN Johannesburg Summit in 2002, ESD was concretized in the Johannesburg Plan of Implementation. The UN Decade Education for Sustainable Development (DESD) was adopted in the resolution 57/254 at the UN General Assembly meeting held in December 2002. The UNESCO defines the objective of the UN Decade ESD from 2005 to 2014 in the following way:

> Everyone has the opportunity to benefit from quality education and learn the values, behaviour and lifestyles required for a sustainable future and for positive societal transformation. (UNESCO 2005: 6).

In the international debate around ESD, the objective is not only access to education, which was the focus of the policy of "Education for All" (UNESCO 2012) or the Millennium Development Goals (MDGs). ESD primarily focuses on the quality of education, and values and behavior that contribute to sustainable development are promoted (UNESCO 2002). Despite these enthusiastic objectives, the *Review of contexts and structures for Education for Sustainable Development* by the UNESCO (2009a) stated:

> a considerable number of countries have committed themselves to integrating ESD into formal education, that there is a remarkable presence of ESD in national policy documents, and that many countries have put in place national coordinating bodies for the DESD. Challenges include the regional unevenness of ESD implementation, awareness of ESD in the wider educational community and in the general public, the reorientation of curricula, and the availability of sufficient budgetary provisions for ESD measures. (UNESCO 2009a)

Bagoly-Simó (2014b) demonstrated in a comparative textbook analysis of the integration of ESD in Bavaria, Mexico, and Romania that approaches vary from ESD being an add-on or organically integrated within subject curricula. Notable is the top-down approach of Mexico, where an interdisciplinary workgroup sets sustainability as a common framework and provides methods for implementation within and across subjects. The positive evaluation of integrating ESD highlights the need of political importance ESD is given at a national level for wide-ranging curriculum integration.

In 2009, the participants of the UNESCO World Conference on Education for Sustainable Development in Bonn adopted the *Bonn Declaration* promoting comprehensive policies, committees, and networks to reorient and strengthen formal and informal education toward sustainability. To integrate sustainable development themes, a systemic approach through the development of effective pedagogical approaches, teacher education, teaching practice, curricula, and learning materials was supported at the practice level. The *Bonn Declaration* encourages:

> a new direction for education and learning [...] based on values, principles and practices necessary to respond effectively to current and future challenges [...emphasizing...] creative and critical approaches, long-term thinking, innovation and empowerment for dealing with uncertainty, and for solving complex problems. (UNESCO 2009b)

In 2012, 20 years after the UNCED in Rio de Janeiro, 192 member state representatives—again including India and Germany—reaffirmed their commitment to the idea of ESD in Agenda 21 beyond the UNDESD, specifically in the form of a non-binding document entitled "The Future We Want." Herein, it was suggested "to prepare people to pursue sustainable development, including through enhanced teacher training [and] the development of curricula around sustainability" (United Nations Conference on Sustainable Development 2012: 230). Currently, the post-2015 agenda with the *Sustainable Development Goal* (SDG) 4 further focuses on the promotion of quality education, whereby the role of Education for Sustainable Development is not emphasized clearly. Nevertheless, prior and ongoing political support for ESD has helped develop the idea into a powerful tool with which to critically challenge educational methods and content in classroom teaching in practice, specifically for the benefit of school education in a globalized, environmentally degrading world that is at present experiencing rapidly changing life conditions.

3.1.3 Objectives of Education for Sustainable Development

The objectives of Education for Sustainable Development (ESD) are broadly framed by UNESCO (2005), but can be substantiated through educational theories and related approaches such as environmental education and global learning. UNESCO (2005) states that ESD postulates environmental knowledge, awareness, and action through education explicitly addressing the global challenge of environmental degradation. Skills to identify environmental problems and to offer opportunities to participate and solve these problems should be promoted (UNESCO 2005). Education can contribute to a better understanding of complex relations of controversial topics and has an effect on personal lifestyle decisions, critically considering the environmental impact. Furthermore, education should provide students not only with skills to take a lead in environmental decision-making, but also with skills to adjust to rapid changes and new requirements, which a complex and globalized world demands (UNESCO 2005). Schools should prepare students for living in a globalized world with environmental problems and participating in a socially and environmentally responsible way in society. Hence, education should not only focus on the teaching of factual knowledge, but also on the promotion of norms, values, and skills, which enable citizens to shape their environments (UNESCO 2005: 7). Textual and didactic adjustments of existing curricula are meant to enable students to participate in society (cf. Sect. 3.1.2).

ESD developed from environmental education (UNESCO 2009a: 28, 29; Bonnett 1999). ESD includes besides economic and environmental aspects also social aspects and interconnects ecological perspectives with broader developmental challenges from local to global scales (Bonnett 1999). Similar to environmental education, ESD includes a normative compound as value education. Environmental education (EE) focuses on raising awareness about nature and action toward environmental pollution and resource degradation (UNESCO 2005). While environmental education looks mainly at educating students on nature and their environment, ESD emphasizes social, political, and cultural perspectives on economic and environmental development and also embraces a governance perspective on current environmental conflicts. However, even if sustainability is not a prevalent paradigm in environmental education, it can be implied indirectly. This was demonstrated in the explorative study by Raschke (2015) who depicted the local, regional, and national concepts and practices of environmental education in China, which are uninfluenced by international sustainability and ESD debates. She used a phenomenological inductive approach to study the notions of environmental education in Chinese green schools. The green school program is the Chinese state program to implement and promote environmental education in Chinese schools, as well as a symbol for the state engagement of China's sustainability and environment activities (Raschke 2015: 63). Raschke (2015: 279) observed a close link between culture and nature that is inherent of Chinese culture and thought and gave some indication of the relevant cultural interpretation of environmental education, which implies sustainability. In contrast, ESD more explicitly formulates optimistic perspectives focusing on opportunities for social, cultural, and economic sustainability.

Apart from the close relation to environment education, ESD also includes aspects of global learning as it sets a new focus on social and developmental issues in north–south relations in the context of globalization. Global learning teaches causes, dynamics, and impacts of globalization as well as opportunities to shape this process and to prepare for a life in a globalized world and thus combines the two fields of ecological and developmental education (Schrüfer 2010). Therefore, ESD merges environmental education and global learning and also leads to a major conceptual shift in both fields dealing with the reciprocity of environmental and developmental learning (de Haan et al. 2010: 202).

While ESD functions as a political and economic instrument to achieve the concrete objectives of sustainable development (Hopkins and McKeown 2002), it also aims at promoting skills in enhancing individual self-determination and the ability to become an agent for change (De Haan 2006: 5). UNESCO (2002: 21) stressed promoting "learning how to learn, how to analyze and solve complex problems, how to think creatively and critically about the future, how to anticipate and make our own histories." ESD does not refer to specific skills for concrete situations or behavior, but promotes skills for a reflective, responsible participation in a sustainable society. A multitude of studies and official international and national documents have defined skills and methods, didactic principles, and educational frameworks for ESD, but there is a continuing "lack of clarity" on ESD principles, which Huckle (1993) already stated more than two decades ago. The *International Implementation Scheme* for the Decade of Education for Sustainable Development claims "a variety of pedagogical techniques that promote participatory learning and higher-order thinking skills" (UNESCO 2005: 31) and aims for "changes in behavior that will create a more sustainable future" (UNESCO 2005: 6). Although empirical evidence on teaching methods is not available, Tilbury (2011: 36–38) cites "commonly adopted ESD pedagogies in higher education" in the UNESCO ESD *Expert Review of Processes and Learning*. In a survey on university educators in the UK, Cotton and Winter (2010) identified that active learning techniques are essential for ESD, e.g., including group discussions, debates, critical reading and writing, and problem-based learning. Tilbury (2011: 25) found a consensus on *critical reflective thinking* and described an educational shift proposed by ESD from "sending messages" to "dialogue, negotiation and action." According to the UNESCO (2005), ESD learning refers to asking critical questions, clarifying one's own values, envisioning more positive and sustainable futures, thinking systematically, responding through applied learning, and exploring the dialectic between tradition and innovation (Tilbury 2011).

Teaching methods in the context of ESD should be student-centered and focus on constructive and cooperative learning principles, address content holistically, and connect to the students' everyday life. Concerning teaching methodology, the focus is on skill development that promotes participation in society, including primarily cooperative learning, discussions, and creativity to solve problems. In the German context, de Haan and Harenberg (1999) coined the term *Gestaltungskompetenz*, literally translated as "shaping competence," or in the *Lucerne Declaration* as "action competence." The German National Commission for Education (BLK-21) divides *Gestaltungskompetenz* into 12 competences including cooperative learning, showing solidarity and empathy as well as anticipatory thinking and action. These competences are oriented to the competence categories of the OECD reference framework (Rychen 2008). Didactic principles stated are, among others, systemic and critical thinking as well as action and communication skills, which are meant to enable the participation in societal decision processes. Hence, the understanding and the willingness of students to act on complex topics should be promoted (UNESCO 2005). In a two-stage Delphi study, ten German educational stakeholders were asked which professional competencies teachers should have to integrate ESD within the German educational system (Hellberg-Rode et al. 2014). The results confirm the competencies mentioned in UNESCO and national documents, whereas the definitions and operationalization of the terminology remain unclear and are not theoretically or empirically grounded.

The principles and skills to be promoted through ESD can be linked to several educational theories. ESD includes a constructive understanding of learning, in which knowledge is constructed and adapted to existing knowledge. This understanding can be grounded with Piaget (1962), who states that knowledge is integrated into existing mental models (assimilation) and changes mental models (accommodation). The strong participatory and communicative component of ESD can be linked to Vygotsky's *model of social constructivism* (1962), which stressed that learning takes place in a social context and that the social environment is especially important for learning. The role of the teacher is to create social contexts for students to learn actively. According to Vygotsky (1978), the zone of proximal development "is the distance between the actual development level as determined by independent problem solving and the level of potential development." This centers learning around instructional scaffolding to language and problem learning, which is achieved through enabling children to learn beyond their abilities with the support of a more experienced teacher. Scaffolding can also be promoted by ESD, for example, when the relevance of environmental problems should be derived from local to global scales. In addition to educational theories, scaffolding and social constructivism have also great support through psychological theories of learning and neurodidactic studies on brain-based learning (Spitzer 2007).

ESD's focus on contents and skills relevant to society can be linked to the German educationist Wolfgang Klafki (2013). He defined epochal-typical key problems ("epochaltypische Schlüsselprobleme"), which include peace, environment, and inequalities in society for meaningful teaching. The role of school education is the analysis of societal key problems and the mediation of key qualifications and orientation benchmarks, including spatial action competences. Because of its similar postulations, ESD aims at addressing epochal-typical key problems. Furthermore, Klafki defined basic competences, which a person with general education should have that also overlap with the postulations of ESD, such as the ability to criticize, to argue, to feel empathy, and to think in networks. Klafki (2013) combined material (educational content that students should know as it is important) and formal (education that is important for students) educational theory to *categorical education*. He defined five principles to decide whether contents are necessary to study: the meaning of the exemplary (local examples of global relevance), the present and the future, the structure of content, as well as the accessibility, the means to approach students. Applying these to ESD may help teachers to structure their teaching content.

Similarities with ESD can also be found in interactive constructivism (Reich 2007). This approach is based on Dewey (1916) in "Education and Democracy" and stresses the importance of "experience" and "learning by doing." These examples demonstrate that the didactic principles named in ESD are not new, but that these can be linked to existing educational theories. These linkages affirm the concept of ESD and underline its importance. Furthermore, this indicates that there is great scope to ground ESD to several theorems. This could result in interesting approaches to conceptualize ESD and develop theoretically informed empirical studies.

3.1.4 Critical Perspectives on Education for Sustainable Development

Both sustainable development and ESD are umbrella phrases, which are applied to various settings, including a range of fields such as environmental, nature, peace, and developmental education. Critiques of ESD range from those of a promising regulative idea to those of a blank compromise (Steiner 2011: 26). In fact, ESD is "poorly researched and weakly evidenced" (UNESCO 2011: 9). The unclear epistemological definition and therefore unclear distinction from environmental education cause confusion and misunderstanding particularly among teachers, which can "deprive Environmental Education of its educational value" (Malatesta and Camuffo 2007: 62). Similarly to sustainable development, ESD has political origins and was not developed from educational theory. Hence, the sources of legitimacy of education have categorically changed, and ESD is legitimized by politics rather than pedagogy and educational research (Hasse 2010: 45). Therefore, it was the international political agenda that pushed ESD forward at a global level although environment education has existed in educational theory for a long time (Rousseau 1979). This resulted in ministries developing the content and concepts for educators to implement (Hopkins and McKeown 2002: 14). Scientific research has to fill in these gaps subsequently and develop concepts, models, and measurements to ground ESD scientifically. This turns around the role of educational research—it aims at underlining educational policies with theoretical and empirical foundations, instead of provoking educational policies in the first place. In the following, critical accounts of ESD from different perspectives are depicted in detail.

The concept of ESD is criticized particularly for its globalist rhetoric, which overshadows the need for locally relevant educational approaches (Manteaw 2012). Whether and how ESD's global developmental alignment is transferable to diverse national, regional, and local levels leads to the question as to whether a normative ethical paradigm is assignable to all socio-cultural contexts. Based on a critical discourse analysis, Manteaw (2012) argued that ESD underlies a hegemonic educational discourse and a Western epistemology that ignores the needs and values of regional contexts. With reference to African countries, Manteaw concluded that the "interpretation of the concept of sustainable development from a 'globalist perspective' is not only theoretical and idealistic, but also an indication of [a] geopolitical power-play which tends to privilege hegemonic understandings of the concept" (Manteaw 2012: 381). Hence, there is a need to "re-conceptualize sustainable development by linking it to the lived experiences and the cultural knowledge systems" (Manteaw 2012: 381). Similarly to other educational reforms, studies of ESD have frequently stressed the importance of contextualizing ESD to the "region's unique social, ecological, economic and political challenges" (Manteaw 2012: 376). Manteaw's critical discourse analysis of ESD showed the high expectation of education as a means for social change, despite its lack of research and implementation in this regard (Manteaw 2012: 378). Linking teaching and learning to community epistemologies and locality-specific complex challenges strives "towards a pedagogical logic" (Manteaw 2012: 382). Particularly, the constructivist understanding of learning with student-centered and competence-based teaching methods contrasts the hierarchically structured classroom interaction that reproduces social inequality, as cited in the literature (Sriprakash 2010). This indicates that ESD needs to be culturally adapted to social and material conditions and institutional systems in India and elsewhere. Manteaw (2012) suggests experiential learning and place-based education "using local resources and knowledge systems to deconstruct local problems to explore alternative possibilities."

As Bonnett (1999) argued, sustainability is historically, geographically, and culturally local. Thus, he questioned ESD as a normative principle serving many countries as global guideline for stakeholders at the international level. He claimed that there are epistemological, semantic, and ethical difficulties with the concept of sustainability, and he questioned how sustainability can be defined while we have an "imperfect state of our current knowledge" (Bonnett 1999: 316) due to the complexity of the term itself. Bernstein (1990) criticized paradigms similar to ESD for their incommensurability, as they cannot be understood as it is talked about one term from different standpoints that are ideologically incompatible (cf. Moore 2013: 4).

Another critique is that the term Education "for" Sustainable Development prescribes a pre-determined mode of thinking and thus education is instrumentalized (Jickling 1994). Educational theory suggests that education is autotelic and not prescriptive; hence, education cannot have the objective to educate "for" a particular view. Barth (2011) argued that the objective to change behavior is not in line with educational theories. Instead, the objective should be to think critically, including the critical reflection of the term "sustainability." In OECD documents such as *The definition and selection of key competencies*, education is seen as tool for "adaptive qualities" that "individuals need in order to function well in society" (OECD 2005: 6). The demand-led approach is also seen critically by educationists (cf. Graupe and Krautz 2014) as people are meant to adapt to changing conditions, which contradicts educational theories that see the sense of education in being able to question and think critically. Competences can be seen as reductionist as they only measure explicit knowledge, leaving out implicit knowledge, which is fundamental for our understanding of the world and ourselves (Hoffmann et al. 2012). Furthermore, Graupe and Krautz (2014) argued that the term competences is neither theoretically nor empirically consolidated, but rather an approach to economize education and adjust educational systems to the interest of the economy and its lobbies, including the OECD. Practically, this should lead to a more scientific understanding and less political indoctrination leading to situations in which students measure water quality without understanding "cultural practices that cause and tolerate multiple forms of pollution" (Gruenewald 2004: 86).

On the basis of a Foucauldian analysis of environmental education, Gruenewald (2004) criticized that environment education is becoming absorbed, institutionalized, and disciplined within general education by being aligned to "conform to the norms, codes and routines of schools" (Gruenewald 2004). In Foucauldian terms, ESD underlies a *panopticon*, being assimilated to an external power and "in the context of the dominant educational discourse, environmental education is easily ignored and can be stripped of its revolutionary political content as it becomes constituted as disciplinary practice" (Gruenewald 2004). Therefore, it is necessary to consider ESD not only as an add-on to the curriculum, but make it in itself relevant. This is also reflected in the Indian discussion whether to infuse environmental education in all subjects and textbooks or whether to create a separate subject for EE (Hoffmann and Bharucha 2013). Despite some justified critiques to consider, with its political support ESD is a powerful tool to rethink contents and methods of environmental education for a variety of subjects. Among many, it offers one interpretation to encourage a transformative action-oriented approach for geography education, tackling environmental challenges of our time. In the following subchapter, the relevance of ESD to geography education is depicted.

3.1.5 Education for Sustainable Development in Geography Education

As the subject of geography integrates natural and human sciences and deals with immanent interdisciplinary topics, it is especially relevant as a subject for ESD. Topics focus on resources and societies with spatial and temporal relations on different scales. As outlined in the prior subchapter, ESD combines environmental education, developmental learning, intercultural and global learning, which all intersect with geography. McKeown-ice (1994) ascribed the link between environment education and geography to the common thematic interest in human–environment interactions and the thinking in different spatial scales (local, regional, global). Thematically, ESD links to the same topics and also promotes skills overlapping with geographical skills.

These links have also been addressed in two relevant international proclamations for geography education: The *International Charter on Geographical Education* (Haubrich 1992) and the *Lucerne Declaration on Geographical Education for Sustainable Development* (Haubrich et al. 2007). Concerning environmental and developmental education, the *International Charter on Geographical Education* (Haubrich 1992) responded to the Preparatory Committee for the UN *Conference on Environment and Development* (UNCED) in a 1992 statement of a need to strengthen educational systems as a prerequisite to environmental and development education:

> Geographical Education contributes to this by ensuring that individuals become aware of the impact of their own behaviour and that of their societies, have access to accurate information and skills to enable them to make environmentally sound decisions, and to develop an environmental ethic to guide their actions. (Haubrich 1992, 1.9)

The *Lucerne Declaration on Geographical Education for Sustainable Development* by the International Geographical Union (Haubrich et al. 2007) supports that "the paradigm of sustainable development should be integrated into the teaching of geography at all levels and in all regions of the world" (Haubrich et al. 2007: 244). As already laid out in the *International Charter on Geographical Education* in 1992, "in the context of problems facing humanity, the right to education includes the right to high quality geographical education that encourages both a balanced regional and national identity and a commitment to international and global perspectives" (Haubrich 1992, 1.4). The *Lucerne Declaration* recommends the following principles, which indicate the overlaps with geography education (Fig. 3.3).

Topics taught in geography, such as globalization, urbanization, climate change, or resource pressure qualify for teaching on sustainability issues in the subject as proclaimed by ESD. The *Lucerne Declaration* (Haubrich et al. 2007) points out that geographical knowledge, skills, values, and attitudes to human–environment relationships are relevant for sustainable development.

ESD principles and objectives can particularly be linked to the latest paradigm change within geography since the spatial turn in the social sciences in the 1970s. Human geography research has moved away from the regional approach to post-structuralist, feminist, post-colonial, discourse analytical, and critical geography

Principles of ESD (Haubrich et al. 2007)

- relevant for everyone at all levels of learning, formal and informal education
- an ongoing, continuous process
- acceptance of processes of societal change
- a cross-sectorial task that has an integrative function
- improving the contexts in which people live
- new opportunities for individuals, society and economic life
- promoting global responsibility
- principles and values that underlie sustainable development
- a healthy balance of all dimensions of sustainability: environmental, social, cultural and economic sustainability
- local and global contexts

Fig. 3.3 Principles of education for sustainable development (Haubrich 2007: 32)

approaches that view knowledge as historically and socially contextual. Societal issues have become increasingly complex, and therefore, comprehensive research designs include various factors and complexities. Everyday productions of space and the occupancy of space by individuals lead to "everyday regionalization" (Werlen 1999); geography doing is seen as a constant process. Therefore, a region is not seen as given or determined, but as constructed. Hence, human geography includes more political conflicts with a focus on examining power relations, for example, dealing with water conflicts in the Indus basin, the Tigris–Euphrates river system, the Nile River Basin, or the Jordan River Basin. On the one hand, the objective is to identify stakeholders, their interests and strategies. On the other hand, an analysis of power relations can help to deconstruct and critically question guiding principles and societal framework conditions. Through this, social and political conflicts about resources can often be understood as conflicts about power and space (Reuber 2012). Meanwhile, they are closely connected to poverty, marginality, and vulnerability.

These geographical approaches highlight how ESD can be included and transferred to geography *teaching*. Political geography, for example, offers a critical analysis and allows the deconstruction of conflicts about resources, e.g., water conflicts using units of household, city, state, country, and beyond. Water conflicts are seen as social, geopolitical conflicts of space and power (Reuber 2012). In a regional teaching approach, the location of resources (e.g., rivers) is important, whereas in a deconstructing teaching approach, different positions in water conflicts are analyzed. To bring ESD to the classroom, the interests of different agents need to be identified and analyzed as well as their power relations and strategies. These can be dismantled argumentatively in everyday products such as newspaper articles, maps, and pictures. Factual knowledge is still important but not sufficient, as causal relationships and network thinking are necessary to understand causes and implications of conflicts. An ESD approach to geography education promotes explaining and relating processes rather than the repetition of definitions and facts. Therefore, geographical teaching methods for ESD should promote thinking in networks and discussions structured by the *structure* and *cause* of a problem, as well as potential *responses* to solve the problem (Haubrich et al. 2007: 33). Concerning the topic of water conflicts, environmental, economic, and societal reasons and consequences of unequal water distribution can be analyzed and integrated. Opportunities to solve the problem can be developed at an individual or community level. Schockemöhle (2009) developed and evaluated a concept for regional learning as strategy for ESD. Her study brought evidence that regional learning can promote communication skills, anticipatory and network thinking as well as cooperative learning, which are demanded by ESD. This approach demonstrates the need of place-based learning for developing ESD skills. As mentioned earlier, Bagoly-Simó (2014b) examined the implementation of ESD in geography education and other subjects in a comparative analysis of 255 German, Mexican, and Romanian curricula. He pointed out differences in the cultural interpretation of sustainable development and the constructivist paradigm in these countries. His textbook analysis demonstrated "missing synergies between subject-specific skill acquisition and ESD" (Bagoly-Simó 2014a: 182). The studies mentioned demonstrate that there are research approaches to ESD in geography education, and the *Lucerne Declaration* on ESD from the umbrella organization IGU provides political support and interest for geographical education research on ESD.

3.2 Argumentation Skills in Geography Education

An overarching orientation for ESD in formal school education can be framed with Freire's postulation for dialogue learning and the promotion of skills to achieve *critical consciousness* (cf. Sect. 2.2.2). Developing critical consciousness is not only necessary in consideration of bringing forward social injustices and political action, as demanded by Freire (1996) and Giroux (2004) and other critical pedagogues, but it is necessary to raise environmental awareness as a prerequisite for environmental action (Mandl and Gerstenmaier 2000). UNESCO (2005) stressed that basic education must be "expanded to include critical-thinking skills," building a "citizenry capable of thinking through some of the more complex sustainability issues that face communities and nations." Learning is a social process in which problems are solved through communication (Vygotsky 1962, 1978). This implies that not only the teacher speaks; in contrast, students should have an equal opportunity to participate in discussions and give their own reasons. I argue that the promotion of argumentation skills is a didactic approach that can contribute to Bernstein's democratic pedagogic rights of enhancement, inclusion, and participation and Freire's concept of *critical consciousness*. In Bernstein's terms, the integration of argumentation leads to addressing, developing, and questioning multiple perspectives in classroom communication. Students have more space to express their own opinions, rather than only reproduce information from textbooks. This implies that strongly framed teaching transfers to weakly framed communication patterns. This indicates a shift from a performance to a competence-oriented teaching style, as content, structure, and purpose of messages are changed. This suggests that the demands of ESD can be translated through the promotion of argumentation skills in the formal educational system.

3.2.1 *The Societal Relevance of Argumentation, Particularly in India*

The oldest discussions on adequate argumentation date back to the Greek philosophers Platon and Aristoteles and the Daoist philosophers Laozi and Zhuangzi. In the Indian subcontinent, the Buddhist emperor Ashoka was the first to establish rules for conducting debates and disputes, with the opponents being "duly honoured in every way on all occasions" (Sen 2005: xiii). The principles of these argumentative traditions still hold relevance in contemporary societies (Sen 2005).

Andrews (2009: 5) argued that argumentation has multiple functions such as "clarification, catharsis, amusement, defense, attach and winning; the discovery of ideas, the creation and resolution of difference." In the "Theory of communicative interaction," Habermas (1984) stated that critical and rational discourse is the foundation of society. Ideally, every individual should be able to participate in discourses and call anyone's opinion, regardless of authority, into question. In a dialogue free of domination and with rational argumentation, norms are justified with the rationality which receives highest consent. In a democracy, the argumentative model of agreement is institutionalized as process for collective decision-making (Kopperschmidt 2000: 26). Habermas' ethical program described the ideal state of a collective use of rationality; however, power relations shape societal discourses, particularly in socially stratified societies as the Indian.

The dissolution of traditional industrial class structures in modern society, increasing globalization, transnational structures, and individualization, has led to a "risk society" (Beck 1986), in which existing life pre-structuring orientation systems, particularly class and caste associations, have become increasingly indistinct. With the growth of opportunities and plural lifestyles, individuals have to make more life choices under specific risks and uncertainties of "reflexive modernity" (Beck et al. 1995). With the plurality of opinions and the need for constant individual decision-making, for example, on lifestyles, the ability to understand and differentiate perspectives is an asset. In these "postmodern times," argumentation helps to orient, to make decisions, and to act deliberately (Wohlrapp 2006). Thus, the ability to argue becomes a power resource to enforce own economic, political, ecological, or social interests and to obtain approval from someone else (Budke and Uhlenwinkel 2011: 114).

The argumentative traditions mentioned at the beginning raise the question if they have any relevance to contemporary societies. Do cultures, or nations, have different contents and forms of argumentation, or even attribute different functions to argumentations? Combs (2004) wanted culturally diverse studies on argumentation. He supported these arguments by pointing out the fundamentally different systems of thought in Daoism as Chinese philosophical tradition and the Greco-Roman, "Western" tradition. The Daoist tradition reflects principles such as unity, harmony and non-contentiousness. These principles are reflected in the "East Asian" argumentative tradition that focuses on contextual, dialectic, and holistic perspectives and uses relational, not adversarial, argumentation (Combs 2004: 69). In contrast, the Greco-

Roman argumentative tradition of the "West" is marked by rationalism, analytical reasoning, objectivity, cause and effect relations, and categorization as heritage of the Greek philosophers.[3] Similarly, the social psychologists Nisbett et al. (2001) took a dualistic approach and differentiated between an "East Asian" holistic and a "Western" analytical cognition and assumed that the origin of these differences is accountable to differing social structures, ecologies, philosophies, and educational systems.

Concerning empirical studies on the question of culture and argumentation, Allen and Preiss (1997) conducted a meta-analysis on whether statistical or anecdotal evidence in support of claims is more persuasive and noted that all experiments were conducted with "Western" participants only. As an attempt to address this research gap, Hornikx and de Best (2011) conducted studies with Indian students, to compare their ratings of the persuasiveness of argumentations with American, German, and Italian students. The results indicate that Indian students rated different evidence types (anecdotal, statistical, causal, and expert evidence) and evidence quality (normatively strong and normatively weak) in the same manner as their "Western" counterparts. However, the methodological study design had multiple weaknesses such as a too small statistical sample of 183 Indian students and too few features on the socio-economic background, which could correlate with perceptions of persuasiveness. These methodological issues indicate how difficult it is to extract the factor *culture* or nationality as the determining factor in judging argumentations on its persuasiveness.

The highly generalizing dualism of "Western" and "Eastern" cultures is constituted by the notion of a culture as demarcated "container." It neither holds any relevance from a cultural geography perspective, particularly enforced since the transcultural turn, nor does it do justice to the complex intersectionalities of any community and society. Neither extreme of a culture-specific and a culture-blind approach to argumentation sufficiently considers the multiple influences of age, gender, caste, class, religion, ethnicity, and other social divides which constitute culture. Hence, I argue that a culture-sensitive approach to argumentation is desirable; an approach that examines the role of argumentative traditions as one among many intersecting factors influencing argumentation. For the Indian context, for example, Oetke (1996) argues linking "commonsense-inferences," or so-called non-monotonic reasoning, to "Indian logic" based on ancient Indian theories of proof and inference. He stated a connection of both on two levels, the subject matter itself, as well as the manner in which the subject matter is reflected. However, the mathematical-astronomical work of Aryabhata in classical India, for example, stands in stark contrast to the characteristics of "Indian logic." This is just one of many examples that could work to contradict reductions to one cultural argumentative tradition, not even speaking of the multiplicity of argumentative characteristics of diverse disciplinary thoughts in contemporary societies. Hence, it is relevant to see argumentation and its messages

[3]Interestingly, the characteristics of the Greco-Roman argumentative traditions are those, which post-structuralism attempts to overcome through deconstructing and destabilizing the meaning of binary oppositions, and pointing out the "situatedness" of knowledge.

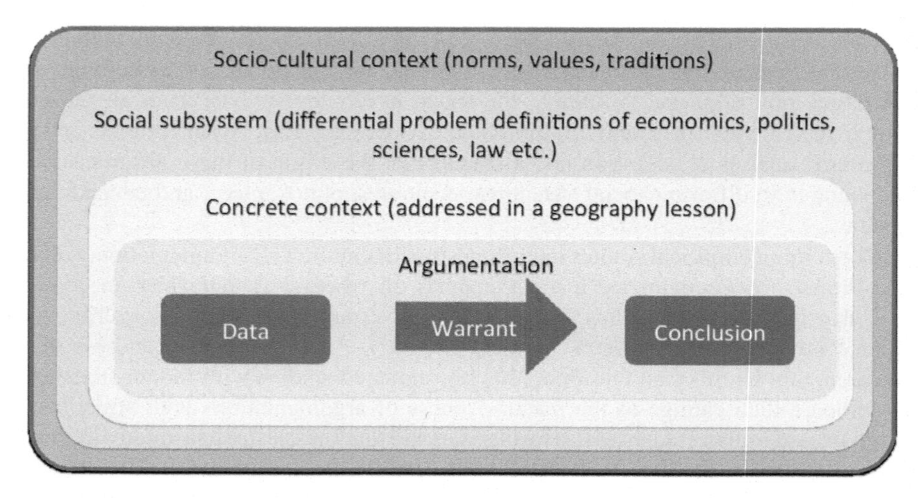

Fig. 3.4 Structure of argumentation embedded in the concrete and societal context (Toulmin (1996) and Budke et al. (2010))

influenced on multiple levels: firstly, on a macro-scale, by underlying socio-cultural norms, values, and traditions, which could be attributed to culture in the broadest sense; secondly, by a social subsystem, which pre-structures the categories of thought through domain-specific problem definitions and approaches; thirdly, the specific context in which the interaction takes place (cf. Fig. 3.4).

As the book's title already suggests, "The Argumentative Indian" by Nobel Prize winner Amartya Sen (2005) depicts India's long argumentative tradition. Sen developed an understanding of the history, culture, and politics of India by examining the country's argumentative traditions. He noted the traditions of the skeptical argument, the acceptance of heterodoxy, and divergent viewpoints as markers of argumentative tradition within the diverse country of India. He also pointed out that while "arguments tend to be biased in the direction of the articulations of the powerful and the well schooled, many of the most interesting accounts of arguments from the past involve members of disadvantaged groups" (Sen 2005: xiii). Hence, he recognized the value of argumentation skills to all sections of societies regardless of class, caste, gender, and other social divisions.

It is relevant to note here that hierarchies are inherent to the Indian society. In the past and to a great extent still today, people with a relative lower status are not allowed to state opinions toward established higher ranks of society determined by caste, economic status, gender, and age. Hierarchies of gender and age are also lived within the household (Mandelbaum 1975: 38). From childhood onwards, children learn that the elder family members, starting with elder siblings, are respected, as they "possess wisdom acquired through life experiences and long moral apprenticeship to their elders" (Marrow 2008). Concerning gender hierarchy, men are considered superior to women, and even younger males are superior to elder women. Despite its official abolition in 1947, the ranks of the caste system are still followed and reproduced through arranged marriages of the same caste. These social structures implicate that argumentation with persons of higher status, which could be in age, gender, or caste, is culturally undesirable. Thus, questioning or arguing with people of higher rank is generally avoided, which holds true for all ranks within the educational system, from student, to teacher, to principal and higher state and national representatives of educational boards.

Hence, argumentation has a particular important value in terms of contributing to social justice in India. Argumentation as part of the formal educational system may contribute to let "the Subaltern speak" (Spivak 2008). This refers not only to individual promotion of argumentation skills of the marginalized and deprived, but also to the inclusion of diverse argumentations and hence multiple perspectives, as well as the deconstruction of discourses, clichés, stereotypes, and narratives on the constitutions of hierarchies. These could be tackled through critical and reflective teaching approaches in textbooks and curricula uncovering attributions to poverty, class, caste, gender, and other social divides inherent in Indian society. These deliberations implicate the important task of schools to prepare students for societal discussions with an argumentative approach to tackle and uncover social injustices. Students' abilities of receptive, productive, as well as to interactive argumentation skills need to be promoted to be able to deconstruct and react in debates.

3.2.2 The Didactic Relevance of Argumentation

In pedagogical contexts, argumentation has two major didactic functions. On the one hand, argumentation skills promote negotiating skills, i.e., understanding and producing written and oral argumentations in different subjects, as well as interdisciplinary, in societal discourse (Budke and Uhlenwinkel 2011). On the other hand, argumentation skills help to construct and consolidate subject-specific knowledge. Firstly, argumentation skills improve communication skills and thus can enhance negotiations within societal participation (Budke and Uhlenwinkel 2011: 114). Argumentation functions as a problem-solving process to reach consent by convincing someone of one's perspective through logical reasoning (Budke 2012: 12). This indicates the functional role of argumentation, which is to negotiate and persuade someone with a specific intention in mind. In a dialogue, reasoning and new perspectives are expressed and analyzed. Disputable conclusions can be confuted or confirmed (Kopperschmidt 1995). The ability to argue leads to critical thinking, opinion formation, the reflection on one's own and societal values, and hence, the ability to take responsibility (Budke and Meyer 2015). This facilitates social and affective competences, such as the ability to find compromises and solutions peacefully, as well as the acceptance of contradictions and diverse perspectives, and not least, self-confidence to express own opinions (Budke and Meyer 2015).

Secondly, the formulation of argumentation helps to produce knowledge and enhances network and systemic thinking. Knowledge is constructed in that different perspectives are understood and information is integrated into individual existing knowledge networks (Wuttke 2005). Students learn subject-specific questions, controversial theses, conclusions, and subject-specific ways of rationality (Budke and Meyer 2015: 13–14). The understanding of argumentation as individual knowledge representation is based on constructive learning theory (Dewey 1916; Reich 2010), which assumes contextual learning and the individual integration of new concepts in existing mental representations. Thus, argumentation demands that different factors are individually prioritized in their causal relations. These two roles of argumentation demonstrate that this skill is a central prerequisite for the willingness to act, which is a central premise for ESD.

Didactic research has led to several conceptions of argumentation skills. In contrast to an *explanation*, which is a statement backed up with reasons in a causal relation, *argumentation* assumes disputable conclusions and reasons (Budke 2012: 12). A further abundant communicative action, which should not be confused with argumentation, is a *description,* which is a characterization of a subject or phenomenon from a specific perspective and thus presents a linguistic construct of reality.

The most prominent argumentation theory was proposed by Toulmin (1996). It includes a model in which an argumentation consists of three elements: a disputable *conclusion* (or statement) supported by *data* (or premises), connected by a *warrant*. This warrant is not valid for every person nor in every context and thus is based on perceptions and standards, which may or may not be known to the interaction partner. Furthermore, different operators can be support warrants by indicating the degree of strength of the warrant. Moreover, conditions and exceptions can be cited under which the warrant is relevant. Argumentation is addressed in a specific situation within a social subsystem, which pre-structures the categories of thought through domain-specific problem definitions and approaches. Argumentation and its messages are influenced by underlying socio-cultural norms, values, and traditions and have to be considered as contextual and relative (cf. Fig. 3.4).

The structure of an argument as described by Toulmin (1996) has been used for empirical data analysis in educational research, including geography didactics studies (Budke et al. 2010; Kuckuck 2014; Hoogen 2016). Argumentation can be evaluated on its quality by analyzing their completeness, factual scope, eligibility and relevance, and appropriateness to the audience (Kopperschmidt 2000). Further, the degree of complexity can be judged by its multi-dimensionality of perspectives (Rhode-Jüchtern 1995). Argumentation can be differentiated in factual and normative argumentation, of which the prior can focus on different facts, e.g., on influences of anthropogenic climate change (Budke et al. 2010). Kienpointer (1983) pointed out that every argumentation has a distinct normative proportion. For example, the factual accuracy of the scientific evidence on climate change indicators can be criticized from a normative perspective. Hence, it is necessary to recognize that ESD as normative educational objective encourages normative argumentation about human–environment relations.

3.2.3 The Promotion of Argumentation Skills in Classrooms

While argumentation with people of a higher status is culturally undesirable in India (cf. Sect. 3.2.1), the latest National Curriculum Framework 2005 supports critical-thinking skills within the educational system of India. Building on this public educational shift of interest, promoting argumentation skills in classrooms presents one approach to contribute to Education for Sustainable Development. As this approach calls for a systemic change of pedagogic practice, the teaching methodology of promoting argumentation challenges not only students, but also teachers, textbook authors, syllabi and curricula designers, and educational policy makers. Because argumentation is a central cultural technique to solve conflicts of interests and should be the basis of democratic decision-making in political, social, economic, and ecological domains, argumentation skills should be promoted at all levels in the formal educational system (Budke et al. 2010).

For subject teaching, several studies have been conducted to analyze and measure argumentation production and reception in several subjects. While in natural sciences, the understanding of concepts and individual knowledge construction is of interest, in language didactics and social science didactics, the focus is on the role of argumentation for individual opinion formation and value orientation (Budke et al. 2010: 180). Several studies have proved deficits in traditional science teaching with regard to students' ability to develop argumentation skills (Aufschnaiter et al. 2008; Kuhn 1992). Aufschnaiter et al. (2008) proved that the level of prior knowledge and the level of argumentation are closely linked for complex argumentation.

In geography education, argumentation skills could be promoted to analyze maps, texts, pictures, etc., and to deconstruct represented perspectives and their intentions. Especially in spatial conflicts concerning the use of resources, stakeholders may have different interests that need to be understood. A study by Budke et al. (2010) showed a lack of tasks in German textbooks that promote argumentation skills as well as a lack of the promotion of argumentation skills in geography lessons. Furthermore, she revealed deficits in students, teachers, and textbooks' abilities to deal with discussions (Budke et al. 2010). Budke et al. (2010) developed a hypothetical model to measure and classify geographical argumentation. She transferred the linguistic dimensions from the Common European Framework of Reference for Languages (CEFR) by the Council of Europe to geographical argumentation skill. The CEFR categorizes dimensions of oral and written argumentation skill into different levels of reception, interaction, and production. The result of a pilot study has shown that the level of written argumentation of students of 7th grade and 11th grade is similarly low (Budke et al. 2010).

The argumentation skill model developed by Budke et al. (2010) was tested and altered by Kuckuck (2014) for receptive argumentation skills in social discourses on local spatial conflicts in geography teaching. Students had to identify stakeholders and their interests by ranking stakeholders' argumentation according to its integrity, complexity, relevance, significance, applicability, and appropriateness to the addressed. Through this, power relations and discourses behind a local spatial conflict are deconstructed. Both studies encourage a more language-sensitive and communicative teaching style and the promotion of argumentation skills in classroom teaching.

3.3 Didactic Framework for Education for Sustainable Development Through Argumentation in Geography Teaching

Education for Sustainable Development aims at raising awareness and action on the ecological, social, and economic implications of resource depletion. Only if the use of resources is critically questioned and different interests and perspectives are understood, can environmental action ultimately be achieved (Mandl and Gerstenmaier 2000). To attain this aim, pedagogic practice, particular in geography education, has to promote relevant knowledge, critical consciousness (Freire 1996), and skills for sustainable environmental action. I argue that the objectives of ESD can be transmitted through the didactic approach of argumentation. The promotion of argumentation skills facilitates understanding and reasoning for environmental action for oneself and toward others in society and can foster critical consciousness concerning questions on sustainable development. The implementation of ESD through argumentation in geography education can encourage seeing resource conflicts as a result of diverging stakeholder interests and socio-cultural power relations. Building on the theories of Bernstein (1975, 1990) and Freire (1996), the link of ESD and argumentation is an opportunity to conceptualize and operationalize ESD for both classroom teaching and didactic research. According to Bernstein (1990), weaker framing and classification of teaching contents and methods change teacher and student communication in classrooms. Such an approach is particularly helpful for contexts in which reproductive pedagogy is prevalent (Leder and Bharucha 2015).

Since argumentation skills can also be instrumental to those who have inhuman or environmentally unsustainable intentions, the normative framework of sustainable development can steer the direction of argumentation. Yet, sustainable development itself also needs to be critically questioned and argued about. The evaluation of practices on their sustainability is dependent on the socio-cultural and concrete context. As pointed out earlier, premises for sustainable development do not count in the same manner in every socio-cultural context, as diverse benchmarks are applied, in how far and why natural resources and ecosystems should be tapped in. In India, for example, this is the discussion of the Bengal tiger, which competes with humans for living space. Depending on the respective sustainability perspective, e.g., an economic one from a tourist guide perspective, a socio-cultural perspective from a farmer, or an environmental perspective from an NGO working for biodiversity, the argumentations for sustainability, and the consequences for actions can be entirely different. Another topic, which can be evaluated from different sustainability perspectives, is deforestation within the same or in different contexts. Through argumentation, one

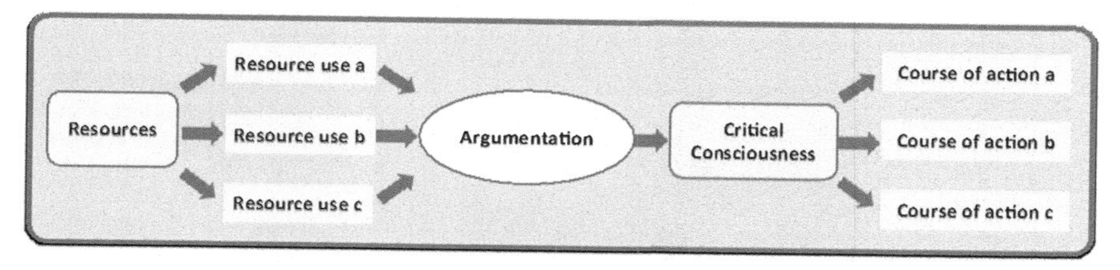

Fig. 3.5 Argumentation on resource conflicts can facilitate environmental action in the sense of Education for Sustainable Development (Leder 2015)

can reflect on different resource uses and then, based on a critical consciousness, decide on different course of actions (Fig. 3.5). Furthermore, argumentations can be evaluated contrarily by discipline, e.g., from a biological or an economic perspective. The understanding of these perspectives can lead to critical awareness, e.g., on resource conflicts. Argumentation can enhance decision-making for sustainable development, and hence, potentially lead to more reflective environmental action as desired by ESD. Consequently, one's own environmental action can be understood and explained through argumentation.

To argue that content needs to be understood in an interrelated manner and under different perspectives needs to be constituted. A critical expression of opinion in conflicts of interests includes critical influencing factors to substantiate the reasons for resource scarcity conclusively. The guiding principles of ESD call for systemic thinking, communication skills, and participation in decision-making; the didactic approach of argumentation fulfills these demands. The promotion of argumentation skills integrates the didactic principles of *network and critical thinking* as well as *student orientation and participation*. Hence, I argue that argumentation can conceptualize ESD principles for teaching methodology. Thus, the didactic approach of argumentation skills is a valuable concept to transfer ESD to the classroom. In this study, I will use the following definition:

> ESD is concerned with pedagogic practice that promotes critical consciousness through argumentation to empower for debates and decision-making on sustainable natural resource use by facilitating learner-centered, problem-posing, and network-thinking teaching approaches.

This definition can be embedded into geography education, in which the three dimensions of sustainable development are thematically integrated within human and physical geography themes. Network and critical thinking on these topics as well as student orientation and participation can be promoted through the didactic approach of argumentation (Fig. 3.6).

Argumentation fosters *student orientation* as it encourages the willingness to act. The aim of participation can only be achieved if students realize the necessity of

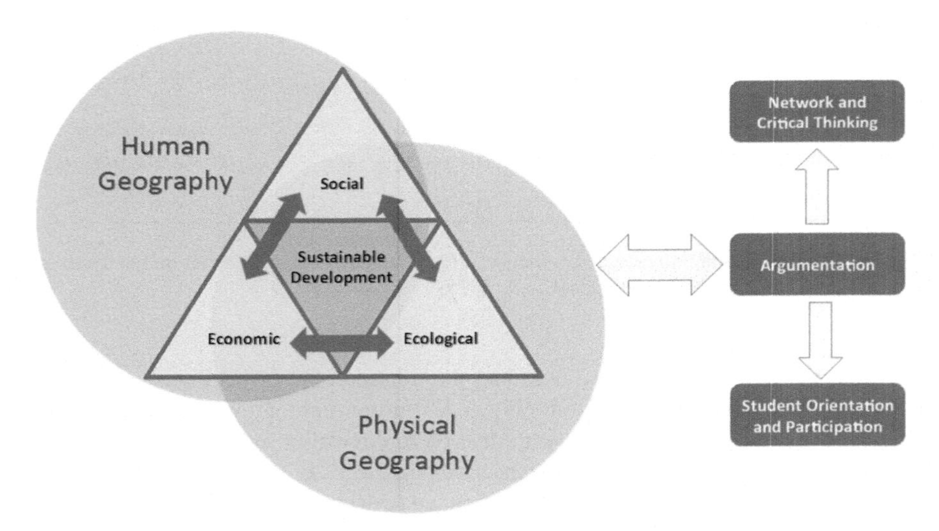

Fig. 3.6 ESD through argumentation in geography education (own draft)

action. This only occurs if a feeling of responsibility persists, for which a direct relation to the living environment of the students is a precondition. Moreover, studies have proven that environmental awareness does not consecutively lead to environmental action and that there is a gap between these two (Mandl and Gerstenmaier 2000). Therefore, the development of argumentation skills is paramount to achieve capacities for sustainable development.

Argumentation skills also facilitate *thinking in networks* by interlinking ecological, economic, and social aspects of human–environment relations. Vester (2002) proclaimed network thinking and transferable knowledge, which he derived from cybernetics and neurobiological research. Interlinked knowledge can be easily consolidated and serves to identify the structure, cause, and solution of environmental or developmental problems of sustainability (Vester 2002). Especially the subject geography can contribute because the interconnectedness of physical and human geography is immanent. Therefore, it enables students to develop a complex understanding of controversial topics along with skill development for a responsible participation in society.

These principles have been proven relevant to learning by empirical educational research, but also neurobiological research. In neurodidactics, evidence demonstrates that teaching content related to the students' direct environment increases motivation, which again has a positive effect on learning success (Spitzer 2007: 195). "Brain-based learning," e.g., network thinking and student-centered learning, has been proven to be supportive as the number of transmitters between synapses is boosted and action potentials are increased in strength and frequency (Spitzer 2007). Thus, the didactic principles of argumentation are neurodidactically grounded, and argumentation can be used as teaching methodology for ESD.

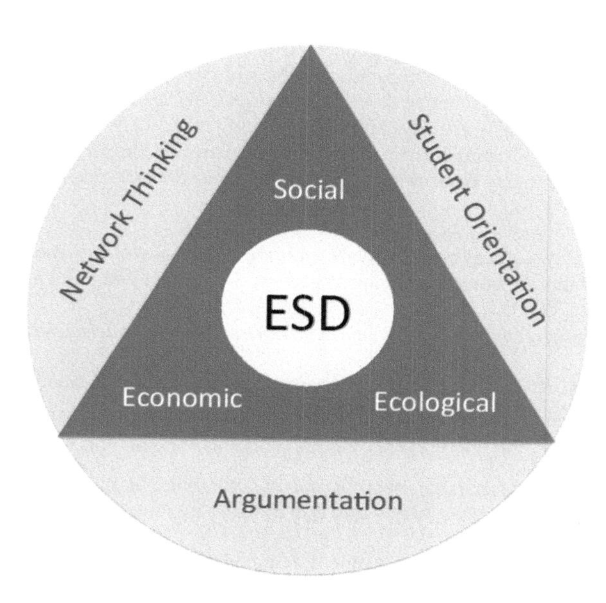

Fig. 3.7 Approach to analyze ESD principles in geography education (own draft)

ESD through argumentation is a promising approach in order to address environmental and social challenges of global change. According to the definition of ESD applied in this study, ESD principles can be analyzed in six fields (Fig. 3.7). Concerning contents, the three dimensions of sustainability are examined: social, economic, and ecological aspects. Concerning teaching methodology, argumentative approaches are studied, as well as the extent to which network thinking and student orientation (synonymously used to learner-centered education) is promoted.

References

Allen, M., & Preiss, R. W. (1997). Comparing the persuasiveness of narrative and statistical evidence using meta-analysis. *Communication Research Reports, 14*(2), 125–131.

Andrews, R. (2009). *The importance of argument in education*. London: Institute of Education.

Aufschnaiter, C., Erduran, S., Osborne, J., & Simon, S. (2008). Arguing to learn and learning to argue: Case studies of how students' argumentation relates to their scientific knowledge. *Journal of Research in Science Teaching, 45*(1), 101–131.

Baccini, P. (2007). Cultural evolution and the concept of sustainable development: From global to local scale and back. In S. Reinfried, Y. Schleicher, & A. Rempfler (Eds.), *Geographical views on education for sustainable development* (pp. 11–26). *Geographiedidaktische Forschungen*.

Bagoly-Simó, P. (2014a). Implementierung von Bildung für nachhaltige Entwicklung in den Fachunterricht im internationalen Vergleich [Implementation of education for sustainable development into formal education: A comparative analysis of German, Mexican and Romanian Curricula]. In M. M. Müller, I. Hemmer, & M. Trappe (Eds.), *Nachhaltigkeit neu denken. Rio + X: Impulse für Bildung und Wissenschaft* München: oekom verlag.

Bagoly-Simó, P. (2014b). Tracing Sustainability: An International Comparison of ESD Implementation into Lower Secondary Education. *Journal of Education for Sustainable Development, 7*(1), 95–112.

Baker, S. (2006). *Sustainable development*. London: Routledge.

Barth, M. (2011). How to deal constructively with the challenges of our times—education for sustainable development as an educational objective. *Sws-Rundschau, 51*(3), 275–291. <Go to ISI>: //000295428700003.

Beck, U. (1986). *Risikogesellschaft. Auf dem Weg in eine andere Moderne*. Frankfurt am Main: Suhrkamp.

Beck, U., Giddens, A., & Lash, S. (1995). Reflexive modernization. Politics, traditions and aesthetics in the modern social order. Cambridge: Polity Press.

Benedict, F. (1999). A systemic approach to sustainable environmental education. *Cambridge Journal of Education, 29*(3), 433.

Bernstein, B. (1975). *Class and pedagogies: visible and invisible*. Paris: OECD.

Bernstein, B. (1990). *Class, codes and control. The structuring of pedagogic discourse* (Vol. IV). London: Routledge.

Bonnett, M. (1999). Education for sustainable development: A coherent philosophy for environmental education? *Cambridge Journal of Education, 29*(3), 313–324.

Brunotte, E. (2005). Nachhaltigkeit. In C. Martin, E. Brunotte, H. Gebhardt, M. Meurer, P. Meusburger, & J. Nipper (Eds.), *Lexikon der Geographie* Wiesbaden: Spektrum Akademischer Verlag.

Budke, A. (2012). Ich argumentiere, also verstehe ich.—Über die Bedeutung von Kommunikation und Argumentation für den Geographieunterricht. In A. Budke (Ed.), *Kommunkation und Argumentation* (pp. 5–18). Braunschweig: Westermann Verlag. Geo Di 14 (KGF), Alexandras einführungsartikel ausgdruckt.

Budke, A., & Meyer, M. (2015). Fachlich argumentieren lernen—Die Bedeutung der Argumentation in den unterschiedlichen Schulfächern. In A. Budke, M. Kuckuck, M. Meyer, F. Schäbitz, K. Schlüter, & G. Weiss (Eds.), *Fachlich argumentieren lernen* Münster: Waxmann.

Budke, A., & Uhlenwinkel, A. (2011). Argumentieren im Geographieunterricht—Theoretische Grundlagen und unterrichtspraktische Umsetzungen. In C. Meyer, R. Henry, & G. Stöber (Eds.), *Geographische Bildung* (pp. 114–129). Braunschweig: Westermann.

Budke, A., Schiefele, U., & Uhlenwinkel, A. (2010). Entwicklung eines Argumentationskompetenzmodells für den Geographieunterricht. *Geographie und ihre Didaktik, 3*, 180–190.

Combs, S. C. (2004). The useless-/usefulness of argumentation: The DAO of disputation. *Argumentation and Advocacy, 41*(2), 58.

Cotton, D. R. R., & Winter, J. (2010). It's not just bits of paper and light bulbs: A review of sustainability pedagogies and their potential for use in higher education. In P. Jones, D. Selb, & S. Sterling (Eds.), *Sustainability Education: Perspectives and Practice Across Higher Education*. London: Earthscan.

de Haan, G. (2006). Bildung für nachhaltige Entwicklung—ein neues Lern- und Handlungsfeld. *UNESCO heute, 1*, 4–8.

de Haan, G., & Harenberg, D. 72 (1999) 'Expertise Förderprogramm Bildung für eine nachhaltige Entwicklung: Gutachten zum Programm' *Bund-Länder-Kommission für Bildungsplanung und Forschungsförderung*.

de Haan, G., Bormann, I., & Leicht, A. (2010). Introduction: the midway point of the UN Decade of education for sustainable development: Current research and practice in ESD. *International Review of Education, 56*(2), 199–206. https://doi.org/10.1007/s11159-010-9162-z.

Dewey, J. (1916). *Democracy and education*. New York: The Macmillan Company.

Ekins, P. (2000). *Economic growth and environmnetal sustainability: The prospects of green growth*. London: Routledge.

Fien, J., & Maclean, R. (2000). Teacher education for sustainability II. Two teacher education projects from Asia and the Pacific. *Journal of Science Education and Technology*, *9*(1), 37–48. http://www.jstor.org/stable/40188539.

Freire, P. (1996). *Pedagogy of the oppressed*. London: Penguin Books Ltd.

Giroux, H. A. (2004). Critical pedagogy and the postmodern/modern divide: towards a pedagogy of democratization. *Teacher Education Quarterly, 31*(1), 31–47.

Graupe, S., & Krautz, J. (2014). Die Macht der Messung. Wie die OECD mit PISA ein neues Bildungskonzept durchsetzt. *COINCIDENTIA—Zeitschrift für europäische Geistesgeschichte*, 4, 139–146.

Gruenewald, D. A. (2004). A foucauldian analysis of environmental education: Toward the socioecological challenge of the earth charter. *Curriculum Inquiry, 34*(1), 71–107. https://doi.org/10.1111/j.1467-873X.2004.00281.x.

Habermas, J. (1984). *The theory of communicative action*. Boston, Mass: Beacon Press.

Hasse, J. (2010). Globales Lernen. Zum ideologischen Gehalt einer Leer-Programmatik. In G. Schrüfer, & I. Schwarz (Eds.), *Globales Lernen. Ein geographischer Diskursbeitrag*. Münster: Waxmann.

Haubrich, H. (1992). *International charter on geographical education for sustainable development*. Nürnberg: Hochschulverband für Geographie und seine Didaktik.

Haubrich, H. (2007). Geography education for sustainable development. In S. Reinfried, Y. Schleicher, & A. Rempfler (Eds.), *Geographical Views on Education for Sustainable Development. Geographiedidaktische Forschungen* (pp. 27–38).

Haubrich, H., Reinfried, S., & Schleicher, Y. (2007). *Lucerne declaration of geographical education for sustainable development*. Switzerland: Geographiedidaktische Forschungen.

Hawkes, J. (2001). *The fourth pillar of sustainability: Culture's essential role in public planning*. Victoria, Australia: Cultural Development Network.

Hellberg-Rode, G., Schrüfer, G., & Hemmer, M. (2014). Brauchen Lehrkräfte für die Umsetzung von Bildung für nachhaltige Entwicklung (BNE) spezifische professionelle Handlungskompetenzen? *Zeitschrift für Geographiedidaktik, 4,* 257–281.

Hoffmann, T., & Bharucha, E. (2013). Education for Sustainable Development (ESD) as modern education—a glimpse on India. Accessed August 27, 2014.

Hoffmann, K. W., Dickel, M., Gryl, I., & Hemmer, M. (2012). Bildung und Unterricht im Fokus der Kompetenzorientierung. *Geographie und Schule, 195*(34), 4–14.

Hoogen, A. (2016). Didaktische Rekonstruktion des Themas illegale Migration. Argumentationsanalytische Untersuchung von Schüler*innenvorstellungen im Fach Geographie. Münster: MV-Verlag.

Hopkins, C., & McKeown, R. (2002). Education for Sustainable Development: an international perspective. In D. Tilbury, R. B. Stevenson, J. Fien, & D. Schreuder (Eds.), *Education and Sustainability. Responding to the Global Challenge* (pp. 13–24). Cambridge: IUCN.

Hornikx, J., & de Best, J. (2011). Persuasive evidence in India: An investigation of the impact of evidence types and evidence quality. *Argumentation and Advocacy, 47,* 246–257.

Huckle, J. (1993). Environmental education and sustainability: A view from critical theory. In J. Fien (Ed.), *Environmental education: A pathway to sustainability* (pp. 43–68). Geelong, Vic: Deakin University.

Jickling, B. (1994). Why i don't want my children to be educated for sustainable development: Sustainable belief. *Trumpeter*, *11*(3). http://trumpeter.athabascau.ca/index.php/trumpet/article/viewArticle/325/497.

Kienpointer, M. (1983). *Argumentationsanalyse*. Innsbruck: Verlag des Instituts für Sprachwissenschaft.

Klafki, W. (2013). *Kategoriale Bildung: Konzeption und Praxis reformpädagogischer Schularbeit zwischen 1948 und 1952*. Bad Heilbrunn: Klinkhardt.

Kopperschmidt, J. (1995). Grundfragen einer allgemeinen Argumentationstheorie unter besonderer Berücksichtigung formaler Argumentationsmuster. In H. Wohlrapp (Ed.), *Wege der Argumentationsforschung* (pp. 50–73). Stuttgart: Frommann-Holzboog.

Kopperschmidt, J. (2000). *Argumentationstheorie. Zur Einführung*. Hamburg: Junius.

Kuckuck, M. (2014). *Konflikte im Raum—Verständnis von gesellschaftlichen Diskursen durch Argumentation im Geographieunterricht*. Münster: MV-Verlag.

Kuhn, D. (1992). Thinking as argument. *Harvard Educational Review, 62*(2), 155–178.

Leder, S. (2015). Bildung für nachhaltige Entwicklung durch Argumentation im Geographieunterricht. In A. Budke, M. Kuckuck, M. Meyer, F. Schäbitz, K. Schlüter, & G. Weiss (Eds.), *Fachlich argumentieren lernen* Münster: Waxmann.

Leder, S., & Bharucha, E. (2015). Changing the educational landscape in india by transnational policies: New perspectives promoted through education for sustainable development (ESD). *ASIEN, 134,* 167–192.

Leng, M. (2009). *Bildung für nachhaltige Entwicklung in europäischen Großschutzgebieten. Möglichkeiten und Grenzen von Bildungskonzepten*. Hamburg: Verlag Dr. Kovac.

Malatesta, S., & Camuffo, M. (2007). Geography, education for sustainable development and primary school curricula: A complex triangle. In S. Reinfried, Y. Schleicher, & A. Rempfler (Eds.), *Geographical Views on Education for Sustainable Development. Geographiedidaktische Forschungen* (pp. 58–65).

Mandelbaum, D. G. (1975). *Society in India*. Noida: Popular Prakashan.

Mandl, H., & Gerstenmaier, J. (2000). *Die Kluft zwischen Wissen und Handeln—Empirische und theoretische Lösungsansätze*. Göttingen: Hogrefe-Verlag.

Manteaw, O. O. (2012). Education for Sustainable Development in Africa: The search for pedagogical logic. *International Journal of Educational Development, 32*(3), 376–383. <Go to ISI>://000301698300003 http://ac.els-cdn.com/S0738059311001301/1-s2.0-S0738059311001301-main.pdf?_tid=e247c694-1df6-11e2-a864-00000aab0f6b&acdnat=1351095877_200e77592eca28a76d7de4643351e023.

Marrow, J. (2008). *Psychiatry, modernity and family values: clenched teeth illness in North India*. Chicago: ProQuest.

McKeown-ice, R. (1994). Environmental education: A geographical perspective. *Journal of Geography, 93*(1), 40–42. https://doi.org/10.1080/00221349408979684.

Meadows, D. H., Meadows, D. L., Randers, J., & Behrens, W. W., III. (1972). *The limits to growth*. New York: Universe Books.

Moore, R. (2013). *Basil Bernstein: The thinker and the field*. London: Routeledge.

Mulà, I., & Tilbury, D. (2009). A United Nations Decade of education for sustainable development (2005–14). *Journal of Education for Sustainable Development, 3*(1), 87–97. https://doi.org/10.1177/097340820900300116.

Neumayer, E. (2003). *Weak versus strong sustainability: Exploring the limits of two opposing paradigms*. Bodmin: MPG Books Ltd.

Nisbett, R. E., Peng, K., Choi, I., & Norenzayan, A. (2001). Culture and systems of thought: Holistic versus analytic cognition. *Psychological Review, 108*(2), 291–310.

OECD. (2005). *The definition and selection of key competencies*. Paris: OECD.

Oetke, C. (1996). Ancient Indian logic as a theory of non-monotonic reasoning. *Journal of Indian Philosophy, 24*(5), 447–539.

Piaget, J. (1962). *The language and thought of the child*. London: Routledge & Kegan.

Raschke, N. P. (2015). *Umweltbildung in China*. Geographiedidaktische Forschungen: Explorative Studien an Grünen Schulen. Münster.

Reich, K. (2007). Interactive constructivism in education. *Education and Culture, 23*(1), 7–26.

Reich, K. (2010). *Systemisch-konstruktivistische Pädagogik—Einführung in die Grundlagen einer interaktionistisch-konstruktivistischen Pädagogik*. Weinheim: Beltz Verlag.

Reuber, P. (2012). *Politische Geographie*. Münster: Schöningh UTB.

Rhode-Jüchtern, T. (1995). *Raum als Text*. Wien: Perspektiven einer konstruktiven Erdkunde.

Rousseau, J.-J. (1979). *Emil or on education*. New York: Basic Books.

Rychen, D. S. (2008). OECD Referenzrahmen für Schlüsselkompetenzen—ein Überblick. Kompetenzen der Bildung für nachhaltige Entwicklung. In I. Bormann, & G. Haan (Eds.), *Kompetenzen der Bildung für nachhaltige Entwicklung. Operationalisierung, Messung, Rahmenbedingungen, Befunde* (pp. 15–22). Wiesbaden: VS Verlag für Sozialwissenschaften. https://doi.org/10.1007/978-3-531-90832-8_3.

Schockemöhle, J. (2009). Außerschulisches regionales Lernen als Bildungsstrategie für eine nachhaltige Entwicklung. Entwicklung und Evaluierung des Konzeptes Regionales Lernen 21+. Münster: Geographiedidaktische Forschungen.

Schrüfer, G. (2010). Förderung interkultureller Kompetenz im Geographieunterricht—Ein Beitrag zum Globalen Lernen. In G. Schrüfer & I. Schwarz (Eds.), *Globales Lernen* (pp. 101–110). Münster: Waxmann.

Sen, A. (2005). *The argumentative Indian*. Noida: Penguin.

Spitzer, M. (2007). *Lernen. Gehirnforschung und die Schule des Lebens*. Berlin: Spektrum Akademischer Verlag.

Spivak, G. (2008). *Can the subaltern speak?*. Wien: Turia+Kan.

Sriprakash, A. (2010). Child-centered education and the promise of democratic learning: pedagogic messages in rural Indian primary schools. *International Journal of Educational Development, 30*(3), 297–304.

Steiner, R. (2011). Kompetenzorientierte Lehrer/innenbildung für Bildung für Nachhaltige Entwicklung. Kompetenzmodell, Fallstudien und Empfehlungen. Münster: MV-Verlag.

Tilbury, D. (2007). Asia-pacific contributions to the UN Decade of education for sustainable development. *Journal for Education for Sustainable Development, 1*, 133–141.

Tilbury, D. (2011). *Education for sustainable development. An expert review of processes and learning*. Paris: UNESCO.

Toulmin, S. E. (1996). *The uses of argument*. New York: Cambridge University Press.

UNESCO. (2002). Education for Sustainability. From Rio to Johannesburg: Lessons learnt from a decade of commitment. Paris: UNESCO.

UNESCO. (2005). United Nations Decade of education for sustainable development (2005–2014): International Implementation Scheme. Paris.

UNESCO. (2006). Water. A shared responsibility. The United Nations World Water Development Report 2. Barcelona: UNESCO, Berghahn Books.

UNESCO. (2009a). *Review of contexts and structures for education for sustainable development 2009*. Paris: UNESCO.

UNESCO (2009b). UNESCO World Conference on ESD: Bonn Declaration.

UNESCO. (2011). *Education for Sustainable Development*. An expert review of processes and learning. Paris: UNESCO.

UNESCO (2012). EFA Global Monitoring Report. Youth and skills-putting education to work.

UNESCO, & UNEP (1977). Tbilisi Declaration.

United Nations Conference on Sustainable Development (2012). The future we want. http://daccess-dds-ny.un.org/doc/UNDOC/GEN/N11/476/10/PDF/N1147610.pdf?OpenElement. Accessed March 10, 2013.

Vester, F. (2002). *Unsere Welt—ein vernetztes System*. München: Deutscher Taschenbuchverlag.

Voß, J.-P., Bauknecht, D., & Kemp, R. (2006). Reflexive Governance for Sustainable Development. Edward Elgar Publishing.

Vygotsky, L. S. (1962). *Thought and Language*. Cambridge, Mass.: MIT Press.

Vygotsky, L. S. (1978). *Mind in society. The development of higher psychological processes*. Cambridge, Massachusetts.: Harvard University Press.

Wals, A. E. J. (2009). A Mid-DESD Review. *Journal of Education for Sustainable Development, 3*(2), 195–204. https://doi.org/10.1177/097340820900300216.

Werlen, B. (1999). Zur Ontologie von Gesellschaft und Raum. Sozialgeographie alltäglicher Regionalisierungen Bd. 1. [On the Ontology of Society and Space. Vol. 1. Social Geography of Everyday Regionalizations]. Stuttgart: Franz Steiner Verlag.

Wohlrapp, H. (2006). Was heißt und zu welchem Ende sollte Argumentationsforschung betrieben werden? In E. Grundler, & R. Vogt (Eds.), *Argumentieren in der Schule und Hochschule. Interdisziplinäre Studien* (pp. 29–40). Tübingen: Stauffenburg Verlag Brigitte Narrr.

World Commission on Environment and Development (1987). Report of the World Commission on Environment and Development: Our Common Future. http://www.un-documents.net/our-common-future.pdf. Accessed March 10, 2014.

Wuttke, E. (2005). Unterrichtskommunikation und Wissenserwerb. Zum Einfluss von Kummunikation auf den Prozess der Wissensgenerierung. Frankfurt a. M: Peter Lang.

Chapter 4
Education and Water Conflicts in Pune, India

Abstract This chapter reflects on India's educational system and environmental challenges that have the potential to be addressed through Education for Sustainable Development (ESD). I examine the development, reforms, and structure of India's educational system with a particular focus on environmental education and the principal role of textbooks in classroom teaching. I discuss the relevance of ESD in India using the example of water resource conflicts. The importance of geography as central subject for ESD-oriented water education in schools is highlighted. To demonstrate the relevance of the topic "water conflicts" as a challenge to sustainable development, I outline the growing water demands as well as intra-urban disparities of water access and controversies on causes of insufficient water supply. I focus in particular on the example of the emerging megacity of Pune.

As an emerging megacity of the newly industrialized country India, Pune is marked by rapid urbanization, population, and economic growth. Located 150 km south-east of the megacity Mumbai, Pune's population is 3.1 million, while the urban agglomeration around Pune with Pimpri and Chinchwad (Pune Urban Agglomeration) accounts for 5 million inhabitants (Government of India 2011a). Threshold values for quantitative definitions of megacities vary between 5 million (Bronger and Trettin 2011) and 20 million inhabitants (UN-Habitat 2013). Therefore, Pune and its urban agglomeration can be called an emerging megacity. Megacities are hubs for globalization processes and marked by great dynamics of social, economic, and ecological development (Kraas and Mertins 2008). The high concentration of population, infrastructure, economic activity, capital and political decisions can lead to uncontrollability of development processes in megacities as well as challenges for governability (Kraas and Mertins 2008). On the one hand, megacities are influenced by global ecological, socio-economic, and political changes, while on the other hand, megacities bring about change and "offer a multitude of potentials for global transformation" (Kraas 2007: 81). Megacities are also global risk areas in which natural disasters or the shortage of natural resources such as water can result in devastating effects (Keck et al. 2008). Competing uses and users increase the pressure on natural resources and infrastructure. Intra-urban socio-economic disparities mark access to

S. Leder, *Transformative Pedagogic Practice*, Education for Sustainability,
https://doi.org/10.1007/978-981-13-2369-0_4

and control over resources, particularly water, as well as services, such as quality education. Recognizing the role of education for social transformation, this study links the thematic focus of water conflicts in Pune to opportunities and challenges for Education for Sustainable Development in Indian pedagogic practice.

This chapter frames the research context within which the study took place. At first, the structure and development of the Indian educational system, educational reforms, and the current educational situation in Pune, the state of Maharashtra, as well as in India are outlined. As it is relevant to this study, the role of environmental education and ESD and environmental education in India are depicted. Secondly, the relevance of urban water conflicts, particularly in Indian megacities, and water supply conflicts in Pune is reviewed. These are important as the empirical analysis centers around the example of water conflicts in Pune. The chapter ends with outlining the relevance of ESD-oriented geographical education to water conflicts in schools in general and in Pune in particular.

4.1 The Educational System of India

4.1.1 Development and Reforms of the Indian Educational System

Education in India has historically been influenced by traditional Hindu education, British colonialism, nationalist and Hinduist interests, and lately, transnational educational policies. This subchapter will give a brief chronological overview of these influences on the Indian educational system.

Education in India was traditionally based on a guru–shiksha relationship,[1] which has highly influenced the choice of knowledge recital as preferred teaching methodology up to today. The ancient ideal of absolute teacher authority "superimposed the values and administrative structures of British colonialism and Indian nationalism" (Alexander 2001: 211). India's educational system has historically been elitist, as in traditional Hindu education only Brahmin boys were taught within a well-developed network of indigenous schools, usually under the open sky near local temples in the center of the village (Indian Institute of Education 2002: 11). Schooling hours and the curriculum depended on the teacher's ability and knowledge and were often adapted to the local environment (Indian Institute of Education 2002: 11). Segregation within education was reinforced under British colonialism (1700–1947), as education remained a privilege for upper classes and excluded girls, scheduled castes, and tribal castes. With the *Charter Act of 1813*, the so-called modern elementary educational system was introduced under British rule (Indian Institute of Education

[1] In Hinduism, a guru is a spiritual leader guiding disciples, shikshas, to knowledge. The ancient guru–shiksha relationship is marked by a close relation of the teacher and the learner, in which students learned to cite, e.g., the Ramakrishna, by word of mouth. Traditionally, thoughts from the guru were transferred through oral communication to the shiksha.

2002: 1). The East India Company was appointed by the British Parliament to invest in and implement elementary education, with English as the medium of instruction. However, education continued to be available only to a limited number of people who were to enter government services, and this provided social upward mobility (National Center on Education and the Economy 2006: 14). The *Wood's Despatch of 1854* transferred the responsibility for education to the Government of India and allowed education in vernacular languages. The *Zakir Hussain Report of 1937* under Mahatma Gandhi promoted more effective education for the masses. In the post-independence era, the structure of the educational system in India remained strongly influenced by the British educational system and English, as a medium of instruction, remained relevant in many schools despite the development of vernacular language schools. Kumar (2005) studied how principles of the British educational system were imposed on the indigenous system of education in India under colonial rule. Due to a highly centralized system, decision-making of curricula, rules of recruiting teachers, textbooks and exams in all regions of India were controlled. A teacher became a "sub-ordinate functionary of the superior officers of the education department" (Kumar 2005: 196). Teaching methods for the masses were "a means of teaching them a morality of obedience while education for children of propertied classes was a means of acquiring the symbolic repertoire consistent with leadership roles" (Kumar 2005: 198). The underlying dogma of the colonial government's agenda was the utilitarian thought of establishing social order. Post-independence challenges were the infusion of nationalist, Hindu values in which the ordinary teacher had to "perform his ancient duties of transferring legitimate truths to the students without the interference of the student's questioning spirit" (Kumar 2005: 200). Thus, pedagogic practice in India was strongly influenced by colonial and Hindu nationalist and ideas.

In the last three decades, the Indian educational system has undergone many reforms (cf. Table 4.1). The *Kothari Commission Report*, which introduced 12 years of schooling, and the *National Policy of Education* (NPE 1986) called for quality education irrespective of caste, location, and gender in upper primary schools. Following the aims of universalization of elementary education in the NPE 1986, the scheme *Operation Blackboard* started in 1987 with the aim to equip elementary schools with essential school facilities, such as additional classrooms, the provision of separate toilet facilities for boys and girls, and learning materials like blackboards, maps, and charts. One aim was also to appoint two teachers per school, whereas more female teachers were appointed to balance gender equality in schools.

Transnational policies such as the MDGs have yielded an increased number of students enrolling in Indian schools by providing basic classroom infrastructure and equipment, especially in rural areas. As a result, members of the scheduled castes and girls were able to surmount the deep social divides that had previously precluded them from receiving an education. National policies by the Government of India (GoI)—such as "Sarva Shiksha Abhiyan" (SSA), the Indian "Education for All" movement—have also been aimed at the universalization of elementary education and the reduction of school dropout rates. In training programs, teachers were sensitized to tribal language, culture, and knowledge systems as well as to activity-based learning methods.

Table 4.1 Educational policies in India since 1986

Year	Policy/report	Focus	Details
1986	National Policy of Education (NPE)	Creation of national system of education capable of responding to India's diversity of geographical and cultural milieus, while also ensuring common values and competencies along with shared academic components	Common core component in school curriculum throughout the country, achieved through the National Curriculum Framework (NCF) formulated by NCERT
1987	Operation Blackboard	Program for blackboards and other crucial facilities in government schools	Equipment for schools and teacher training programs on how to use materials
1991	Supreme Court (SC) accepts the Public Interest Litigation PIL by the environmental lawyer M. C. Mehta	Decision to strengthen Environmental Education in the national education system	Introduction of Environmental Education in all subjects or as a separate subject
1993	"Learning without burden" report of the National Advisory Committee under the chairmanship of Prof. Yash Pal, appointed by the Ministry of Human Resource Development	Critically analyzing the problem of curriculum load for students and the convention of "teaching the text"	Criticizes performance character in examinations, fact-based textbooks, and poorly equipped teachers - as well as societal spirit focusing on the elite
2000	Sarva Shiksha Abhiyan (SSA)	"Education for All" (EFA) movement through central and state governments, funded by UNICEF, World Bank etc.	Focus on universal primary education, especially for girls, scheduled castes, minority, and tribal children
2002	Study on the "Status of infusion of environmental concepts in school curricula and the effectiveness of its delivery" by BVIEER	The Bharati Vidyapeeth Institute of Environment Education and Research (BVIEER) revealed that textbooks do not promote pro-environmental learning	BVIEER analyzed 1845 textbooks in 22 languages, finding that environmental information is too limited and general, inaccurate, fragmented, and biased
2005	National Curriculum Framework (NCF), Position Paper 1.6: Habitat and Learning	Constructivist pedagogical approach, environmental education infused in all subjects and as separate subject	Promotion of life skills and protection of the environment through projects
2009	Right of Children to Free and Compulsory Education Act (RTE)	Anti-discriminatory clauses ensuring access to schools in walking distance, free textbooks, and equal learning opportunities for all	Enacted by the parliament, providing legal support for the implementation of SSA/EFA

The *Right of Children to Free and Compulsory Education Act* (RTE) was passed by the Indian Parliament in 2009 "to provide free and compulsory education to all children of the age of six to fourteen years" (Ministry of Law and Justice 2009). The act meant to reinforce the fundamental right of education as prescribed under Article 45 in the Constitution of India. According to the Government of India (2011b), approximately 230 million children are included in this age cohort. Among its many anti-discrimination clauses, the RTE Act also includes the stipulation that state governments must provide elementary schools within a walking distance of 1 km from every home and secondary schools within 3 km thereof. In addition, free textbooks, writing materials and uniforms, free special learning, and supporting materials for children with disabilities are a requirement, and 25% of seats need to be reserved in private schools for scheduled caste children from lower economic classes.

As an attempt to individualize the existing strict examination structures and to reduce the stress experienced in the course of final board exams, continuous and comprehensive examination (CCE) was introduced for classes from six to ten in 2009. For this, teachers need to report on every student after each lesson. Consequently, students are evaluated not only by summative but also by formative assessment. In practice, the implementation of CCE is difficult because of the high number of students per class, often amounting to around 60 learners, which means placing extraordinary high time burden on the teachers.

These educational reforms show that over the last three decades or so, a number of radical changes have been introduced at the policy level in India. Nevertheless, their actual implementation is difficult, as policies are often detached in their considerations from the available financial, infrastructural, and human resources. The change of behavioral patterns is also slow, particularly as teacher identities and methods are influenced by own experiences as students (cf. Eick and Reed 2002). Institutional change is slow, with reforms often being confronted with a lack of implementation strategies and of genuine will to implement proposed schemes. For example, according to SSA, each school should have at least one computer—even though many schools do not even have access to electricity.

For ESD to be successfully implemented, these challenges have to be faced and overcome. Aiming at enhancing learning, understanding, and taking action for a sustainable future sets a strong focus on the quality of education in a setting where at the same time serious infrastructural, financial, managerial, and human resource challenges remain to be tackled.

4.1.2 Structure of the Indian Educational System

Due to the existence of interrelated administrative bodies, each with multiple responsibilities at the district, state, and national levels, the Indian educational system is a complex and highly diverse entity (cf. Fig. 4.1). Since 1976, education has been on the so-called Concurrent List, which means that it is of concern to both national and state governments in India. To strengthen a uniform orientation for education, the

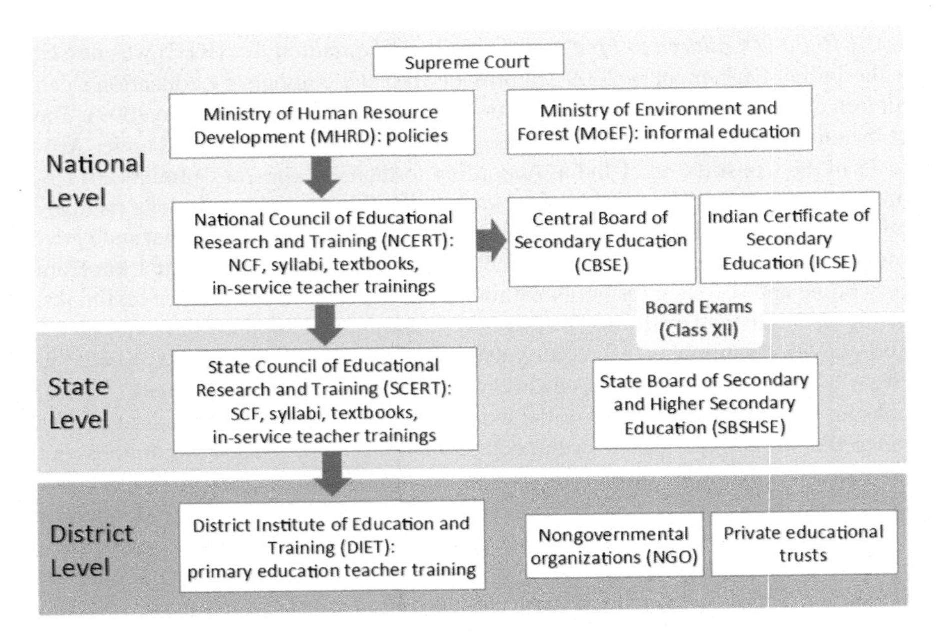

Fig. 4.1 Structure of the Indian educational system (own draft)

National Policy on Education (NPE 1986) suggested a common core component for the school curriculum in every state. The national government ministry responsible for education is the Ministry of Human Resource Development (MHRD), whereas environmental education falls under the mandate of the Ministry of Environment, Forests and Climate Change (MoEFCC). This division of labor leads to unclear responsibilities for environmental education in schools, because the MHRD does not feel it is responsible for environment education while the MoEF is not concerned with formal school education.[2] The National Council for Educational Research and Training (NCERT), meanwhile, is the executive authority charged with overseeing the development and implementation of educational policies and programs. Within the most recent (2005) National Curriculum Framework (NCF) and in other publications, the NCERT has provided educational guidelines and developed and revised national curricula and textbooks for each subject. State Councils for Education, Research, and Training (SCERTs) as well as the State Textbook Bureaus orient their syllabi and textbooks toward NCERT samples. In fact, the SCERTs and the State Textbook Bureaus have full autonomy on public educational matters. However, educational policies are not necessarily effective at the state level, as the Maharashtra state board, for example, has not taken any initiatives to explicitly include ESD principles in their textbooks whereas NCERT has taken it up, as the analysis of academic frameworks (cf. Chap. 6) will demonstrate.

[2]This statement is based on interviews with national educational stakeholders in March 2012.

NCERT textbooks are based on the content of the board examinations taken in class 12, which are developed by the Central Board of Secondary Education (CBSE). These board examinations have to be passed if one is to gain entrance to the country's universities. A number of educational boards each having their own curricula, textbooks, and examinations currently coexist, being funded from either public or private sources (educational trusts, non-governmental organizations, etc.). Schools in India can be run by the government (public) or by private or semiprivate entities. The most common school type is the so-called government school at the municipal level, which is under the authority of the District Institutes of Education and Training (DIETs) and Municipal Corporations (MCs). The DIETs are particularly important as they conduct pre- and in-service teacher trainings. Government schools usually follow public state board curricula and textbooks that are oriented along NCERT textbooks. They can be vernacular and use the state language, or they can be in English or Hindi.

Based on the *Kothari Commission Report* of 1966, the structure of the educational system was divided into 12 years of schooling, which remains valid until today. There are eight years of compulsory primary school (seven years in the state Maharashtra) and four years of secondary school (Fig. 4.2). These are further subdivided into lower primary (5 years) and upper primary (class 6 to 7/8), and lower secondary (grade 9 and 10) and upper secondary (class 11 and 12). Students start school at the age of 6 and end their schooling with respective board exams around the age of 17. After school, bachelor degrees can be obtained from various colleges and universities. The education of a secondary teacher covers a three-year Bachelor of Science (B.Sc.) or Art (B.A.) and is followed by a one-year Bachelor of Education (B. Ed.). Primary teachers are trained for three years at the respective District Institute of Education and Training (DIET), covering in-class services.

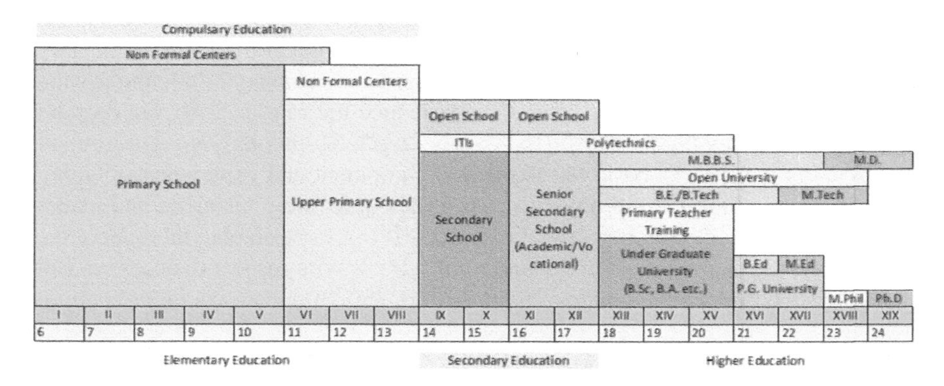

Fig. 4.2 Sequence of the educational system in India (own draft based on MHRD (1993) and Alexander (2001: 85))

4.1.3 The Educational Situation in India

India's educational system is influenced by the country's stark socio-economic disparities, a remarkable urban–rural divide, and the complex urban locality in which children live. The multiple different school systems (such as public/private, English-medium/local vernacular) also contribute their own idiosyncrasies to the country's teaching and learning processes. This produces a great educational heterogeneity within India.

After the end of British rule, the national literacy rate was 12% and improved from 65.0% in 2001 to 74.4% by 2011 (Government of India 2011a). This data leaves aside considerable regional variations in urban and rural areas. According to the latest national census in 2011, the state of Maharashtra, for example, has a literacy rate of 82.32%, Kerala has the highest literacy rate with 94.0%, and Bihar the lowest with 61.8%. Sex-disaggregated data indicates strong gender differences, as the male literacy rate is 82.1% and the female literacy rate is 65.5%. In Maharashtra, male literacy is 88.4% and female literacy is 69.9% (Government of India 2011a). This data indicates a prevailing strong preference for boys to be schooled, while education for girls is compromised: Marry them early and with a lower dowry, as higher education raises the amount of dowry required (Chandramukhee and Leder 2013). In 2011, the expenditure on education was 3.9% of the GDP, in comparison to Germany (4.8%), Nepal (4.1%), or Pakistan (2.2%) moderate, considering that it includes transfers from international sources to the government (UNESCO Institute for Statistics 2016). In 2005, there were one million schools and 202 million students who were taught by 5.5 million teachers (with a high national student–teacher ratio of 37:1) (National Council of Educational Research and Training 2005).

Although the Indian educational system has achieved almost universal enrollment in primary schools (99.3%), it faces multiple development challenges as low student attendance, low secondary school enrollment rates (50.3%), and teacher absenteeism are high (Kingdon 2007). In 2005, 53% of students dropped out from elementary schools (National Council of Educational Research and Training 2005), and despite the objectives of the 11th Five-Year Plan to reduce the rate to 20%, the dropout rate in 2013 was as high as 40% (Indiatoday 2013). To this day, the government school system still lacks basic infrastructural equipment and experiences ongoing administrative issues. Core problems are insufficient learning materials and inadequate teacher education (Government of India 2011a). In particular, there are great discrepancies between Indian urban and rural schools with respect to access to and quality of schooling (Government of India 2011a). 75% of Indian schools are in rural areas, which often have multi-grade classrooms (National Council of Educational Research and Training 2005). There are also stark disparities within cities: Private schools generally provide a higher level of education, while government schools lack financial and human resources—resulting in overcrowded classrooms, insufficient teaching materials, and inadequate pre- and in-service teacher training programs. This results in intra-urban disparities in terms of access to schooling and a close correlation between the quality of school attended and the socio-economic background of the student's parents (Government of India 2011a).

The Indian Institute of Education (2002) noted, based on its study on the educational system of the state Maharashtra, in which Pune is located, that "lacunae are not at the policy level, but at the implementation level." The authors recommend improving infrastructure such as providing sufficient classrooms, drinking water, toilet facilities, playgrounds, libraries and laboratories, as well as speedy appointments of teachers and headmasters, subject-specific training opportunities, and in-service teacher training. Schemes such as free mid-day meals, free provision of textbooks to increase attendance need to be enforced through local "ombudsmen [...] to check for delay, corruption, lethargy and inefficiency" (Indian Institute of Education 2002: 9). The authors further recommend stronger community involvement in school improvement, as they noted better functioning of schools when Village Education Committees were active.

Similar to the majority of empirical studies, government surveys of national, state, and district bodies overemphasize quantitative indicators such as gender-, caste-, and age-disaggregated enrollment, attendance rates, dropout rates, literacy levels, student–teacher ratios, and infrastructure expenditures. Despite the importance of these, context-specific research and awareness to understand the underlying processes are needed to develop more targeted policies to increase the quality of subject-specific teaching methodologies. Berndt (2010: 286) argues for double truths: on the one hand, the public truth of enrollment and the idealist objectives of national programs, and on the other hand, the collective truth of low quality of public education, discrimination, exclusion, and corruption within the educational system. The author of this book identified a contradiction between the national notion of egalitarian education in national government documents and programs and the lived reality of a socially stratified and institutionally hierarchized educational system. She observed a Brahmanocentric and Anglicist notion of classroom practice marked by the authoritarian role of teachers, high amount of private debts, central role of English teaching, utilitarian approach of teaching content, and the applied teaching methods (Berndt 2010: 291). Berndt (2010: 294) argues that these educational realities reproduce stratified inequalities and a dual educational system. Socio-cultural causes of educational disadvantage and emancipatory contents remain untouched within educational objectives, which could address historically rooted power inequalities. The teaching contents of teacher trainings were not critically reviewed and adapted to classrooms and thus, given the small space of teachers' agency, were simply transferred.

Given that the educational discourse in India has been centered on improving access to education, a paradigm change toward more qualitative studies investigating processes of schooling is greatly needed. The importance of educational quality, that is subject-specific learning outputs and wider skill development, has just started to gain attention. Geography teaching methodology, for example, is not an established research area in India. The exclusion of geography education as a scientific field is obvious in the country report on *Progress in Indian Geography 2004–2008* by the Indian National Science Academy (Singh 2012) as it does not include geography education among the 19 subdisciplines reviewed. The Department of Education at the University of Delhi is one of the few academic institutes that has a professor position that covers geography education research.

This overview of India's educational situation implies that the goals of ESD pose a challenge to the current realities of geography education in India. Of particular focus to address is the approach of rote learning prevalent in most of the country's schools. The most important governmental report addressing these issues is "Learning Without Burden," which was submitted by the National Advisory Committee to the Ministry of Human Resource Development of the Government of India in 1993. The lead author Pal (1993) revealed the existence of the convention of "teaching the text" and "an examination system which focuses on information rather than on skills" (1993: 19–20). This indicates the need of cooperatively designing, implementing, and evaluating new teaching approaches in partnership with teachers and other educational stakeholders to understand how to empower students with skills for democratic participation.

4.1.4 The Principal Role of Textbooks for Educational Quality

Textbook provision is one fundamental way to improve the quality of education, if textbook contents are well designed, correct, and meaningful. Textbooks structure teaching content and tasks as well as prescribe the organization and manner of knowledge acquisition. Studies of the World Bank and other international agencies have shown that the provision of textbooks presents a cost-effective strategy for improving overall educational quality in emerging countries (Crossley and Murby 1994: 101). Textbook use has a particularly strong impact on educational achievement in low-income settings where other learning resources are less available (UNESCO 2005: 48). A meta-analysis of 79 studies on the factors that influence student learning in emerging countries demonstrated that the availability of textbooks has a significant effect (Glewwe et al. 2011: 19). Yet, provision by governments and national and international agencies alone is not sufficient; textbooks need to be "pedagogically sound, culturally relevant and physically durable" (World Bank 1990).

In India, the majority of available textbooks are those developed by state boards and national authorities (SCERT and NCERT), while a private textbook market (e.g., Morning Star, Orient Black Swan, Evergreen Publications) is growing, especially for English-medium schools. However, SCERT and NCERT have a greater outreach (also because of their lower costs, which are around 20–50 INR). As the textbook is the central, and often only teaching resource in Indian classrooms, the textbook has been reported as the "sole arbiter of what is worth knowing" (Advani 2004: 103) and "the only authoritative resource" (Pal 1993: 8). Teachers have little choice on teaching content, and Kumar (1988: 453) refers to a prevailing "textbook culture" in the Indian educational system. The dense presentation of textbooks limits the autonomy of teachers and narrows their teaching space.

Since textbooks are steered and legitimized by current national and state authorities, they are political instruments that frame the contents and methods students have to learn. Textbook analyses can reveal the power that state authorities impose on teachers. Kumar (2005) conducted a study on textbooks before and after independence that showed how moral, political, economic, and cultural control had been imposed on teachers and students through textbooks. Advani (2004: 110) supports this argument by demonstrating how Indian textbooks were used for nationalist (Hindu) interests. By awarding a specific national identity privileges over diversified backgrounds and the interests of children, a particular ideology is prescribed. This is well described through the presentation of the ideal national child. This ideal national child obeys authorities, knows correct answers, does not make any mistakes, and is enthusiastic about national symbols (Advani 2004: 106). When the government changed in 2004, the new edition of the National Curriculum Framework (NCF) set a new focus on constructivist learning approaches in 2005. These focused on the discovering child and the learning environment, rather than imposing prescribed answers. This innovative approach broke with prior methods of transmitting nationalist ideologies via the curricula. History textbooks, for example, should build identity through critical reflection on historical events and hence encourage all people to contribute to public discourse in a democratic society (Gottlob 2007). The new direction of the National Curriculum Framework 2005 demonstrates how the values transmitted in the educational system are dependent and affected by the ruling government's interests. This demonstrates how politically driven the choice of textbook contents is.

The choice of English over vernacular textbooks represents another powerful tool for enforcing political interests. The intentional preference of English as a working language enforces social inequality (Advani 2004: 110; Kumar 1989). Although knowledge of English improves access to the economic market, under-privileged students who do not speak English and, in Bourdieu's terms, do not have the cultural capital are excluded from educational success, as English is essential but cannot be adequately taught through textbooks (Advani 2004: 110).

The important role of quality textbooks for improving education in India and elsewhere justifies the analyses of ESD principles in national and state government textbooks of geography (NCERT, SCERT) in this study. The analyses will particularly build on the insights of a national textbook study conducted by Bharati Vidyapeeth Institute of Environment Education and Research (BVIEER) (2002). The study by BVIEER (2002) on "The status of infusion of environmental concepts in school curricula and the effectiveness of its delivery" reveals a lack of both in-depth information and links to students' local environments in textbooks of different subjects and grade. Especially since the environmental examples given in textbooks do not promote pro-environmental activities, it is difficult for students to link classroom knowledge to their own lives (BVIEER 2002: 4). There is no growing complexity in content over the school years, and environmental concepts are either absent or inadequate in disconnected lessons (BVIEER 2002: 5). Thus, the introduction of environmental education has not been covered in Indian schools to a great extent.

4.1.5 The Educational Landscape of Pune, Maharashtra

Due to its numerous universities and partly excellent tertiary education, Pune is considered the "educational capital of India" (e.g., Adhyapak 2016). A general trend regarding primary and secondary education can be noticed, as in other Indian cities: intra-urban disparities with respect to access to schooling and quality of teaching, which is closely connected to the socio-economic background of the students' parents. Government primary schools in Maharashtra are the responsibility of the Zilla Parishads in the rural areas. In Pune, the Pune Municipal Corporation (PMC) manages primary education. Secondary schools are affiliated either with the Maharashtra State Board of Secondary and Higher Secondary Education Board (MSBSHSE) or the national Central Board of Secondary Education (CBSE) or the Indian Certificate of Secondary Education (ICSE). Consequently, state textbooks are from the Maharashtra State Bureau of Textbook Production and Curriculum Research under the Maharashtra State Council Research and Training (MSCERT). The national CBSE uses textbooks from the National Council of Education and Research (NCERT), and the ICSE designs its own textbooks. Similarly, state and national boards have their separate formats and examination schedules. The national ICSE and CBSE board exams in the class 12 are weighted higher over the Maharashtra state board exams due to advanced syllabi. Private schools often adopt the CBSE or ICSE textbooks and exams, or use textbooks by private publishers that comply with the CBSE or ICSE syllabi.

In the city of Pune, there were 603 primary schools with 268,744 students and 6,148 teachers and 460 secondary schools with 376,860 students and 10,994 teachers (Government of India 2001)[3] in 2001. The student dropout rate was 8.0% for primary schools and 17.6% for class 10. Student–teacher ratios are 42:1 in municipal schools (Pune Municipal Corporation 2005). Discrimination of girls in regard to educational access is low and is also reflected in the sex ratio of 945:1000 in Pune, which is the fourth highest sex ratio in cities of 2 million inhabitants. The literacy rate of 91.6% is the second highest literacy in cities over 2 million inhabitants, with 92.5% for males and 86.1% for females (Government of India 2011a).

The Indian Institute of Education (2002: 6) found in its report *A Status and Evaluation Study of the Upper Primary Education of The Elementary Education System* that classrooms in Maharashtra are, similar to national surveys, "highly deficient in basic infrastructure [… and have an] insufficient number of teachers," as well as the lack of subject-specific and in-service training. However, as additional expenses need to be covered, only those families who can afford these expenses can send their children to private schools. For the report, headmasters in Maharashtra were interviewed on their perspectives of the difficulties they face in managing upper primary schools. Their answers included posts of headmasters and teachers kept vacant and not filled in time, lack of trained teachers, inadequate teaching materials, lack of classrooms, educational aids, laboratories, and interference by political leaders (Indian Institute of Education 2002: 76). According to the report of the Indian Institute of Education

[3]The census data for 2011 were not available free of charge.

(2002), enrollment rates and accessibility in terms of distance, connecting roads, and availability of public transport have increased. Dropout rates are lower than in other states; e.g., the dropout rate up to grade 7 declined from 49% in 1990/1991 to 32% in 1999/2000. Over the last three decades, privately run secondary schools have been increasing as government schools are generally perceived by parents as not efficiently managed (Indian Institute of Education 2002: 4). The study on the *Status of Elementary Education in Maharashtra* (Government of Maharashtra 2012) notes that 32% of all Maharashtrian schools are without female teachers, whereas the majority of these schools are primary government schools in rural areas. The percentage of female teachers is 40% in government and local bodies schools, and private aided schools, and as high as 74% in private unaided schools. These infrastructural characteristics of the educational landscape in Pune and Maharashtra demonstrate that Pune can serve as a good example for this study investigating ESD in India's urban educational settings emerging to improve the quality of education.

4.1.6 Environmental Education and ESD in India

Environmental education is the predecessor of ESD, with the former having a stronger focus on nature education (Lee and Chung 2003; Steiner 2011). In India, ESD can build on the attempts of national enforcement to make environmental education a compulsory subject. For two and a half decades, different stakeholders have tried to introduce environmental education to Indian schools at the national level. In 1991, the environmental lawyer Shri M. C. Mehta made use of Public Interest Litigation (PIL) and filed a remarkable case in favor of mainstreaming environmental education in all schools in India. Ruling on this Civil Writ Petition No. 860/1991 titled M. C. Mehta v. Union of India, the Honorable Supreme Court of India held that

> through the medium of education, awareness of the environment and its problems related to pollution should be taught as a compulsory subject [...] we would require every State Government and every Education Board [...] to immediately take steps to enforce compulsory education on environment in a graded way. [...] as by now there is a general acceptance throughout the world as also in our country that protection of environment and keeping it free of pollution is an indispensable necessity for life to survive on this earth. If that be the situation, everyone must turn immediate attention to the proper care to sustain environment in a decent way. (Direction IV of the Hon'ble Supreme Court of India judgment dated 22nd November, 1991)

This decision of the Supreme Court as the highest constitutional court remarkably endorses the principle of environmental education. However, as states and educational authorities did not implement the order, M. C. Mehta filed another case, which led to a second order on the enforcement of environmental education on December 18, 2003:

> we direct all the respondents- States and other authorities concerned to take steps to see that all educational institutions under their control implement respective steps taken by them and as reflected in their affidavits fully, starting from the next academic year, viz. 2004-05 at

> least, if not already implemented. The authorities so concerned shall duly supervise such implementation in every educational institution and non-compliance of the same by any of the institutions should be treated as a disobedience calling for instituting disciplinary action against such institutions.

Following this order, the University Grants Commission (UGC), as India's statutory body for maintaining standards in higher education, introduced environmental education as a compulsory subject in college education and developed a common syllabus and a textbook for environmental sciences. State ministries for education are obliged either to infuse environmental education into each school subject or to introduce environmental studies as a separate subject for students. However, when weakness became clear, for example, that some states were not taking up the order to include environmental education as a subject in colleges, or that teachers were not qualified to teach environmental education, M. C. Mehta redressed his grievances and filed a case arguing that the "lack of education in environment science would prejudicially affect the spirit of [... two] Articles" of the Constitution of India, namely Article 48A providing that "the States should endure to protect and improve the environment and safeguard the forests and wildlife of the country" and Article 51A(g) imposing "duties on every citizen to protect and improve the natural environment, including forests, rivers, lakes, and wildlife and to have compassion for the living creatures." However, his grievance was rejected for procedural reasons and the court asked "the applicant to approach the appropriate forum/court for enforcement of that direction" (National Green Tribunal 2014).

In CBSE schools today, the subject of environmental studies is taught in standard 3 to 5, with the contents of it focusing on the child's immediate environment. Environmental studies' textbooks for classes 3, 4, and 5 provide examples of good practice for ESD in textbooks, addressing the themes of natural and man-made changes in the environment, fauna and flora, food, water, air, shelter, clothing, functions and festivals, health and hygiene, transport and communication, and environmental pollution and protection. At the state level in Maharashtra, however, policy initiatives, curricula, and textbooks have not yet been orientated toward explicitly meeting the UGC's guidelines on environmental education. The state Maharashtra introduced "environmental sciences" as a separate subject, which was, however, substituted in 2013 by "information and technology communication" (ICT). This demonstrates the limited and fading importance given to environmental education.

Despite India's Supreme Court order to introduce environmental education modules to the country's syllabi and textbooks, the introduction of environmental education at the school level has not yet been achieved nationwide (BVIEER 2002). For formal education, the study on the "Status of infusion of environmental concepts in school curricula and the effectiveness of its delivery" by the BVIEER (2002) revealed that environmental examples given in textbooks do not promote pro-environmental activities. Because of a lack of in-depth information alongside the absence of references made to specific localities, it is difficult for many students to make the link between the knowledge they acquired at school and their everyday lives. Teaching content does not clearly progress in complexity with increasing standard levels, lessons are disconnected from daily realities in terms of content, and environmental concepts are either absent or inadequate. While environmental education topics are woven into science, geography, and languages textbooks, they are not identified as being related to the immediate environment of every student. As environmental education is not considered a core curricular subject, it is thought of as being of only minor relevance. Therefore, environmental education does not sufficiently find its way into the Indian classroom at present. The study's findings suggest that the inclusion of an interactive group work will situate environmental education as a distinct subject (BVIEER 2002). Despite its openness and adaptability, the integration of environmental education into Indian syllabi, textbooks, pre- and in-service teacher training programs, and classrooms is currently rather limited—which adds to the enormity of the task faced if ESD is to be successfully integrated into the country's educational system.

There have been several noteworthy environmental education programs in India. The Ministry of Environment and Forest (MoEF) of the GoI launched the flagship scheme "Environmental Education, Awareness and Training (EEAT)" in 1983–1984 to promote environmental awareness and skills to protect the environment among all sections of society in the formal education sector and the non-formal sector. In 1984, the Center for Environment Education (CEE) was established with the mandate to promote environmental awareness nationwide. Besides the National Environment Awareness Campaign (NEAC) and other awareness programs, seminars, conferences, and workshops have been held regularly until today. CEE leads the *National Green Corps* (NGC) as a governmental program for environmental awareness implemented in every district. Every district should have 250 *eco-clubs* in schools for class 9th and higher. Every school can apply for 2500 Rupees (approximately 30.00 €) per year from the central government. Additional funds of the state government can be applied for ranging from 12,500 Rupees (approximately 150.00 €) in Delhi to 3500 Rupees (approximately 43.00 €) in Himachal Pradesh. For the application, schools hand in a utilization report with bills to the state governments. This program provides incentives for schools to develop own environment activities. However, it is criticized for focusing only on environmental awareness, while guidance on implementing environmental action is lacking (Prof 7). In contrast to the extracurricular eco-club approach of the NGC, CEE's nationwide initiative Paryavaran Mitra (Environment's friend) follows a curricular approach in linking action projects to classroom activities. Paryavaran Mitra is a sustainability and climate change education project aimed at the 20 million students in classes 6–8. A curriculum-linked teacher's handbook provides activities to reduce the footprint and increase the handprint on the environment. Despite the envisioned great coverage of the programs, the outreach and effectiveness in schools vary according to stakeholder interviews (cf. Chap. 7).

Non-governmental organizations lead a number of extracurricular green school projects, in which environmental activities follow ESD principles. Non-governmental programs, such as the green school program of the Center for Science and Environment (CSE), cover a total of 21,000 schools in 16 states, of which 1000 schools are active and hand in their annual reports (NGO 2). Each year, 35% of the participating schools are new,[4] which demonstrates a great fluctuation. There are links to government programs in five states,[5] and teacher trainings are conducted in collaboration. These environmental education programs demonstrate the great outreach and leading role both government as well as non-governmental programs play in reaching out to students.

4.2 Water Conflicts in the Emerging Megacity of Pune

4.2.1 Water Supply as Global Challenge for Sustainable Urban Development

Under the continuous pressure of urban population growth, especially in megacities, the availability and access to drinking water are not sufficiently meeting the demands. Municipalities account for 10% of the global freshwater withdrawals, while agriculture accounts for 70% and industries for 20% (Comprehensive Assessment of Water Management in Agriculture 2007). From 1990 to 2008, 1.8 billion people gained access to improved drinking water, of those 59% live in urban areas (UN World Water Assessment Programme 2009: 5). The use of improved drinking water in urban areas has not increased as much as the population growth (World Health Organization and UNICEF 2010: 18).

Global data conveys that the human right of access to safe drinking water and sanitation does not apply to a great part of the global population. The MDG to halve the population without safe drinking water was officially met in 2010, with 91% of the global population now using improved[6] drinking water sources, but that means that 663 million people still lacked access to safe drinking water in 2015 (World Health Organization and UNICEF 2015: 4). Similarly, the MDG target on sanitation was not obtained: 2.4 billion people still lack improved sanitation facilities; 72% of these people live in Asia (World Health Organization and UNICEF 2015: 5).

This data hides the great urban–rural divide, the gendered division of who collects water and suffers from water insecurity and pollution as well as the politics of exclusionary water governance. Particularly, the urban poor who live in informal settlements are excluded from water development schemes. More inclusive water governance is needed (UN-Habitat 2006: 88). The rapid increase of the slum populations in developing contexts is linked to the increase of urban population without access to water and sanitation due to "the inability or (or unwillingness) of local

[4] Information on CSE was given by a staff member during an expert interview.

[5] Andhra Pradesh, Chattisgarh, Delhi, Himachal Pradesh, and Sikkim.

[6] However, improved drinking water source merely means "one where human use is kept separate from use by animals and fecal contamination. In many cases the water from such sources is not of good enough quality to be safe for human consumption" (UN World Water Assessment Programme 2015).

and national governments to provide adequate water and sanitation facilities in these communities" (UN World Water Assessment Programme 2015: 3).

The attention of international and national development bodies strongly focuses on the role of water for sustainable rural and urban development. In 2015, the annually published *UN World Water Development Report (WWDR)*, also decisive for the theme of the annual World Water Week (WWW), stressed the centrality of water by linking it to the three dimensions of sustainable development (the didactic framework for ESD of this study does so likewise). The report highlights the important role of water for poverty elimination, economic growth, as well as increasing social equity and environmental sustainability, while envisioning participatory water governance (UN World Water Assessment Programme 2015: 2). The report theme in 2016 is *Water and Jobs* and emphasizes, next to the human right of safe drinking water and sanitation, the multi-sectorial importance of sustainable water management to employment, particularly for marginalized communities. The relevance of water for the global economy becomes clear when it is considered that 42% of the global workforce is heavily water-dependent, and another 36% is water-related (UN World Water Assessment Programme 2016: 2). If not effectively addressed in policy and planning, water scarcity and conflicts can lead to (forced) migration.

The supply of water in megacities of newly industrialized countries depends on climatic, hydrological, geomorphological as well as social, infrastructural, financial, and political factors. The interdependency of these factors contributes to unequal access to drinking water by different groups who live close together. These groups can be of different ethnic, socio-economic, and educational backgrounds who share and compete for access to water. Through economic development and social differentiation, socio-economic disparities are enhanced by inequalities in access to water and education. Especially, the coexistence of "gated communities" of the globalized middle and upper class and informal settlements shows the socio-economic disparities in society (Borsdorf and Coy 2009). The lack of public regulations and development planning of water supply contributes to informal water access. People living in slums with temporary housings, limited access to clean drinking water, food and energy supply, limited public space as well as low standards of education and healthcare services are prone to poverty, unemployment, and economic insecurities (Hackenbroch et al. 2009; Mertins and Müller 2010; Wehrmann 2008).

The strong dynamics of development and the high complexity of different local, regional, national, and global processes lead to major water governance struggles in megacities of newly industrialized countries. The "urban water crisis" (Herrle et al. 2006) requires new approaches to (drinking) water and wastewater management. These could include building reservoir dams as well as ecological sustainable methods such as storing rainwater for harvesting in private households and the use of recycled water for irrigation and toilet flushing. While water demand between different rural household uses (domestic/agricultural) may compete, several approaches have been introduced to rethink the sectorial approach to water use. In agriculture, for example, the approach of "Multiple Use Water Systems" (MUS) takes multiple water needs of households into account for planning new water systems that are more cost-effective and enhance household well-being and women's productivity (van Koppen et al. 2008). Such an integrated approach to water provision and planning proved more sustainable than single-use water systems (Clement et al. 2015). Similar innovative approaches are needed for urban water management with robust institutional frameworks and participatory and inclusive water governance aiming at socially, economically, and environmentally sustainable resource use. The need for changing behavioral patterns in terms of water use and hence the role of environmental education for raising environmental awareness is particularly relevant. Awareness on sustainable water use for domestic as well as productive use can influence self-reflection and decision-making on water practices (cf. Mandl and Gerstenmaier 2000).

4.2.2 Water Conflicts in Pune

The city of Pune is a good example for an Indian city at the nexus of rapid urbanization, economic growth, socio-economic fragmentation, environmental degradation, and weak water governance. Because of its increasing urbanization and population growth, Pune is on its way to becoming a megacity, which entails great infrastructural, ecological, and social challenges related to water resources.

Located in the Western Ghats of India, Pune has a monsoonal climate with seasonal rainfall from June to September. Rainfall occurring within these four months of the year is stored in six dams west of Pune's urban agglomeration (cf. Fig. 4.3). In the remaining months, the dams provide drinking water for the population and supply water for irrigation agriculture, industries, and energy generation (the Mulshi Dam provides water for energy generation in Mumbai). Rising from the dams, the confluence of the rivers Mula and Mutha is in the city center of Pune and the river then flows to the Gulf of Bengal, which emphasizes the river's transregional role. Upstream and downstream conflicts occur and need to be addressed through supraregional river management approaches (Wagner et al. 2015: 52).

Wagner et al. (2013) modeled land use changes induced by urbanization and deforestation and their impacts on the water balance in the Mula and Mutha rivers catchment upstream of Pune. With the Soil and Water Assessment Tool (SWAT), the authors developed a multi-temporal land use classification and modeled future hydrological scenarios (cf. Fig. 4.3 for a current land use map of Pune). Between 1989 and 2009, the urban area increased from +5.1% to +10.1% and the water yield increased by up to +7.6% (Wagner et al. 2013). Forest, shrubland and grassland decreased in this time period by −9.1%. Future scenarios predict a further expansion of the urban area of +7.9% until 2028, with decreases of −3.6% in agricultural land and −4.3% for natural land. The demand for water for the urban population and industrial use is increasing, while overall water availability is decreasing due to these land use changes, and predicted climatic changes which bring higher variability in precipitation. This indicates a continuous decrease of water availability during the dry season as well as exacerbated competing user interests.

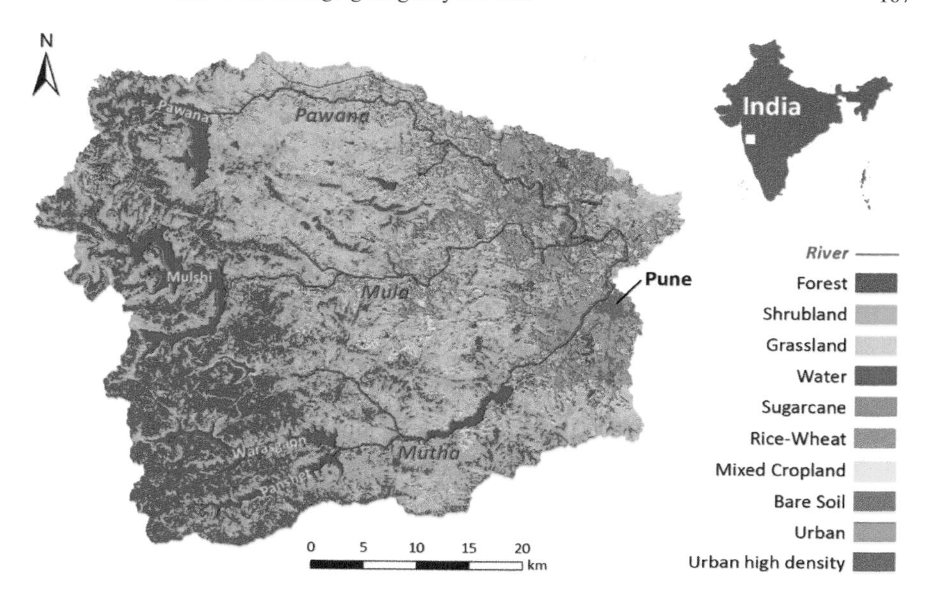

Fig. 4.3 Land use map 2009/2010 of Pune with the six dams in the west, which supply water to sprawling urban Pune in the east (Wagner et al. 2013)

Besides climatic, topographical, infrastructural, and demographic challenges, competing user groups, socio-economic disparities, and exclusive water governance are the key factors to understanding water supply conflicts in Pune. Particularly in the dry season, in the months from March to May before the monsoon, intra-urban conflicts occur due to the distribution of the limited resource of water between public and private stakeholders, competing user groups and different socio-economic groups (The Times of India 2010). This should arouse even more attention considering that the Pune Municipal Corporation (2010) stated that the water supplied to the city in 2009–2010 (14.5 TMC) exceeded *twice* the amount the city required (7–7.8 TMC) according to the Central Public Health and Environmental Engineering Organization (CPHEEO). One reason for this is the high water loss of 40% (Rode 2009: 54) due to infrastructural leakages during the transfer from dams to water end users in Pune. Infrastructural issues include outdated and leaking water pipes, which partly stem from the period of British colonialism, with the oldest water supply system dating back to 1750 (Rode 2009: 49). Further infrastructural challenges are insufficient storage capacities for drinking water and undersupplied city cantons in higher topographical locations because of the inadequate delivery rate of pumps. Apart from infrastructural challenges, unaccounted for water accounts, which range from 25 to 40% in Indian cities (McKenzie and Ray 2009: 447), express the inefficiency of current water governance. Zérah (2000: 22) states in her extensive study on water supply unreliability in Delhi that the "distribution problem [...] lies more in inefficient management of the supply network rather than in a scarcity of water resources," a statement that holds particularly true for Pune.

While the Pune Municipal Corporation (2011: 11) states a daily per capita water supply of 194 L "which is more than 1.5 times compared to other cities," the case study of Kroll (2013) in six different residential areas in Pune indicates high variations in water access related to the socio-economic status of communities. Privileged gated communities and exclusive residential areas have unlimited water access due to community water tanks, while most residential areas have intermittent water supply; i.e., in the mornings and in the evenings a total of four to six hours of water are supplied (Pune Municipal Corporation 2011). Water availability parallels seasonal variations and in dry seasons, water shortages can last up to 48 h, as for example in the "2009 water crisis" (Jadhav 2012). Especially in informal settlements, in which 40% of the city's population lives, several households share one water connection and particularly women have to spend a great deal of their time fetching water daily (Pune Municipal Corporation 2010). Unauthorized settlements are not connected to the municipal water supply network. Therefore, these areas have neither a regulated water supply nor hygienic wastewater disposal. Two in-depth studies on water supply in Delhi point out insights relevant to urban water conflicts. The Foucauldian analysis on the production of Delhi's waste waterscapes in Delhi's informal settlements by Zimmer (2011) revealed that power-laden processes produce stigmatizing discourses and "Othering" of groups and individuals (Spivak 1985: 252). These are manifested in state representatives' practices of labeling slum settlements as "less clean, less ritually pure, or less hygienic" (Zimmer 2011: vi). Excluding these inhabitants in (waste) water governance leads to invisibility of everyday wastewater struggles in informal settlements. Selbach (2009) demonstrated how even in planned urban areas of the capital Delhi, insufficiently constructed, overstretched, and inefficiently managed water supply cause different types of vulnerability. Although public water connections, manually operated pumps, or water tankers are supposed to guarantee a minimum of water supply, multiple measures are needed to improve the water supply situation and to decrease vulnerabilities related to both water scarcity and pollution.

Not only the limited and unequal water availability, but also the degradation of water is an increasing challenge in India and Pune. Insufficient wastewater disposal and purification and inadequate waste management are manifested through institutional under-investment and inadequate management capacities of India's municipal corporations (Nandi and Gamkhar 2013). In Pune, only 55% of the total generated sewerage in the city is treated, while the remaining 45% is released untreated into the river (Pune Municipal Corporation 2010). This leads to exceeding levels of biochemical oxygen demand (BOD) and chemical oxygen demand (COD). This deteriorating water quality is in line with the national trend of 49% untreated municipal sewage within India's metropolitan cities (>1 mio. inhabitants), which flows into water bodies and leads to organic and bacterial contamination (Central Pollution Control Board 2016). Leaking or broken connections and depressions in wastewater treatment plants and their limited number and working capacity lead to water contamination. Similarly, illegal access to groundwater facilitates the rapid subsidence of the groundwater level and increasing contamination through intruding bacteria and germs.

Obvious impacts of insufficient and unsafe drinking water supply are water-induced diseases. Bringing along great financial costs and loss of economic activities, waterborne diseases are India's biggest public health risk (McKenzie and Ray 2009: 445). Pune's rapid urbanization and inadequate provision of water supply infrastructure lead to intra-urban differences in exposure to health risks as well as differential access to healthcare services due to awareness and affordability (Butsch et al. 2015; Butsch 2011; Kroll 2013). Water-induced diseases represent 20–30% of all occurring diseases in Pune (Patwardhan et al. 2003: 6). Particularly during the rainy season, fluctuating water quality promotes waterborne and vector-borne diseases such as gastroenteritis, hepatitis, dengue, typhoid, and malaria (Bork et al. 2009).

As in other (mega)cities with rapid and mostly uncontrolled settlement growth, the Pune Municipal Corporation (PMC) cannot keep up with infrastructure provision due to a lack of finance, human resource, and management capacities (Nandi and Gamkhar 2013). With untransparent and exclusive water governance, increasingly informal and also illegal processes are evolving, which on the one hand temporarily reduce the state of emergency but also do not provide long-term security as these are unregulated processes (Kraas and Nitschke 2006: 23). The urban government is responsible for water supply, but does not have sufficient control over water use and cannot control illegal tapping (Rode 2009: 55). The ratio of metered water connections is only 29.71% (Pune Municipal Corporation 2011: 21). Unpaid or irregularly paid water supply bills (for domestic use 3 Rupees per 1000 L) promote a lacking awareness of the economic value of water and lead to water waste (Jadhav 2012). Vague or unexecuted laws on water supply lead to infractions of the law (Centre for Environment Education 2012).

To improve the current problematic state of Pune's water governance, PMC intends to implement a comprehensive water management system. This includes a controlled tariff structure with metered continuous water supply ("24/7"), which will also reduce water waste during distribution and transmission (Jadhav 2012). Since 2011, the Italian company Studio Galli Ingegneria (SGI) has worked on leak detection, data elicitation, and amount charging. A pilot study on hydraulic defects and water demand of consumers was carried out to yield a long-term improvement of the water supply in Pune. Léautier (2006) highlights how infrastructural provisions are interlinked with technological advances and, possibly, globalization, which is an opportunity for new and more complex multi-level institutions to increase urban performance through robust, transparent, and inclusive water governance systems.

The complexities around water use and supply conflicts in Pune indicate the great relevance ESD could play in addressing multiple perspectives on the causes and consequences of these. The role of ESD on water conflicts in geography education is outlined in the following chapter.

4.3 Relevance of ESD-Oriented Geographical Education on Water Conflicts

Geographical education on water conflicts is relevant due to increasing water scarcity as a problem of global scale with complex social, political, economic, and ecological causes and consequences that can also affect students individually. The UNESCO (2009) calls water education "a key dimension of the international response to the world's water crises" and promotes water education as an integral part of all educational systems. Water education is also meant to be related to wider objectives of the international community, such as "poverty eradication, adaptation to climate change, provision of basic human rights, gender equality and indigenous cultures" (UNESCO 2009). Therefore, the topic of water conflicts is closely related to the objectives and principles of Education for Sustainable Development (ESD) and is, due to its complexity and controversy, a suitable theme to promote network-thinking and argumentation skills.

ESD provides a framework for water education through argumentation as it promotes approaching water conflicts—or other natural resource conflicts—from three perspectives, namely social (including the cultural and political), environmental, and economic dimensions of sustainability. Reflecting and arguing on intra-urban water conflicts from these three perspectives of sustainable development is a novel and required approach for introducing ESD to geography education. ESD-oriented geographical water education indicates one comprehensive problem-, process-, and people-oriented approach that is required in order to inform about environmental and social challenges of sustainable urban development (cf. Kraas 2007). Such an approach works out the controversy of sustainable development in regard to natural resource use and prepares students to become critical citizens sensitized to these present and future challenges. Based on comprehensive knowledge, argumentation skills on water conflicts can influence students' decision-making on sustainable lifestyles.

The example of water conflicts in Pune displays natural resource management challenges in an emerging megacity in the newly industrialized country of India. As an example at the human–environment interface in a complex and globalizing market-driven urban setting, students in Pune can understand the interrelation of multiple human and physical geography processes and factors in their own environment. Since this topic is set in the students' socio-cultural context, it makes it meaningful and increases students' interest and motivation to learn, to engage in, and to value geography education for their community's sustainable development. For this purpose, it is important to understand Pune's water scarcity not only as a natural, technical, or even immutable challenge, but also as a result of human–environment interrelations. Processes contributing to aggravated water scarcity are itself complex and need to be understood in a greater coherence to increase decision-making on lifestyle choice and sensitize for behavioral patterns. This also holds true for related global challenges such as population and economic growth, urbanization-induced land use change, as well as climatic changes. The unequal water resource access of differential groups should be realized as a social problem, which is linked to the relationship between people's socio-economic background and their exclusion in local water governance. Hence, students should be able to argumentatively depict local controversies on the causes for insufficient water supply as well as the conflict potential because of different interests.

References

Adhyapak. (2016). *Pune education*. http://adhyapak.com/schools/pune-education/polytechnic-colleges.html?q=pune-education/polytechnic-colleges.html. Accessed.

Advani, S. (2004). Pedagogy and politics: The case of English textbooks. In A. Vaugier-Chatterjee (Ed.), *Education and Democracy in India*. New Delhi: Manohar Publishers.

Alexander, R. (2001). *Culture and pedagogy. International comparisons in primary education*. Singapore: Blackwell Publishing.

Berndt, C. (2010). *Elementarbildung in Indien im Spannungsverhältnis von Macht und Kultur. Eine Mikrostudie in Andhra Pradesh und West Bengalen*. Berlin: Logos Verlag.

Bharati Vidyapeeth Institute of Environment Education and Research. (2002). *Study of status of infusion of environmental concepts in school curricula and the effectiveness of its delivery*.

Bork, T., Butsch, C., Kraas, F., & Kroll, M. (2009). Megastädte: Neue Risiken für die Gesundheit. *Deutsches Ärzteblatt, 106*(39), 1877–1881.

Borsdorf, A., & Coy, M. (2009). Megacities and global change: Case studies from Latin America. *Die Erde, 140*(4), 341–353.

Bronger, D., & Trettin, L. (2011). *Megastädte - Global Cities HEUTE: Das Zeitalter Asiens?*. Münster: Lit Verlag.

Butsch, C. (2011). *Zugang zu Gesundheitsdienstleistungen. Barrieren und Anreize in Pune, Indien*. Stuttgart: Franz Steiner Verlag.

Butsch, C., Kroll, M., Kraas, F., & Bharucha, E. (2015). How is rapid urbanization in India affecting human health? Findings from a case study in Pune. *ASIEN, 134*, 73–93.

Central Pollution Control Board. (2016). *Water quality trend*. http://www.cpcb.nic.in/water.php. Accessed March 27, 2016.

Centre for Environment Education. (2012). *Drinking water in Pune*. unpublished.

Chandramukhee, & Leder, S. (2013). Dowry practices and gendered space in urban Patna/India. *Gender Forum*, 42. http://www.genderforum.org/issues/gender-and-urban-space/dowry-practices-and-gendered-space-in-urban-patnaindia/.

Clement, F., Pokhrel, P., & Yang Chung Sherpa, T. (2015). Sustainability and replicability of multiple-use water systems (MUS). *Study for the Market Access and Water Technology for Women project (MAWTW)*.

Comprehensive Assessment of Water Management in Agriculture. (2007). *Water for food, water for life: A comprehensive assessment of water management in agriculture*. London, Colombo: Earthscan, International Water Management Institute.

Crossley, M., & Murby, M. (1994). Textbook provision and the quality of the school curriculum in developing countries: Issues and policy options. *Comparative Education, 30*(2), 99–114. http://www.jstor.org/stable/3099059.

Eick, C. J., & Reed, C. J. (2002). What makes an inquiry-oriented science teacher? The influence of learning histories on student teacher role identity and practice. *Science Education, 86*(3), 401–416.

Glewwe, P., Hanushek, E. A., Humpage, S., & Ravina, R. (2011). School resources and educational outcomes in developing countries: A review of the literature from 1990 to 2010. *NBER working paper 17554*.

Gottlob, M. (2007). Changing concepts of identity in the Indian textbook controversy. *Internationale Schulbuchforschung - International Textbook Research, 29*(4), 341–353.
Government of India. (2001). *Census of India*. New Delhi: GOI.
Government of India. (2011a). *Census of India: Literacy in India*. http://www.census2011.co.in/literacy.php. Accessed March 4, 2014.
Government of India. (2011b). *Census of India: Population enumeration data*. Single Year Age Data—C13 Table. http://www.censusindia.gov.in/2011census/population_enumeration.aspx. Accessed August 8, 2014.
Government of Maharashtra. (2012). *Status of elementary education in Maharashtra state and municipal corporation profiles. District information System for Education (DISE), 2011–12*. Mumbai: Government of India, School Education and Sports Department.
Hackenbroch, K., Baumgart, S., & Kreibich, V. (2009). The spatiality of livelihoods: Urban public space as an asset for the livelihoods of the urban poor in Dhaka, Bangladesh. *Die Erde, 140*(1), 47–68.
Herrle, P., Jachno, A., & Ley, A. (2006). Die Metropolen des Südens: Labor für Innovationen? Mit neuen Allianzen zu besserem Stadtmanagement. *Stiftung Entwicklung und Frieden, Policy Paper 25. Bonn*.
Indian Institute of Education. (2002). *A status and evaluation study of the elementary education system*. Pune: Indian Institute of Education.
Indiatoday. (2013). India has 40 per cent drop-out rate in elementary schools: Report. http://indiatoday.intoday.in/story/india-has-40-per-cent-drop-out-rate-in-elementary-schools/1/324717.html. Accessed.
Jadhav, R. (2012, February 29). PMC's tunnel vision blocks water supply plans. *The Times of India*. Retrieved from http://articles.timesofindia.indiatimes.com/2012-02-29/pune/31109975_1_water-supply-parvati-water-works-total-water-consumption.
Keck, M., Etzold, B., Bohle, H.-G., & Zingel, W.-P. (2008). Reis für die Megacity. Nahrungsversorgung von Dhaka zwischen globalen Risiken und lokalen Verwundbarkeiten. *Geographische Rundschau, 60*(11), 28–37.
Kingdon, G. G. (2007). *The progress of school education in India*. Oxford: ESRC Global Poverty Research Group.
Kraas, F. (2007). Megacities and global change: Key priorities. *Geographical Journal, 173*(1), 79–82. https://doi.org/10.1111/J.1475-4959.2007.232_2.X.
Kraas, F., & Mertins, G. (2008). Megastädte in Entwicklungsländern: Vulnerabilität, Informalität, Regier- und Steuerbarkeit. *Geographische Rundschau, 60*(11), 4–10.
Kraas, F., & Nitschke, U. (2006). Megastädte als Motoren globalen Wandels - Neue Herausforderungen weltweiter Urbanisierung. *Internationale Politik, 61*(11), 18–28.
Kroll, M. (2013). *Gesundheitliche Disparitäten im urbanen Indien. Auswirkungen des sozioökonomischen Status auf die Gesundheit in Pune*. Stuttgart: Franz Steiner Verlag.
Kumar, K. (1988). Origins of India's "textbook culture". *Comparative Education Review, 32*(4), 452–464. http://www.jstor.org/stable/1188251.
Kumar, K. (1989). Learning to be backward. In K. Kumar (Ed.), *Social character of learning*. New Delhi: Sage.
Kumar, K. (2005). *Political agenda of education. A study of colonialist and nationalist ideas*. New Delhi: Sage.
Léautier, F. A. (2006). Cities in a globalizing world. Governance, performance and sustainability. Washington D.C: The World Bank Institute.
Lee, J. C. K., & Chung, Y. P. (2003). *Knowledge foundation: Education for sustainable development*. Oxford: EOLSS.
Mandl, H., & Gerstenmaier, J. (2000). *Die Kluft zwischen Wissen und Handeln - Empirische und theoretische Lösungsansätze*. Göttingen: Hogrefe-Verlag.
McKenzie, D., & Ray, I. (2009). Urban water supply in India: Status, reform options and possible lessons. *Water Policy, 11*(4), 442. https://doi.org/10.2166/wp.2009.056.

Mertins, G., & Müller, U. (2010). Gewalt und Unsicherheit in lateinamerikanischen Megastädten. Auswirkungen auf politische Fragmentierung, sozialräumliche Segregation und Regierbarkeit. *Geographische Rundschau, 60*(11), 48–55.

Ministry of Law and Justice. (2009). *The Right of Children to Free and Compulsory Education Act 2009*. New Delhi: Government of India.

Nandi, S., & Gamkhar, S. (2013). Urban challenges in India: A review of recent policy measures. *Habitat International, 39,* 55–61.

National Center on Education and the Economy. (2006). *A profile of the Indian education system*. Washington D.C.: National Center on Education and the Economy.

National Council of Educational Research and Training. (2005). *National curriculum framework 2005*. New Delhi: NCERT.

National Green Tribunal. (2014). M. C. Mehta vs. University Grants Commission Ors on 17 July, 2014. https://indiankanoon.org/doc/155218083/. Accessed March 29, 2016.

Pal, Y. (1993). *Learning without burden*. New Delhi: Ministry of Human Resource Development, Government of India.

Patwardhan, A., Sahasrabuddhe, K., Mahabaleshwarkar, M., Joshi, J., Kanade, R., Goturkar, S., et al. (2003). Changing status of urban water bodies and associated health concerns in Pune, India. In M. J. Bunch, V. M. Suresh, & T. V. Kumaran (Eds.), *Proceedings of the Third International Conference on Environment and Health, Chennai, India, 15–17 December, 2003* (p. 6). York University.

Pune Municipal Corporation. (2005). *Education in Pune*. Pune: PMC.

Pune Municipal Corporation. (2010). *Environmental status report 2009–10*. Pune: PMC.

Pune Municipal Corporation. (2011). *Environmental status report*. Pune: PMC.

Rode, S. (2009). Sustainable drinking water supply in Pune metropolitan region: Alternative policies. *Theoretical and Empirical Researches in Urban Management, Special Number 15/April: Urban Issues in Asia* (pp. 48–59).

Selbach, V. (2009). *Wasserversorgung und Verwundbarkeit in der Megastadt Delhi/Indien*. Köln.

Singh, R. B. (2012). Progress in Indian geography, country report 2008–2012 und 2004–2008 (IGC 2012). *Indian National Science Academy*.

Spivak, G. (1985). The Rani of Sirmur: An essay in reading the archives. *History and Theory, 24*(3), 247–272.

Steiner, R. (2011). *Kompetenzorientierte Lehrer/innenbildung für Bildung für Nachhaltige Entwicklung. Kompetenzmodell, Fallstudien und Empfehlungen*. Münster: MV-Verlag.

The Times of India. (2010). *Two reasons why planners must look for options*. July 27, 2010. http://articles.timesofindia.indiatimes.com/2010-07-27/pune/28297240_1_dams-that-supply-water-water-quota-severe-water-crisis.

UN World Water Assessment Programme. (2009). *Water for sustainable urban human settlements*. Paris: UNESCO.

UN World Water Assessment Programme. (2015). *The United Nations world water development report 2015: Water for a sustainable world*. Paris: UNESCO.

UN World Water Assessment Programme. (2016). *The United Nations world water development report 2016: Water and jobs*. Paris: UNESCO.

UNESCO. (2005). *EFA Global monitoring report 2005*. Paris: UNESCO.

UNESCO. (2009). *Water education for sustainable development*. Paris: UNESCO.

UNESCO Institute for Statistics. (2016). *Government expenditure on education as % of GDP (%)*. The World Bank. http://data.worldbank.org/indicator/SE.XPD.TOTL.GD.ZS/countries/1W?display=default. Accessed March 22, 2016.

UN-Habitat. (2006). Water and human settlements in an urbanizing world. *The 2nd UN World Water Development Report: Water, a shared responsibility*. Paris: UNESCO. http://webworld.unesco.org/water/wwap/wwdr/wwdr2/table_contents.shtml.

UN-Habitat. (2013). *State of the world's cities 2012/2013. Prosperity of cities*. New York: Routledge.

van Koppen, B., Smits, S., Moriarty, P., & de Vries, F. P. (2008). Community-scale multiple-use water services: MUS to climb the water ladder. In *International Symposium on multiple-use water services*, Addis Ababa, Ethiopia, 4–6 November 2008.

Wagner, P. D., Fiener, P., & Schneider, K. (2015). Hydrologische Auswirkungen des Globalen Wandels in den Westghats. *Geographische Rundschau, 1,* 48–53.

Wagner, P. D., Kumar, S., & Schneider, K. (2013). An assessment of land use change impacts on the water resources of the Mula and Mutha Rivers catchment upstream of Pune, India. *Hydrology and Earth System Sciences, 17,* 2233–2246.

Wehrmann, B. (2008). Existenzstrategien von Kinderhaushalten in Marginalvierteln Nairobis. *Geographische Rundschau, 60*(11), 20–27.

World Bank. (1990). *Papua New Guinea, Primary education project completion report*. Washington D.C.: World Bank.

World Health Organization, & UNICEF. (2010). *Progress on sanitation and drinking water*. Geneva: WHO Press.

World Health Organization, & UNICEF. (2015). *Progress on sanitation and drinking water. 2015 Update and MDG Assessment*. Paris: UNESCO.

Zérah, M.-H. (2000). *Water—Unreliable supply in Delhi*. Delhi: Manohar.

Zimmer, A. (2011). *Everyday governance of the waste waterscapes: A Foucauldian analysis in Delhi's informal settlements*. Bonn: ULB.

Chapter 5
A Framework for Qualitative Geographical Education Research

Abstract This chapter introduces a methodological framework for qualitative geographical education research for international contexts. I use this framework to explore the status quo and opportunities of Education for Sustainable Development through fieldwork, document analyses and an intervention study in India. I argue that geographical education research needs to be embedded in its socio-cultural context by considering how power relations and cultural norms reflected in educational systems shape pedagogic practice. Against this background, I discuss the applied research methods of participant and classroom observations, qualitative interviews, focus group discussions, action research, and questionnaires and I critically reflected in their application to geographical education research. The process of selecting and accessing schools for extensive fieldwork is described. Finally, I reflect on the role and the perspective of the researcher against the setting of the cultural context.

5.1 Methodological Considerations

To assess geographical education research in a foreign context, a didactic-methodological analysis of geography teaching has to be considered against the background of the institutional and structural framework, as well as the respective socio-economic and socio-cultural context. The structures of the educational system, the development of curricula, syllabi and textbooks, as well as power relations in teacher–student communication influence classroom interactions on several levels. For empirical geographical research in developing contexts, Bronger (1977) states it is necessary to collect and analyze empirical data related to endogenous and exogenous factors. Endogenous factors are the ecological, infrastructural, social, and historical context, and exogenous factors are the international political and economic contexts that determine the current development processes of a region (ibid.). Through the understanding of the influence of these factors, research in a developing country is embedded in a larger context of development processes. Thus, a contextual, comprehensive methodological research design integrating these different perspectives has to be developed and analyzed.

S. Leder, *Transformative Pedagogic Practice*, Education for Sustainability,
https://doi.org/10.1007/978-981-13-2369-0_5

The theoretical framework, as outlined in Chaps. 2 and 3, promotes the link of endogenous and exogenous factors to the contents and methods of geography education in India. Educational geography research combined with concepts from the *Sociological Theory of Education* put forward by Bernstein (1975, 1990) and the concept of a *critical consciousness* for social transformation through education put forward by Paulo Freire (1996) thus embeds pedagogic practice as social practice in its specific socio-cultural context. Bernsteinian concepts are applied to the investigation of Education for Sustainable Development in geography education in order to understand the content of teaching (what is learnt and taught) and the methods of teaching (how are students taught and how do they learn) in the classroom. The Bernsteinian concepts of *framing* and *classification* enable the analysis of monologic and dialogic texts used in textbooks, syllabi, classroom practices, and teacher training. Using these concepts, power relations and principles of control in current pedagogic practice are described and analyzed, as is their relationship to the transnational educational objective of ESD.

Bernstein's concepts provide an external language of description for the internal analysis of societal power relations (Neves and Morais 2001: 190). This perspective offers an understanding not only of current teaching and learning practices, but also of how the social and organizational structures of the educational system, educational boards, recent educational policies, and existing school types influence pedagogic practice. Thus, empirical geography education research is embedded in a broad theoretical approach to assess the link between teaching methodology, and institutional, social, and infrastructural conditions of geography teaching in India.

Based on this guiding theoretical framework, an in-depth analysis of the research questions requires the collection of comprehensive and relevant empirical data on the content and teaching methodology in geography education, as well as on the educational system. To reflect on the contextual social constructions of order and reality from the perspective of the stakeholders (Lamnek 2005: 42), the comprehensive methodological research design is guided by quality criteria for social science research (cf. Sect. 5.2.1).

In the study at hand, I take a social constructivist perspective and observe how social reality and knowledge is constituted in social interactions, subjective meanings and shared artifacts. I developed my understanding of pedagogic practice through the applied methodology heuristically. As an observer from outside, I added to my knowledge and reflected continuously on my observations. I used triangulation to reinterpret my results and also reflected on the influence of my perspective as researcher from a different context. I also discussed my interpretations with several stakeholders to confirm or revise my perspective.

This chapter elucidates, contextualizes, and reflects on the selected qualitative methodological framework (Sect. 5.2), access to and selection of schools (Sect. 5.3), and the research process (Sect. 5.4). Furthermore, the applied empirical research methods for data documentation, processing, and analysis (Sect. 5.5) are summarized, and their use in this study is explained. Limitations of the applied research methods as well as the role and perspective of the research in the socio-cultural context are also discussed (Sect. 5.6).

5.2 Framework for Qualitative Geographical Education Research

To address research questions focusing on classroom teaching in its respective educational system and socio-cultural background, the qualitative research design consists of the combination of extensive fieldwork, document analyses and an intervention study in the form of action research (cf. Fig. 5.1). Fieldwork and document analyses help to examine the status quo of contents and methods in Indian geography lessons, as well as challenges for the implementation of the transnational educational objective of ESD. Based on this, action research can test new methods for geography education and, through close contact with teachers and students, gain further insights into barriers and opportunities for changes in geography education.

This study takes three perspectives into account: first, the institutional perspective, which comprises the perceptions of policy makers, curriculum designers and textbook authors on constraints and opportunities for ESD. Second, the teachers' views on their own and new teaching methods are analyzed. Third, the students' perspectives on how and what they learn and would like to learn are assessed.

The perspective of educational stakeholders contributes to integrating single observations into a complex understanding and to identifying institutional, (infra)structural and administrative barriers to ESD as well as societal, historic, and socio-cultural framework conditions. The analyses of the teachers' perspectives give

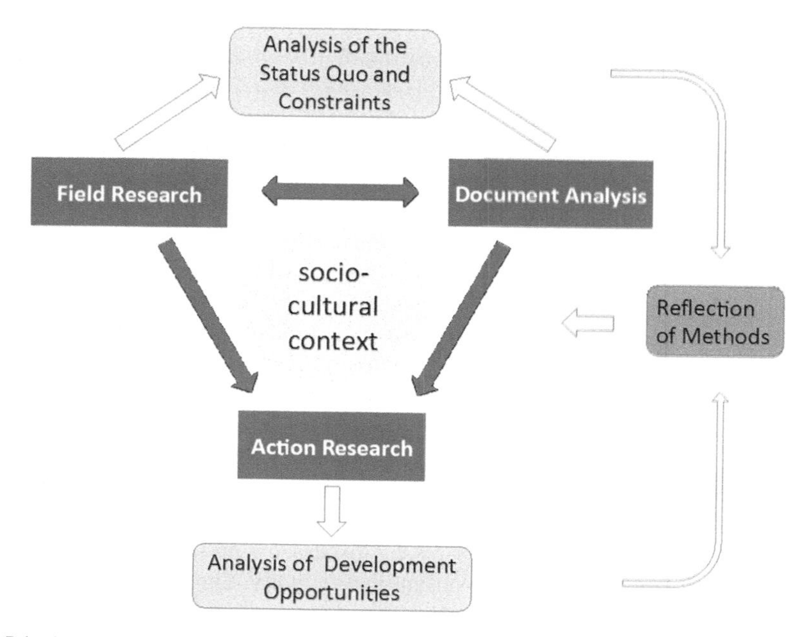

Fig. 5.1 A qualitative research design for geography education research in an international context (own draft)

an insight into their daily teaching practice as well as their understanding of the barriers and opportunities for ESD in their specific teaching contexts. Furthermore, the constant feedback from teachers allows a continuous revision of my own perspective as a researcher as well as specific criticism in the process of teaching module implementation. The students' perspectives reveal the learners' knowledge of water, their attitudes toward ESD as well as the learners' perceptions of existing and new teaching methodologies. Interviews, questionnaires, essays, drawings, filmed group work, and classroom interactions give an insight into learners' knowledge, interests, and attitudes toward ESD, as well as their behaviors in different classroom settings.

The triangulation of these three perspectives enables a comprehensive analysis of pedagogic practice. This multilayered research process requires flexibility toward the applied research methods. To access these perspectives on several levels, research methods from geographical education, qualitative social science, and ethnographic research are combined. During this process, the role and the perspective of the researcher were continually reflected on (cf. Sect. 5.6). The chosen qualitative research framework allows a flexible researcher-object interaction process and openness toward the object of study with redrafts, supplements, and revisions of the theoretical structure, research questions, hypotheses, and method reflection (Mayring 2002: 28) as depicted through feedback arrows in Fig. 5.1.

5.2.1 Quality Criteria for Qualitative Social Research

To ensure the validity and reliability of these qualitative research methods, six quality criteria suggested by Mayring (2002: 144) are observed. These quality criteria are precise documentation, argumentatively constituted interpretations, rule-governed analysis, subject proximity, communicative validity, and triangulation. Firstly, the research process and obtained data are precisely documented and the data analyses and empirical findings are depicted in detail in field notes to make the research process accountable. Secondly, during the data analyses, interpretations are constituted argumentatively, and wherever it is not possible, gaps in knowledge are explained. Third, the overall methodological approaches as well as each analytical step are rule-governed. Although the strength of qualitative research is its theoretical, methodological and analytical openness and flexibility, the data collection, management, processing and analyses are systematically documented and analytical rules are outlined. Fourth, subject proximity is ensured through fieldwork over a period of two years with extensive participant observations in schools and a focus school with close contacts to two classes and three teachers to gain access to their daily life. Fifth, throughout the fieldwork as well as in the final stages of book writing in a separate field visit, results were presented and discussed with state and national educational stakeholders and teachers to ensure their validity. Through the feedback on the results and the discussion of diagrams, the interpretations as well as knowledge gaps due to nonexistent and unreliable literature, e.g., on the educational system, were verified by communicative validity. Sixth, triangulation is the most important methodological quality criteria for this qualitative study.

Triangulation describes the constitution of a research object from at least two angles (Flick 2011b: 11). Triangulation was conceptualized for sociology by Denzin (1970: 300) to overcome "personalistic biases that stem from single methodologies" through the triangulation of data sources, investigator perspectives, theories, and methodologies between or within methods. To answer the research questions of this study, different perspectives in terms of methods and data sources are systematically interrelated and, thus, validate and widen the findings on different levels. Triangulation as the coordinated relation of several methods to each other (between-method triangulation), e.g., text-based interviews and image-based video sequence analysis, enables the validation of obtained data and allows detailed observations to be put into a larger mosaic of knowledge, thus achieving a better understanding of teacher–student interaction. The triangulation of different data sources also validates the material and enables a view from different perspectives, e.g., from teacher and student interviews (within methods). The combination of the interpretations of qualitative interviews with respondents of various positions, document analyses, questionnaires, as well as filmed classroom interaction and group work contributes to a holistic understanding of the methods and contents of geography teaching, as well as the barriers and opportunities for ESD in Indian classroom teaching. Furthermore, data sources are reflected in terms of their validity for their specific context. Thus, multiple triangulations ground this research methodology and lead to an in-depth understanding. Orienting my research methodology along with the quality criteria for qualitative social research has not only helped me to achieve reliable and valid results, but has also provided guidelines for the research processes.

5.3 Access to Schools

The selection of and the access to schools and educational stakeholders requires cultural sensitivity as well as a certain knowledge about the structural set up of the educational system in India. A great variety of school types exist due to the deeply entrenched hierarchical societal structure, multifaceted cultural backgrounds, and the colonial legacy in India. School types are defined according to their respective school board, their funding donors, and whether the schools are government, government-aided or private schools. In particular, the available funding of schools greatly impacts on the availability of infrastructure and teaching material. Both the type and the language used in the school, either English or vernacular languages, give an idea as to the socio-economic backgrounds of students and teachers. In urban areas, parents mostly prefer English-medium schools if they can afford to pay fees (National Center on Education and the Economy 2006: 5). The location of the school, especially whether it is a rural or urban school, reflects the daily opportunities for teachers and students to access information, resources, and institutions.

A total of 25 different schools (government/private, English-medium/Hindi-medium/Marathi-medium, rural/urban, etc.) were visited to observe the school environment and classroom lessons at various school types. Most schools were located in urban areas in Pune, Mumbai, and Delhi. To get a broader overview of the different settings and conditions of schools in rural areas, observations and qualitative interviews were also conducted in rural schools.

During the school visits, social interaction in the school courtyard during breaks and during lessons in classrooms was observed, and informal conversations were initiated with principals, teachers, and students. Observing daily routines in various schools gave an understanding of the diverse Indian school system. It also facilitated the classification of schools according to the existing infrastructure, teaching tools and practices as well as the socio-economic background of the student body and their livelihoods. For example, while many rural schools did not have desks or chairs, private urban schools were equipped with laptops and projectors. Similarly, the class size varied from 16 to 65 students per class.

Of 12 schools initially visited in urban Pune, five English-medium secondary schools were selected for the implementation of teaching modules. In contrast to most rural and vernacular schools, the selected English-medium schools in Pune schools are primarily attended by urban middle-class students, with socio-economic backgrounds comprising of diverse religions (Hindus, Muslims, Christians, Sikhs, and Khasi[1]) and castes (Brahmin, Kshatriya, Maratha, Mali, Shimpi, Gurav, Garo, Dimasa, Nahvi, Sutar, Lingayat Vane, and Sali). The existing memorandum of understanding (MoU) between the Institute of Geography at the University of Cologne and the Bharati Vidyapeeth Institute of Environment Education and Research (BVIEER) in Pune facilitated the access to schools in Pune, as cooperation with schools already existed through different environmental education programs. BVIEER has worked on implementing ESD in textbooks, curricula, and teacher training for several years. Because of the previous environmental education projects with schools through BVIEER, the selection of schools was not randomly sampled and therefore is not representative. Criteria for the five selected schools were as follows:

(1) *English-medium schools* (EMS)[2]: The number of EMS is especially increasing in urban settings and popular among the urban middle class. In 2002, 25% of secondary schools were EMS in India (cf. Meganathan 2011: 26). Therefore, the results of this study are only covering one section of society and are biased especially against the majority of rural government schools. Although the inclusion of Marathi schools would have covered a broader socio-economic background of students, the convenience and efficiency of direct communication in English were the decisive factor. The selection of EMS avoided needing the help of a translator, which might have led to misunderstandings and misinterpretations and, thus, less data reliability as well as additional expenditure of time. As English is usually learned as second language, teachers and students

[1] Khasi is an indigenous matrilineal ethnic group of Meghalaya and Assam in northeastern India.

[2] Although usually not the native language, English is used as medium of instruction. First introduced during colonialism under the British Empire, English plays a central role in education and business.

switch occasionally to Marathi (code-switching). Oftentimes, teachers find it more convenient to explain concepts in their students' mother tongue rather than in their second language, even in EMS.[3]

(2) *Different socio-economic background*: Within the selected EMS, government-aided and unaided schools were chosen to cover different socio-economic backgrounds of students. The student bodies ranged from lower middle-class to upper-class backgrounds.[4] The school equipment as well as the jobs of students' parents as well as their housing situation were indicators of classifying the schools. This was based on triangulating various perceptions gained through discussions and home visits.

(3) *Different organizational types of school boards*: The selected schools cover the two main national boards ICSE and CBSE as well as the Maharashtra state board (SSC) and an educational trust (Deccan Education Society).

(4) *Location in urban Pune*: Urban and rural schools greatly differ in educational equipment, access to information, environmental setting as well as water access and related conflicts. Pune is an emerging megacity with challenging conflicts relating to water access. As this study focuses on local urban water supply and conflicts in Pune, only schools with similar urban living environments, and for which the content of the study was relevant were selected (Fig. 5.2).

(5) *Access to standard nine*: To ensure a similar level of knowledge and age group, one class of standard nine was selected at each secondary school. Class 10 and higher were not possible because of standardized board exams and college entrance examination preparations.

With these criteria, motivated and engaged geography teachers at five schools could be found to conduct interviews and observe geography lessons and to develop school profiles (cf. Table 5.1). During the process of recruiting schools for the project, repeat visits to educational trustees, school board members, principals, head teachers, and senior teachers covering different positions in the hierarchical structure of the school organization promoted the trust in the researcher as well as a deeper insight in decision-making processes in local school structures. The five schools eventually selected for the introduction of ESD modules are briefly described in the following (cf. Table 5.1):

(1) *School 1 (S1)* is a government-aided private school founded in 1992. It offers the Maharashtra state board exams (SSC) after the class 10. Students come from the immediate neighborhood and belong to the upper lower class to lower middle class. The equipment of classrooms is rudimentarily limited to a blackboard and tables and chairs. The walls between the classrooms are not soundproof so the class next door can be heard. The average class size is about 50 students.

---

[3]Code-switching was mentioned in teacher interviews and observed in classrooms in March 2012.

[4]The class background stated for each school is not verified by socio-economic data, but based on the statements of principals and teachers, as well as other educational stakeholders who were familiar with these schools. The heterogenity of middle classes in opposition to simple and idealized images of the new middle class in India was studied in detail by Fernandes (2006).

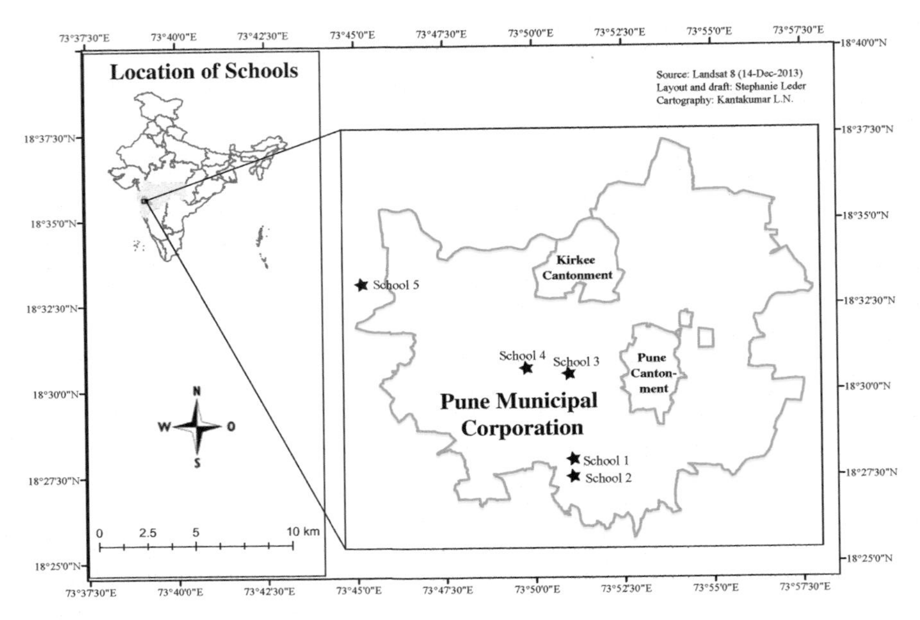

Fig. 5.2 Location of the five selected English-medium secondary schools in Pune (own draft, layout: Lakshmi Kanth Kumar)

(2) *School 2 (S2)* is a government-aided private school affiliated to a university and involved in programs such as an international student exchange, due to which the school has a good reputation. The school follows the national curriculum and NCERT textbooks and offers CBSE exams. Attending students have a middle middle to upper middle-class background. Up to 50 students are in one classroom, whereas in observed lessons, the class sizes were around 35–40 students. Students' homes are located within a radius of approximately 5 km of this school. This school was also chosen as the *focus school*.

(3) *School 3 (S3)* is a government-aided private school in the central part of Pune following the NCERT curriculum and offering CBSE board exams. As this school opened recently in 2009, classes offered are only from 7th to class 10 with fewer than 25 students per class; thus, the students' numbers are low. At the point of research, classrooms used were in a primary school, which are empty in the afternoon. Attending students have an upper middle-class background.

(4) *School 4 (S4)* is an unaided private school in the central part of Pune, with two separate locations for standard 1 and 7 and 8–12. The school covers the state curriculum and offers SSC board examinations. This school is involved in several environmental education projects. The average classroom size is 60 students per classroom with teachers. Teachers teach in two daily shifts.

School 5 (S5) is an unaided private school whose financial expenses are covered by parents and private trustees. The students have an upper-class background and come from different parts of the city with school buses with up to 60 min

Table 5.1 Profiles of the schools in Pune for this study

Code for school	Location	Board	Funding (government, private aided or unaided)	Standards taught	Timings and shifts	Estimated classroom size	Student: teacher ratio	Student background
S1	South of Pune	SSC	private aided	KG–12	7.30–12.00 (I–V); 12.20–5.20 (VI–XII)	50	900: 40 (23/1)	lower middle class
S2	South of Pune	CBSE	private aided	KG–12	7.00–12.20 (V–X); 12.30–5.40 (I–V); 10.30–3.30 (XI–XII)	50	1256: 57 (22/1)	lower, middle and higher class
S3	Central Pune	CBSE	private aided	7–10	1.00–5.00 (VII–X)	25	100: 5 (20/1)	lower middle class
S4	Central Pune	SSC	private unaided	1–12	7.10–12.25 (I–VII); 12.25–5.40 (I –VII)	60	1700: 65 (26/1)	middle and upper middle class
S5	West of Pune	ICSE/CBSE	private unaided	KG–12	8.00–4.00 (I–XII)	30	1093: 59 (19/1)	upper class

> of travel time. The school offers primarily the ICSE board exams, but also the CBSE board exams.[5] The director stressed that the RTE "Right to Education" law which meant including 25% of students of scheduled castes with lower economic backgrounds was followed, but these students were not noticed. The modern building complex is located at the urban fringe and is well equipped with teaching tools, iPads, colorful school uniforms. S5 is a day-boarding school and the only school with a single full day shift in contrast to the other schools with two student shifts a day. Class sizes are relatively small, with about 25 students per class. Students do their homework, revise, and do garden projects during their time at school.

Approaching schools as an outsider brought several constraints from a researcher's perspective as I often encountered empathetic curiosity and approaches in schools. My initial visits were accompanied by curious looks from students and teachers. Groups of students gathered around me to ask questions, to show me something, and to take pictures with me. The visits to schools were usually officially arranged and the presence of a researcher often attracted attention. It interrupted daily routines and made it difficult for teachers and also students to retain their role and their natural behavioral patterns. Therefore, I selected a *focus school* for regular participative observations over a period of two academic years.

Due to regular interactions, students and teachers grew accustomed to the presence of an outsider. Thus, I could observe teacher and student interactions in a natural setting to approximate the perspective of students and teachers at the focus school. Building more informal relationships with teachers and students allowed me to ask more sensitive and critical questions which they answered willingly and deliberately in informal discussions. Through my regular involvement in school, teachers and students felt less distracted or observed, and behaved more comfortably when I was around. I was also invited to visit several teacher and student homes. This enabled the observation of the living environment of students and teachers. In their private homes, students and teachers opened up and were willing to share their perspectives on schooling, even when I was not asking questions. Classroom interactions in general, also in the other schools, seemed to follow the usual pattern, as it did not change over the period of my observations.

S2 was selected as the focus school because the student body represented the broadest range of socio-economic, ethnical, and religious differences (covering Hindus, Sikhs, Jains, Muslims and Christians). This diversity of students' backgrounds shed light on cultural or socio-economic differences in the perception of water, which are important to consider for pedagogic practice in such a diverse country. In the focus school, I observed classes of several standards. Furthermore, I participated in festive events at the school and attended an excursion to the zoo, a temple and a dairy farm, which was organized by ten students. Through these contacts, I gained a greater insight into the daily routine of the school, of the students, and of the teachers. During interactions outside of the school complex, teachers and students often gave different answers to questions previously asked in the school building and

---

[5]CBSE board exams are for the weaker students of the school, according to a teacher of the school.

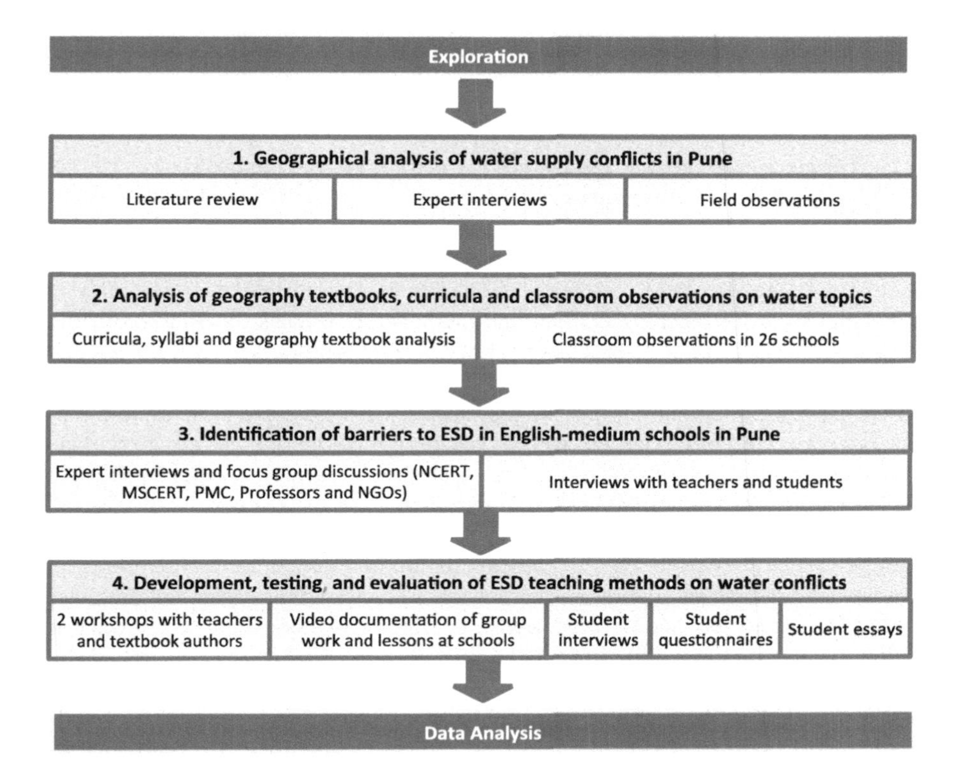

Fig. 5.3 Structure of the research process and methods

were more willing to criticize methodologies of teaching and learning. This enabled a better holistic understanding as well as access to honest thoughts on the current educational situations.

5.4 Structure of the Research Process

The research process was structured in four phases of fieldwork (Fig. 5.3). Over a period of two years, a total of nine months of fieldwork were conducted in Pune and Delhi. An extensive phase of exploration was necessary to access and become acquainted with the field of research, to collect relevant documents, and to establish contacts to possible cooperation schools. The first field visit of three months (February to May 2012) served as an orientation to reframe and adjust the research objectives to the research context.

In phase 1, a geographical analysis of the water supply situation and recent social, environmental, and economic dimensions of urban water conflicts from a sustainability perspective was conducted. This analysis is based on literature research, expert interviews and field observations. In phase 2, documents from institutions and student products such as essays, posters, and drawings were collected for screening. Furthermore, twenty geography textbooks (10 NCERT, 10 MSCERT) were analyzed for their content and teaching approaches to the topic of water and ESD.

In phase 3, qualitative interviews and focus group discussions were conducted with educational stakeholders and teachers on the opportunities and challenges of ESD in Indian geography education. The results of the analyses of explorative interviews, the textbook analyses, and classroom observations all contributed to the analysis of the status quo of teaching content and methods, as well as the analysis of the constraints to ESD in geography teaching in order to build on acquired knowledge for the following research phase.

In phase 4, the intervention study was conducted. On the basis of the textbook analyses and classroom observations, an ESD teaching module on local water conflicts was conceptualized and piloted to explore methodological opportunities for ESD in English-medium schools in Pune. The modified module was introduced, discussed, and reworked at a teacher workshop. ESD teaching modules were introduced through teachers in their classes to test the implementation of a new didactical approach in geography lessons at English-medium secondary schools. The implementation was documented, analyzed, and the results were communicatively validated. During this implementation process, difficulties in the ESD teaching module were identified and modified.

The recorded lessons and interviews and data collected during phases 1 to 4 were thematically organized, literally transcribed, and systematically coded following the principles of qualitative content analysis (Mayring 2010) and grounded theory elements (Glaser and Strauss 2008) using the software Atlas.TI. Through the empirical data analysis, theoretical considerations were systematically triangulated and revised. In a final step, the results of the study as well as the interpretative conclusions were discussed and evaluated with the partner institute BVIEER and with textbook authors from 17 different Indian states in a group discussion during a workshop session. The results were discussed with further educational stakeholders as well as with the teachers and principals of the participating schools to relate to current political educational interests in India and internal institutional barriers. In interviews with educational experts, the results were reflected and opportunities and barriers for the modified implementation of these teaching modules in teacher trainings, curricula, and textbooks were discussed. An affiliation with the Centre de Sciences Humaines (CSH) in Delhi enabled academic exchange and further discussions about the results. These discussions and interviews served as communicative validation adding to a better understanding of content and methodology in ESD-oriented geography teaching in India in general and the existing structural and institutional framework conditions given in English-medium schools in Pune in particular.

Table 5.2 Applied empirical research methods for geographical education research in Pune

Investigation method	Subject or object of research	Documentation method	Data preparation and processing	Data analysis
Fieldwork participant and classroom observation	School routines school environments geography lessons	Field notes videography photography	Literal transcription summarizing protocols	Qualitative content analysis by Mayring (2002) and grounded theory (Glaser and Strauss 2008): inductive and deductive categories
Qualitative interviews	Educational stakeholders, teachers, students	Voice recorder field notes	Literal transcription	
Focus group discussions	Educational stakeholders, teachers, students	Voice recorder field notes	Literal transcription summarizing protocols	
Questionnaires	Teachers students	Tables field notes	Tables diagrams	
Document analyses	Curricula, syllabi, textbooks, student drawings and essays, posters, newspaper articles, (non-) governmental documents, gray literature	Tables field notes	Summaries tables diagrams	
Action research	Teacher workshop classroom intervention	Videography voice recorder photography field notes	Literal transcription summarizing protocols	

5.5 Research Methods for Data Documentation, Preparation, and Analysis

Using the above-outlined research design combining methods of fieldwork, document analyses and action research, different perspectives, such as those of politicians, curriculum designers, textbook authors as well as student and teacher perceptions can be documented and integrated. In the following, the applied qualitative research methods which were used for data collection, documentation, preparation, and analyses (cf. Table 5.2) are introduced.

5.5.1 Participant Observations and Classroom Observations

To analyze the structuring principles of social order and the mechanisms producing structures in daily pedagogic practice and social interaction within the schooling context, I conducted classroom observations and participant observations in schools. Participant observation originally developed in anthropology and ethnology and the social reform movements at the end of the nineteenth century and in the beginning of the twentieth century in Great Britain and the USA. In the 1920s and 1930s, ethnographic field methods for urban environments developed in the context of the Chicago School of Sociology as relevant social science methodology to describe social reality (Lüders 2012: 385). In methodological discussions, the term *ethnography* increasingly substituted participant observation, which also resulted in conceptual changes (Lüders 2012). Ethnography indicates long periods of observations with a flexible application of methods to situational conditions (Flick 2011a: 299).

> The ethnographer participates, overtly or covertly, in people's daily lives for an extended period of time, watching what happens, listening to what is said, asking questions; in fact collecting whatever data are available to throw light on the issues with which he or she is concerned. (Hammersly and Atkinson, 1983: 2)

To gain a deeper understanding of the power structures and communication patterns in educational contexts, ethnographic research and participant observations in schools and classrooms provide the central principles and techniques for this study. I followed the two axioms of Bruyn (1966: 14): I participated in regular school life, and, in my role as researcher in the field, I balanced involvement and impartiality. To come closer to an internal view of agents in the school context and to lower the "reactivity" (Bernard 2013: 317) to the disturbing presence of a researcher, I spent three time periods of two months each observing daily processes in schools and classrooms (cf. photograph in appendix). These observations served to understand how pedagogic practice is embedded and functions in the social institution of a school. My presence was neither neutral nor uninvolved, as an important part of accessing information was "gaining rapport" (Bernard 2013: 327), in terms of relationship building to students and teachers. As participative observer, this methodology implied flexible balancing between research interest and situational challenges (Lüders 2012: 393). To ensure objectivity and plausibility, observations were documented in four books of field notes in the form of reconstructive listings, descriptions of single situations, direct speech, and summarizing protocols. Separately, I periodically produced reflecting reviews and interpretations of the field notes aimed at differentiating protocols and interpretations.

Intense participant observations were conducted at the focus school enabling a deeper insight into daily processes and structures of teachers and students. In geography lessons, a focus was set on the constitution of teacher–student communication and the role of teaching material, especially the textbook. On the content level, the discourse on the topic of water was carefully screened in terms of the interrelation of environmental, social, and economic dimensions of sustainability as well as across scales of time and space. On the method level, the form of communication was examined with regard to its argumentative content. Further, categories for observation were student participation, network thinking and the use of media and teaching tools. The role of the textbook was especially observed: How are the contents and methods in textbooks integrated into the classroom? How does the textbook structure teaching process? Social interaction in the classroom was viewed through a Bernsteinian lens and interpreted with his terms of visible and invisible pedagogy, classification and framing, performance and competence.

Besides regular school observations, environmental education projects of non-governmental organizations (Don Bosco, Aarohi, School without Borders) were visited to observe how non-formal education initiatives implement their own projects. These were usually extracurricular and not bound by rigid schedules, but offered innovative initiatives and opened the perspective for educational formats for ESD beyond classroom structures.

All observations were documented in field notes through summarizing protocols and structuring tables, and the role as researcher in school environments was repeatedly reflected. While observations were of high interpretative value, additional analytical depth was achieved through detailed video sequences of geography lessons.

Videography Documentation and Analysis

To document authentic behavior and become familiar with the pedagogic practice, interaction processes between teachers and students have to be assessed in their naturalistic settings (Goffman 1971). Since the 1970s, classroom research has been conducted by videography to document and analyze didactic processes and structures of teaching as well as social interaction patterns of teachers and students (Kocher and Wyss 2008: 81–82). For example, Wagner-Willi (2007) identified how rituals in the classroom constitute the social identities of students or teachers and how institutional expectations are expressed through postures. For the analyses of pedagogic practice, especially in foreign socio-cultural contexts, Goffman (1959) theoretically grounded empiricism is suitable as it describes and alienates social reality with dramaturgical elements as metaphors to describe self-presentation in daily life. Actions in front of the camera are structured in terms of Goffman's dramaturgical perspective on social interactions (cf. Reichertz and Englert 2011: 38) in terms of stage, requisites, actors, symbolic interaction, and agency. On the basis of audio–visual recordings, a *micro-functional analysis* can offer patterns of natural behavior and embed single behavioral acts in sequential behavioral patterns (cf. Willems, 2012: 43–44). Following Goffman, the analysis of sequences is one strategy to reconstruct the process of interaction from different perspectives which all have their own relativity in their interpretations (Willems 2012: 45, 48). Furthermore, implicit patterns of interpretation need to be interpreted, as defined by Giddens (1984: 12) as *double hermeneutic*. After receiving the permission of the principals, teachers, and students, I sat with the handheld video camera at the side of the classroom, positioned to film the interaction of teachers and students. The methodological procedure of analyzing filmed geography lessons was transcribed and analyzed in multiple layers (Reichertz and Englert 2011: 32):

1. open coding of single fixed images in two-minute intervals
2. *literal transcription* of sequences relevant to the research questions[6]
3. protocol and coding of *takes* (=camera positions)
4. open coding of *moves* (action processes) as central unit for analysis
5. selection of relevant sequences for selective coding based on *theoretical sampling* (Strauss 1998: 70)
6. *selective coding* with Bernstein's thematic concepts of hierarchical, criterial, and pacing rules.

The final transcript includes all layers of transcribed observation and interpretation, producing a highly inferential analysis basis (Kocher and Wyss 2008: 84). Based on the documentary method developed by Bohnsack (2009), the reconstruction of a cultural product's meaning detaches the interpretation from actual processes. The interpretation of selected scenes focuses on *formulating* thematic sequences (what is happening?), *reflecting* framing processes (how is it happening?—the *modus operandi*) and a final generalization. *Formulating interpretations* is both *pre-iconographic* in that objects, phenomena, and moves are described ("putting one's hand up"), and *iconographic* in that the intention to do something (indicating the request to speak) is analyzed. Iconographic interpretation allows the identification of role-specific actions, whereas the *pre-iconic* level is more relevant to the analysis than the plot on the *iconographic* level. *Reflecting interpretation* focuses on the framing of orientation, the specific *habitus* (Bourdieu 1986) documented in the audio–visualization of a group. Primary expressions such as gestures or statements are related to following secondary expressions and not considered in isolation. The relation of gesture and reaction constitutes regularity, which has to be deduced to analyze the significance of the first expression (Bohnsack 2009: 20). The deeper semantics of a gesture can only be interpreted in relation to the full body and the interaction scene. In addition, photographs as snapshots in time served to analyze the arrangement of classroom order and the position, gestures, and facial expressions of teachers and students during interaction. Open codes focus on teaching strategy, teacher's message, student's message, and interaction pace, whereas selective codes focused on Bernstein's concepts (Sect. 2.2.1).

Video sequences offer a great data record of audio–visual material which are selectively analyzed. However, they are only a single time and space bound excerpt of reality and are not representative. The selective coding of chosen video sequences offers in-depth analyses of pedagogic practice, especially with the use of abstract terms describing social interaction by Bernstein. Although students' attention was drawn to the camera initially and they seemed amused or intimidated during the lesson, most of the students focused on the teacher and participated in the lesson. Other video studies in classroom settings, such as TIMSS 1999, have shown that the behavior of teachers and students hardly changes and that camera effects were less substantial than expected (cf. Kocher and Wyss 2008: 81–82). Yet, altered behavior

[6]Sequences for transcription were introductory and concluding lesson sections, scenes of increasing density of interaction or changing interaction and position patterns (e.g., discontinuities in teacher–student communication, change of teaching methods).

due to the camera and the presence of the researcher cannot be excluded. Nevertheless, videography accompanied by photography documentations qualifies for a multi-level analysis of authentic classroom interaction.

5.5.2 *Qualitative Interviews*

To access the knowledge, perspectives, and attitudes of educational stakeholders, teachers and students, qualitative interviews were conducted. These varied from initially explorative interviews with narrative elements, to problem-centered, semi-structured, and guided interviews. Narrative interviews access subjective perspectives in an atmosphere of trust (Meier Kruker and Rauh 2005: 67). Guided interviews include prepared questions on essential research interests and can also include ad hoc questions (Mayring 2002: 70). The choice of questions depends on the position of the interview partner, the interview situation, and the point of time in the research process.

The selection of interview partners is based on theoretical sampling (Strauss 1998: 70) and thus aims to explore and triangulate student, teacher, and institutional perspectives on pedagogic practice in geography education (cf. list of interview partners in the appendix). Interviews were conducted with students and teachers in selected schools and covered institutional perspectives through interviews with principals in schools, officials at governmental institutions and academic institutions on the district, state, and national level. Interviews with educational stakeholders representing the institutional perspective included policy makers, curriculum designers, textbook writers, and professors. Central institutions on the national level were the National Council for Educational Research and Training (NCERT) responsible for the development of national guidelines such as the National Curriculum Framework (NCF), syllabi, and national textbooks.

On the state level, the Maharashtra State Council of Educational Research and Training (MSCERT), and the Balbharti Textbook Bureau were approached. On the district level, the environmental officer and engineers of the Pune Municipal Corporation (PMC) were interviewed. In addition to these institutional perspectives, non-governmental organizations (NGOs) gave a rather critical view on the formal school system and provided insights into their extracurricular programs that embraced progressive and critical environmental teaching approaches. Leading NGOs and think tanks such as the Center of Environment Education (CEE) and Center for Science and Environment (CSE) as well as staff of local NGOs were interviewed. Most of these contacts were established via snowball sampling.

The guided interviews included open questions and were semi-structured in three parts. At first, questions on the background and the interviewee's subjective idea of relevance which they would give to the topic were asked. During the interview, several topics which were pre-drafted for the guided interview were covered. Additionally, ad hoc questions were formulated in between and after the preliminary developed guiding questions (cf. Mayring 2002: 70). Interviews with educational stakeholders focused on opportunities and constraints for ESD and argumentative pedagogic practice for geography education within the formal educational system. Teacher interviews focused on possible links between ESD relating to water and their own geography teaching, and their perception of students' geographical knowledge and skills. Student interviews represented the learner perspective and focused on their knowledge of and interest in the subject of geography and water, and related teaching methodology. The interviews were thematically structured along the following aspects (cf. detailed guiding questions in appendix):

Educational Stakeholders

- status of the Indian educational system; understanding and perspective on current educational policies, specifically ESD
- infrastructural, institutional, content-wise and methodological constraints to ESD in (geography) education
- aspects of the topic of water covered in formal and non-formal educational institutions
- relevance of different topics in geography education and their methodological realization
- role and interest in the subject of geography
- the geographical knowledge and skills of students
- attitudes to challenges, opportunities and visions for the development of argumentative pedagogic practice in the educational system.

Teachers

- water supply situation in Pune (relevance, access, conflicts)
- teaching content and methods relating to water
- role of the subject geography, environmental education, and ESD
- students' geographical knowledge and skills on water
- opportunities and barriers of ESD in classroom teaching.

Students

- knowledge, attitudes, and experiences with water (use, access, diseases, problems)
- information on water learned in school
- interest in the subject geography
- teaching methods and communication in geography class (debates, etc.)
- knowledge and attitude toward ESD.

Interview situations are artificial situations, in which the validation of investigation can be negatively influenced (Meier Kruker and Rauh 2005: 99). Being a researcher from a different country may influence the answers that are given in terms of social desirability and the expectation toward their role in the respective institution. As experienced during the research process, responses varied depending on the place and settings (e.g., in the school or at home), the available time, the presence of third parties, and the previous relation to and interaction with the researcher.

Interview questions were discussed with researchers and piloted with teachers which resulted in repeated changes to achieve the most clear and precise formulation. To reduce biases during the interviews, the order of interview questions was purposely arranged according to their complexity and thematic focus, but also flexibly re-arranged when needed. During the analysis, similar statements were compared and if biases existed, statements were re-presented in a second interview or to another interview partner with a differing perspective for the purpose of communicative validation.

5.5.3 Focus Group Discussions

Focus group discussion is a method to create a lively, social situation in which a group of equal hierarchical background becomes involved in a thematic discussion developing its own dynamic (cf. Flick 2011a: 248). Focus group discussions uncover subjective perceptions, experiences, and visions. Furthermore, collective attitudes, agreements and disagreements can be identified when individual opinions are stated, affirmed, or contradicted. These communicative corrections through the group function as a means of validation (Flick 2011a: 251). Attitudes shaping thinking and acting can be uncovered in focus group discussions as they resemble social situations from which they usually emerge (Mayring 2002: 77).

Focus group discussions were a central element in teacher and textbook writer workshops and also in relation to students and curriculum designers (cf. photograph in appendix). Student group discussions were conducted more frequently, as some students were more encouraged to speak up in a group than in a one-to-one interview. A group discussion with nine teachers was initiated after some results from the classroom observations and textbook analysis were presented. Research results as well as evaluated and modified teaching methods were introduced and discussed. A group discussion with textbook authors from 17 different states was part of a session at a workshop on how to infuse Education for Sustainable Development in textbooks. The discussion was initiated by presenting best practice examples of textbook pages on ESD topics and tasks focusing on skill development. This initiated a lively discussion on difficulties and opportunities for ESD as well as the role of textbooks.

Student group interviews were conducted informally or after the application of a teaching module to obtain access to the students' perceptions of traditional and new teaching methods, their constraints as students, their views on the subject of geography, and their water-related knowledge, attitudes and behavior. Furthermore, group interviews functioned as a central element to obtain honest feedback on the developed teaching methods.

5.5.4 Action Research

To develop guiding principles for social action, changes in social practice can be tested, observed and analyzed in an intervention study when researcher and subject start a discourse. To overcome the gap between the theory and practice of educational research, the relation of theoretical logic to practical logic research (Bourdieu 1976) can take place during social action. Lewin (1946) coined the term action research as "a comparative research on the conditions and effects of various forms of social action and research leading to social action." In Lewin's view, action research is a process of planning, action, and effect analysis. It begins with a defined social, practical problem with the objective being a change of praxis during the process. The project process changes between systematic collection of information, discourse, and practical action. In this process, practitioners are not the experimental subject, but are seen as equal partners to the researcher in communicative with the goal of an exchange without domination (Mayring 2002: 50–51). The process is recorded with field notes, summarizing protocols, a video camera, and photography as documentation of the observations.

Action research was a common research design during educational reform processes in the 1970s in Germany (Mayring 2002: 50). After explorative interviews, observations, and document analyses for this study, the need for examining changes in pedagogic practice through the introduction of ESD modules in Indian geography education became apparent. Changes in pedagogic practice could only be assessed if educational stakeholders, teachers, and students were involved in the process. Their participation served to modify and adapt research tools, to gain access to their perspective, as well as to reflect and validate interpretations throughout the implementation process. Hence, this study is not entirely ethnographically framed as it does not only rely on internal valuation standards (Flick 2006: 439).

After the selection of five schools for this study I decided to conceptualize three ESD teaching modules, of which one, the "Visual Network," was piloted in 14 group works in class 9 at seven English-medium schools. To overcome the prior identified constraints to ESD, the teaching concept was developed in such a way that teachers could introduce it in their lessons under the National Curriculum Framework without great additional effort and expense. Through this teaching module, student-oriented and communicative group work on local resource conflicts were assessed. The module piloting helped to understand whether the students could work with the material and the tasks, and whether modifications were necessary. This process was documented and analyzed on the basis of worksheets, transcriptions of audio recordings, videography, and photography. A limitation on the intervention in group work was that although I asked for equally sized groups and students of different academic strengths, the groups' sizes varied between 3 and 8 students and usually students with good grades were sent, so that the selection was not representative for students of the school. Single students showing competitive behavior often dominated group work, thus, the term "group work" has to be carefully reflected in terms of group interaction.

After the group work, individual and group student interviews were conducted to reflect on the methodological approach of this teaching module. 68 student questionnaires relating to their knowledge of and interest in the topic of water in geography teaching, their position on ESD and their perspective of teaching methodologies in geography education were filled out. Additionally, teacher interviews were conducted with those interested in participating at the planned teacher workshop and the planned intervention study a few months later.

At a teacher workshop with nine geography teachers from five schools, the conception and the results of the teaching module pilot in group work were presented, discussed, and modified (cf. teacher workshop schedule in appendix). Furthermore, the general acceptance, opportunities, and barriers for ESD in geography teaching were examined with teachers. The workshop was organized in two parts: The first part was a vivid discussion about the preliminary results of the research project. In the second part, teachers tested three teaching modules in groups to modify and evaluate these. These modifications were discussed in plenum with professors and PhD students from BVIEER. This workshop encouraged the teachers to introduce one teaching module in their own lessons.

This workshop was documented and served three purposes: to assess the teachers' perceptions of and attitudes toward changes in their pedagogic practice, to modify and adapt the teaching modules, as well as to introduce the teaching modules to teachers for the implementation in a lesson of their own.

The classroom intervention, which took place a few weeks after the workshop, was observed and documented with videography, voice recorder, photography, and student worksheets. After the lesson, 148 student questionnaires for the evaluation of the respective teaching modules were filled out. Student and teacher interviews were conducted to discuss, reflect, and evaluate the new methodological approach (cf. table of participating teachers and teacher evaluation of the methods as well as list of interviewed students and the student questionnaire in appendix). This process was discussed with and evaluated by textbook writers from different states in a workshop session to validate the results (cf. textbook author evaluation of textbooks in appendix).

Although action research is marked by close cooperation with practitioners, the presence of a researcher influenced the behavior of teachers and students and thus maintained an artificial situation. Due to the rigid timetable, exceptional time and spatial arrangements in the classroom had to be made. Yet, the data obtained during the workshop and the classroom intervention proved to be of high explanatory value to explain changes in pedagogic practice and sharpened the analysis in regard to regular classroom interaction between teachers and students.

5.5.5 *Questionnaires*

To collect a greater number of responses and to reduce possible biases due to the interview situation and tendencies of social desirability, questionnaires were used as a standardized instrument to access facts, opinions, knowledge, and attitudes (Meier Kruker and Rauh 2005: 90). This was particularly relevant for understanding, and collecting information about the student body. The design of the student questionnaires included closed format question giving multiple choice answers and verbal rating scales, as well as open questions to which answers could be freely formulated. Students filled out questionnaires before and after interventions. The categories of questions and topics covered in student questionnaires were as follows (Meier Kruker and Rauh 2005: 91):

- factual questions (students' backgrounds in terms of age, household size, religion/caste, access to water)
- questions on knowledge (water sources, water problems on a local to global scale)
- questions on attitudes (toward geography as a subject, the topic of water, teaching methods, discussions)
- questions on behavior (e.g., regarding the relation of knowledge learned in school and environmental action at home)
- questions on suggestions (e.g., visions of how geography lessons should be).

The questionnaires were piloted with five students after which some questions were reformulated, with the aim of achieving a clear and precise formulation. This was useful because some questions had to be simplified, different terms had to be used to clarify aspects, and some were found to be redundant. Subsequently, the revised questionnaire was filled out by 68 students from the schools (cf. student questionnaire in appendix). Changes in the introductory explanation and instructions for filling out the questionnaires were minimized. The analysis centered on quantification of factual data of the students' background as well as thematic similarities and differences. The data obtained through questionnaires gave an overview of students' perspectives on geography education and also highlighted disparities between different schools.

5.5.6 Data Analysis

The interview, focus group discussion and intervention study data was analyzed following the principles of qualitative content analysis (Mayring 2010) and grounded theory elements (Glaser and Strauss 2008) with the software Atlas.TI. The combination of these two analytical methods allowed both inductive and deductive coding (Fig. 5.4). Inductive categories were formed through summarizing and open coding as suggested by the grounded theory (Glaser and Strauss 2008). The steps of paraphrasing, generalization, and reduction are repeated.

In a second reading, the data was coded deductively through structuring coding (Mayring 2010: 67, 92) with terms derived from the theoretical framework based on Bernstein, Freire, ESD, and argumentation. Because of the different data types and sources, different coding systems evolve, which partly overlap, however. The relation of inductive and deductive category formation varied according to the analyzed data type. After the indication of all citations, coextensive concepts were summarized

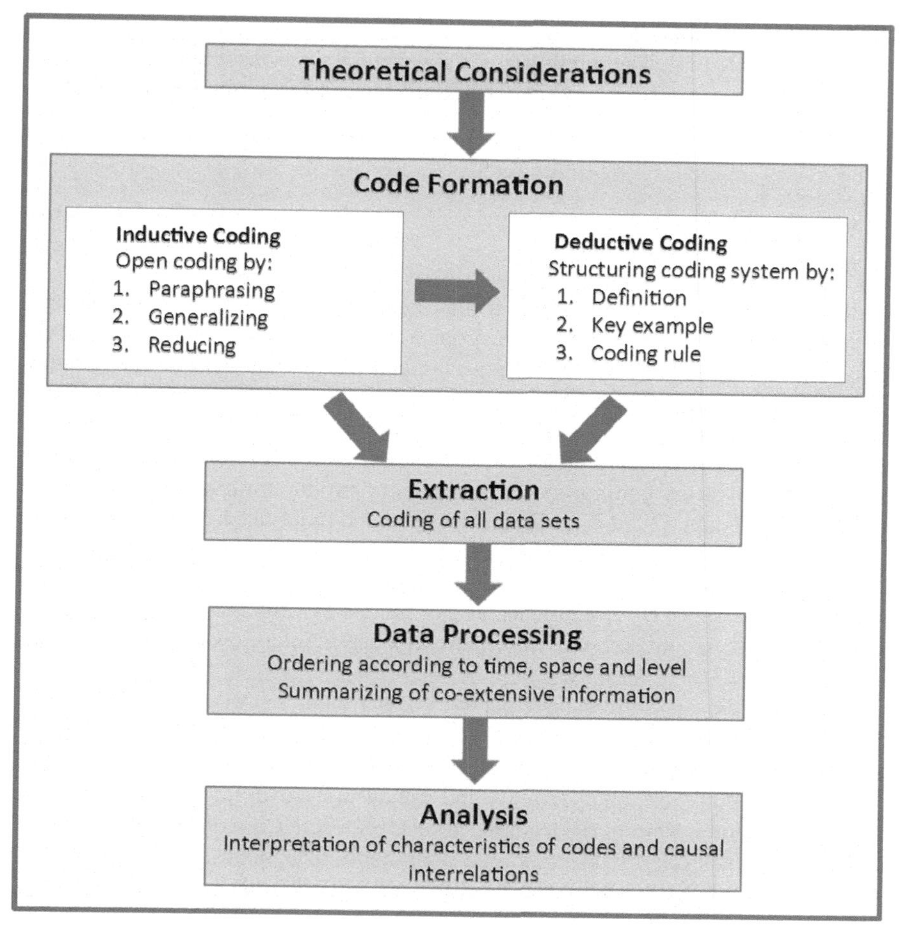

Fig. 5.4 Process model of qualitative content analysis with inductive–deductive code formation (combined and altered from Mayring 2010: 68, 93)

and ordered in dimensions referring to time and space. In a final step, characteristics of the extracted codes and citations were described and analyzed on basis of their causal interrelations.

Due to the different types of texts, such as interviews, questionnaires, observations and documents, for each type of text, a different set of categories, although partly overlapping, was developed. For example, for the textbook analyses, three textual categories related to the triangle of sustainability, namely ecological, economic, and social aspects, and three didactic categories deducted from theoretical considerations, namely argumentation, network thinking, and student orientation, were analyzed. If the respective perspective or level, e.g., of the student, teacher, government, was relevant, it was included in the category name, and subcategories were defined. Each category was clearly defined, governed by a coding rule and depicted with a model example (cf. Table 5.3). The reliability of the categories was tested through the help of an interrater. After the coding, networks and mind maps were developed which structure the results.

5.5.7 Document Analysis

Documents are artifacts of different formats and produced for communicating a particular purpose. If considered in the context of their production and use, documents are a medium for the construction of social realities in institutional contexts (Flick 2011a: 331). The guiding question for this method is: Who produced what in which form for whom and for which purpose? Documents are differentiated by their form (texts, drawings, photographs), producers (e.g., students, teachers, professors), purpose (e.g., to get a good grade), conditions of production (already existent or purposefully produced?), and the intended addressee (e.g., teachers). The selection of documents does not present a representative sample, but documents were purposefully selected according to the criteria of their authenticity, representativity and relevance (Scott 1990: 6). Documents are not treated as validation for factual statements, but rather to contextualize information gained in interviews and observations (Flick 2011b: 327).

To analyze the institutional perspective, educational reports and policy documents of the Indian government, particularly of the Ministry of Human Resources and Development (MHRD), and the National Council for Educational Research and Training (NCERT) were analyzed. To understand non-governmental perspectives, educational reports, directives, teaching tool kits, teacher handbooks and teacher training guidelines (cf. list of examined documents in appendix) of international agencies such as UNESCO and the World Bank, international and national NGOs were inspected. Furthermore, articles of national English newspapers such as *The Times of India*, as well as gray literature were screened and analyzed. Limited access to secondary data was challenging due to the reservation of government authorities with information in interviews and with issuance of documents and data.

To access differential subjective student realities, student products such as posters, essays, worksheets, and drawings were collected during school visits and classroom observations. Analyzing the relation of explicit content, implicit meaning and the context of function and use of the documents were challenging (Flick 2011a: 331). Constraints of document analysis were the bias of purposefully produced documents such as student essays and drawings intended to please the teacher or researcher, incomplete documents as well as not accessible documents, for example, the Maharashtra state syllabus. Nonetheless, documents proved to be an important source of information to construct current perspectives on pedagogic practice.

Table 5.3 Categories for Indian geography textbook and curricula analyses on ESD and water

Category	Definition	Coding Rule	Example
General concept of sustainability	Explicit reference to the sustainability concept or implicit reference to future concerns of water supply	Is the sustainability concept explained or mentioned? Are sustainability principles or references to Agenda 21 or other related documents and literature given? Is the conservation of resources mentioned?	Definition of sustainable development with reference to the Rio Summit 1992 (NCERT 2009: 3)
Water content	Topics covered with a main focus on water	Which aspects of water topics are covered (e.g., hydrological cycle, water resources, supply, use, access, etc.)?	In NCERT V, the only topic covered is irrigation
Water and sustainability	Reference to the sustainability of the use of water resources	(How) are availability, depletion, pollution or environmental effects of water covered?	"We face a shortage of water" (NCERT 2006a: 34)
Environmental issues	Effects on the environment due to deteriorating water quality	How are natural resources, especially water, seen in regard to pollution, depletion and future supply?	"The sea gets polluted near the big cities on the coast and near the areas at the mouth of the polluted rivers" (NCERT 2006: 58)
Social issues	Causes, effects, and conflicts of anthropogenic water depletion are addressed	(How) are multi-perspectives, poverty/intra-generational equity, governance, gender, health effects, sanitation availability and efficiency, groundwater, population growth, irrigation, impact of human behavior addressed?	"Resources are unevenly distributed" (NCERT 2007: 30)
Economic issues	Economic effects of water scarcity are addressed	(How) are water costs, water markets, industrial and agricultural use, and economic implications addressed?	"Large-scale death of cattle and other animals, migration of humans and livestock are the most common sight to be seen in the drought-affected areas" (NCERT 2006c: 90)

(continued)

Table 5.3 (continued)

Category	Definition	Coding Rule	Example
Network thinking	More than two factors are mentioned to explain issues	Are factors explained in one coherent context so that integration becomes obvious?	"What is water scarcity and what are its main causes?" (NCERT 2009: 33)
Argumentation	Exercises and tasks which encourage communication on controversial topics	(How) do exercises promote discussions on controversial topics or encourage critical thinking about (dis-) advantages, personal and societal values and multi-perspectives?	"Discuss the factors responsible for depletion of water resources?" (NCERT 2007: 71)
Learner focus	Students are encouraged to act	(How) do texts and tasks promote solidarity, values, motivation, participation and local references?	"It is the duty of man to protect all the components of the environment and make judicial use of natural resources" (NCERT 2009: 32)

Curricula, syllabi and textbooks were particularly important for the analysis as they formulate the frame for the institutional conditions for teaching and learning contents and methods. Therefore, their analysis is described in detail in the following two subchapters.

Curricula and Syllabi Analysis

The curriculum is a message system transmitting the aims of education. Its analysis is a "means of access to the public forms of thought" (Bernstein 1975: 87). To understand the legitimized knowledge and methodology of geography education, national curricula, syllabi and educational government documents were analyzed from an environmental awareness and action-oriented perspective. The objective of these analyses was the identification of sustainability contents and pedagogic principles on the topic of water. Therefore, the structure, objectives, contents, and methods suggested were evaluated. A particular focus was set on whether curricula include a transformative vision for ESD in Freire's (1996) sense of social transformation.

Curricula were analyzed on their selection, sequencing, pacing and evaluation rules, and curricula types in Bernstein's terms were also analyzed. According to Bernstein (1975), the prescribed educational standards can be classified into two modes of curricula concerning their framing and classification: strongly framed curricula include precise objectives, contents and structure teaching through prescribed methods and detailed learning control benchmarks and are thus in a *closed mode*. In contrast, weakly framed curricula are rather orientations and include guidelines as recommendations, thus in an *open mode*. Concerning classification, Bernstein (1975) distinguishes between collection and integrated curricula. If the degree of insulation between subjects is high, classification is strong and thus subjects have strong boundaries and their own specialized rules, their own voice and identity. This is called a *collection curriculum*. In contrast, the *integrated curricula* type has weak boundaries and thus subject boundaries become indistinct (Bernstein 1975: 87–88).

For the curriculum analysis, the current Indian National Curriculum Framework (2005) and relevant position papers of 21 national focus groups, e.g., "Habitat and Learning," were analyzed. Further, the prior National Curriculum Framework for School Education 2000 and the National Policy on Education (NPE) 1986 were analyzed relating to references to the environment. Since on a national level the geography syllabus exists only for classes 9–12, this as well as the environmental studies syllabus for classes 3–5 and the social science and science syllabus for classes 6–8 were analyzed. State syllabi were excluded as they contain similar information and were only available in Marathi.

Textbook Analysis

Textbooks structure subject knowledge prescribed by the curriculum to be taught within one school year. A textbook substantiates and is oriented toward the textual requirements of the curriculum (Rinschede 2005: 350 pp.). As printed publications, textbooks contain teaching and learning contents through texts and functionally integrated visual presentations such as diagrams, maps, pictures, drawings, and tables. The objective is to examine on the one hand the content, namely how ecological, economic, and social aspects of sustainability are covered and on the other hand, whether the didactic textbook presentation supports the development of competences as demanded by ESD.

To analyze the teaching approach and identify possible linkages for the implementation of ESD, a total corpora of 33 textbooks for classes 1–12 were analyzed in terms of content and task structures on the topic of water (cf. list of analyzed textbooks in appendix). Public national (NCERT) and Maharashtra state textbooks (MSCERT) as well as private textbooks used by the schools involved in this study were selected. The textbooks analyzed are the English translation of the original Hindi or Marathi version. Every sentence is literally translated, but may contain linguistically differently nuances, which show restrictions of English textbook use.

The subject geography is introduced at CBSE schools in class V. In total, 12 NCERT textbooks were analyzed from classes V–XII, with additional, practical textbooks in classes XI and XII. Maharashtra State Board introduces geography as a subject in class III, with different textbooks for each district in class III, resulting in a total of 11 analyzed MSCERT textbooks. Furthermore, NCERT textbooks of Environmental Studies, available for standard III, IV, and V were analyzed to cover this adjacent subject to ESD. As private textbook production is a growing market in India, 10 additionally selected textbooks from private providers (IBAIE, CANDID,

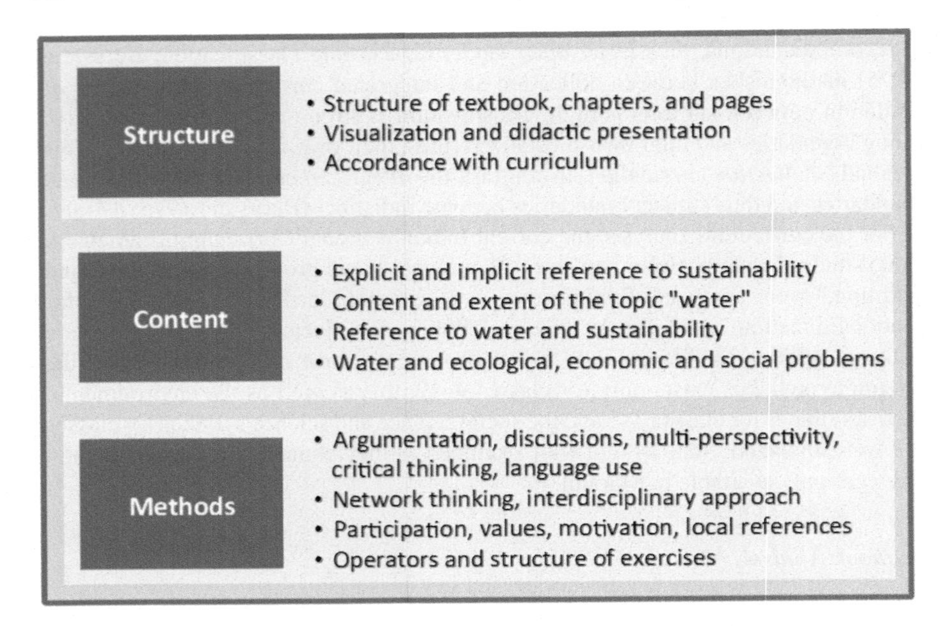

Fig. 5.5 Structure of textbook and curricula analyses

PrimGeo, ICSE) were included in the analysis to reflect a wide range of textbook types.

Since a generally accepted theoretical framework and a methodology for curricula as well as textbook analyses do not exist, a set of categories from the theoretical approach to ESD in geography education was deducted, and the principles and processes of the qualitative content analyses of Mayring (2010) were applied.

The analysis is divided into three parts (cf. Fig. 5.5). The first part examines the outer appearance as well as the accordance with the curriculum, while in the second section, content is analyzed on the basis of the three dimensions of sustainability, and the third focuses on the methods and principles covered regarding the demands of ESD (argumentation, network thinking, student orientation).

The content analysis identified explicit and implicit references to sustainability, and the perspective on and approach to the topic water. The texts were scanned for the coverage, scope and profoundness of ecological, economic, and social dimensions of sustainability in regard to factors influencing the availability of and access to water.

The didactic analysis covered categories which enhance content learning and competence development as demanded by ESD. In exercises, a focus was the promotion of argumentations, discussions, multi-perspectivity, and critical thinking through the use of language. For this purpose, the structure and requirements of exercises were identified as well as the use of operators, for example, "discuss." In texts, content was examined whether a better understanding of complex systems and interdisciplinary or holistic thinking is promoted or discouraged. This was defined through the interconnectedness of several aspects and factors explaining reasons or effects of water-related issues.

In a final step, the texts and exercises were analyzed to understand whether they support students' initiatives and convey values, as well as whether local references are given. While some tasks were meant to motivate students, the actual impact is valued in that the focus on locality and concrete and doable activities is examined. In Table 5.3, each category is defined and has a question as coding rule and an example. The coding rule was developed in a second cycle of categorization. Whether aspects were included and how they were specifically treated is analyzed separately.

5.6 Methodological Reflections on the Role and Limited Perspective of the Researcher in the Cultural Context

In a project with fieldwork in a foreign country, it is of major importance to reflect on the investigation process critically during fieldwork, during the analysis, and during the writing process of the obtained data. The heuristic research methodology of this study is an open process in which theory, methodology, interpretations, the perspective and the role of the researcher are revised. This study focuses specifically on the cultural background of the research subjects. Cultural differences due to being a European who was not socialized in the Indian context were detected throughout the fieldwork.

A central methodological objective was the reflection and adjustment of methods to the researched subjects during the research process. As a researcher with a different socio-cultural background, I had to understand my own perspective on the observed. Thus, I explicitly describe and explain the categories of thinking which pre-structure the empirical data analyses. For example, the theoretical concept of the translation of policies (cf. Sect. 2.4), and the process and gradual reinterpretations of concepts reveal how pedagogic practice is dominated by Eurocentric perspectives, but also how this deconstruction relativizes educational concepts.

Language barriers because of translation difficulties and linguistic misunderstandings interfered with communication during interviews. Abstract concepts provide an external language of precise description and analysis and enable a nonjudgmental perspective. Thus, this study attempted emic research, deriving categories for data collection after explorative field investigations, observations, and interaction with Indian partners and their beliefs, values, and behavior. In a later process of explanation and interpretation, also existing categories were applied for data analysis, relating to etic research (Alexander 1982: 64). During investigations in the field, I reflected on the mingling of perspectives and differentiating viewpoints in field notes, and the data analyses were split into inductive and deducting coding. I reflected on my thoughts and behavior critically during the investigations and was aware that dealing with different norms and values as well as socio-cultural interpretations requires cultural sensitivity. The following example illustrates culturally differing perspectives I encountered during the investigation:

> During the group work, students describe a picture of a buffalo in the river as "the buffalo pollutes the water", while my intention was to demonstrate the possible effect of water pollution on animals. I explicated the underlying norms of these interpretations – the buffalo in his function to work for agricultural purposes in villages is perceived as "dirty". In contrast, my "Western" perspective viewed the buffalo as "victim" of environmental impacts on ecosystems. (Field notes, 15/09/2012)

This example shows that I had to critically reflect on my subjective motivation and my perception as well my decisions that influenced the research process. Field visits were a learning process as I continuously developed deeper contextual knowledge which influenced not only my methodology, e.g., by re-formulating interview questions, but also my behavior in approaching people, as I became acquainted with the field, developed explicit expectations of my interviewees and partly developed relationships with those studied. Certain research decisions were bound to my subjective estimation, and my motives, assumptions and agenda influenced the research process and interpretations (Maxwell (2002: 21). Thus, biases and reactivity influence the validity of qualitative research, but may also strengthen the access to information and thus the depth of interpretation.

The dual relationship between researcher and participants also underlies critical ethical and political issues as an unequal power relation marks researcher and researched interactions. The British anthropologist E. E. Evans-Pritchard clarified already in 1937 how manipulative ethnography is (cf. Bernard 2013: 327): On the one hand, the researcher asks for information and withdraws reflections which might be unsettling and uncomfortable to the teachers or students, while on the other hand, the participants hardly benefit from the researcher. Especially in the hierarchically structured Indian context, I tried to overcome my position and the resulting power imbalances by clarifying my research interests at the beginning and through participatory strategies, like the workshop, from which teachers benefitted as they reported. As in guided interviews, the researcher structures the conversation (Maxwell 2002: 26), I also tried to develop dialogues and included narrative elements in the interviews. Further, in discussions I tried to relativize expectations of interviewees labeling "Western" educational concepts as "modern, progressive, and good ideas." On the contrary, the presentation of my data analysis for communicative validation once also led to contradiction from a public educational stakeholder defending her method of teaching. This demonstrated the relevance but also sensitivity concerning the study of pedagogic practice.

Conversations and interviews may also be deceiving, as the conversation partner is not an equal "friend" who listens to problems, but analyzes problems as relevant for research and objectivizes and rationalizes these from a researcher's perspective (Bernard 2013: 327). Especially, action research can develop dynamics which were not intended:

> One teacher, dedicated to my research, interfered with her principal when asking for more time to prepare students for my intervention. Her emotionally disgruntled call brought my

research interests into conflict: while feeling guilty about having caused alienation and frustration, I also found a further argument for my research concerning the challenges of the introduction of student-oriented pedagogic practice: the agreement of the principal for more time to prepare teaching methodologies. (Field notes 10/02/2013)

In another case, I was frequently contacted via phone by a student after I visited her home challenging me to set limits to personal interaction and the expectations of a student. Wolcott (1995) described gaining rapport as the "darker arts" of fieldwork. Despite these difficulties, the research design comprising of fieldwork, document analyses, and action research assess pedagogic practice comprehensively in its respective socio-economic, cultural, institutional, and structural contexts.

References

Alexander, R. (1982). Participant observation, ethnography, and their use in educational evaluation: A review of selected works. *Studies in Art Education, 24*(1), 63–69.

Bernard, H. R. (2013). *Social research methods. Qualitative and quantitative approaches*. London: Sage.

Bernstein, B. (1975). Class, codes and control. Towards a theory of educational transmission. London: Routledge.

Bernstein, B. (1990). *Class, codes and control. The structuring of pedagogic discourse* (Vol. IV). London: Routledge.

Bohnsack, R. (2009). *Qualitative Bild- und Videointerpretation. Die dokumentarische Methode*. Leverkusen: Verlag Barbara Burich Opladen & Farmington Hills.

Bourdieu, P. (1976). *Entwurf einer Theorie der Praxis auf der ethnologischen Grundlage der kabylischen Gesellschaft*. Frankfurt: Suhrkamp.

Bourdieu, P. (1986). The forms of capital. In J. Richardson (Ed.), *Handbook of theory and research for the sociology of education* (pp. 241–258). Westport, CT: Greenwood.

Bronger, D. (1977). Methodical problems of empirical developing country research: The concept of comparative regional research. *GeoJournal, 1*(6), 49–64. http://www.jstor.org/stable/41142051.

Bruyn, S. T. (1966). *The human perspective of sociology. The methodology of participant observation*. Englewood Cliffs, N.J.: Prentice Hall.

Denzin, N. K. (1970). *The research act*. Chicago: Aldine.

Fernandes, L. (2006). *India's new middle class: Democratic politics in an era of economic reform*. Minneapolis: University of Minnesota Press.

Flick, U. (2006). *Qualität der qualitativen Evaluationsforschung*. Qualitative Evaluationsforschung Hamburg: Rowohlt.

Flick, U. (2011a). *Qualitative Sozialforschung*. Berlin: Rowohlt Taschenbuch Verlag.

Flick, U. (2011b). *Triangulation - Eine Einführung*. Wiesbaden: VS Verlag für Sozialwissenschaften, Springer Fachmedien.

Freire, P. (1996). *Pedagogy of the oppressed*. London: Penguin Books Ltd.

Giddens, A. (1984). *The constitution of society: Outline of the theory of structuration*. Cambridge: Polity Press.

Glaser, B., & Strauss, A. (2008). *Grounded theory: Strategien qualitativer Forschung*. Bern: Huber.

Goffman, E. (1959). *The presentation of self in everyday life*. New York: Doubleday.

Goffman, E. (1971). *Relations in public. Microstudies of the public order*. New York: Harper and Row.

Hammersly, M., & Atkinson, P. (1983). *Ethnography—Principles in practice*. London: Tavistock.

Kocher, M., & Wyss, C. (2008). Unterrichtsbezogene Kompetenzen in der Lehrerinnen- und Lehrerausbildung. Eine Videoanalyse. Neuried: ars et unitas.

Lamnek, S. (2005). *Qualitative Sozialforschung*. Weinheim: Beltz Verlag.

Lewin, K. J. S. (1946). Action research and minority problems. *Journal of Social Issues, 2*(4), 34–46.

Lüders, C. (2012). Beobachten im Feld und Ethnographie. In U. Flick, E. von Kardorff, & I. Steinke (Eds.), *Qualitative Forschung*. Ein Handbuch Reinbek bei Hamburg: Rowohlt Taschenbuch Verlag.

Maxwell, J. A. (2002). *Realism and the roles of the researcher in qualitative psychology*. Tübingen: Verlag Ingeborg Huber.

Mayring, P. (2002). *Einführung in die qualitative Sozialforschung*. Weinheim: Beltz.

Mayring, P. (2010). *Qualitative Inhaltsanalyse: Grundlagen und Techniken*. Weinheim: Beltz Verlag.

Meganathan, R. (2011). Dreams and realities: Developing countries and the English language. In H. Coleman (Ed.), *Language policy in education and the role of English in India: From library language to language of empowerment*. London: British Council.

Meier Kruker, V., & Rauh, J. (2005). *Arbeitsmethoden der Humangeographie*. Darmstadt: Wissenschaftliche Buchgesellschaft.

National Center on Education and the Economy. (2006). *A profile of the Indian education system*. Washington D.C.: National Center on Education and the Economy.

National Council of Educational Research and Training (2006). *National Curriculum Framework 2005. Syllabus for classes at the elementary level* (Vol. I). New Delhi: NCERT.

National Council of Educational Research and Training (2009). *National Curriculum Framework 2005. Position papers on National Focus Groups on Systemic Reform* (Vol. II). New Delhi: NCERT.

Neves, I., & Morais, A. M. (2001). Texts and contexts in educational systems: Studies of recontextualising spaces. In A. M. Morais, I. Neves, B. Davies, & H. Daniels (Eds.), *Towards a sociology of pedagogy. The contribution of Basil Bernstein to research* (pp. 223–249). New York: Peter Lang.

Reichertz, J., & Englert, C. J. (2011). *Einführung in die qualitative Videoanalyse*. Eine hermeneutisch.wissens-soziologische Fallanalyse. Wiesbaden: VS-Verlag.

Rinschede, G. (2005). *Geographiedidaktik*. Paderborn: Schöningh.

Scott, J. (1990). *A matter of record—Documentary sources in social research*. Cambridge: Polity.

Strauss, A. L. (1998). *Grundlagen qualitativer Sozialforschung*. München: UTB für Wissenschaft.

Wagner-Willi, M. (2007). Videoanalysen des Schulalltags. Die dokumentarische Interpretation schulischer Übergangsrituale. In R. Bohnsack, I. Nentwig-Gesemann, & A.-M. Nohl (Eds.), *Die dokumentarische Methode und ihre Forschungspraxis. Grundlagen qualitative Sozialforschung* Wiesbaden: VS Verlag für Sozialwissenschaften.

Willems, H. (2012). Erving Goffmans Forschungsstil. In U. Flick, U. von Kardorff, & I. Steinke (Eds.), *Qualitative Forschung*. Ein Handbuch Reinbek bei Hamburg: Rowohlt Taschenbuch Verlag.

Wolcott, H. (1995). *The art of fieldwork*. New York: AltaMira Press US.

Chapter 6
Academic Frameworks for Formal School Education and ESD Principles

Abstract This chapter uses a didactic framework to explore the extent to which India's educational frameworks reflect the principles of Education for Sustainable Development (ESD). Syllabi and textbook analyses serve as an important backdrop for understanding how classroom teaching is structured and for outlining the limitations to achieve ESD objectives. The results show that the pedagogic principles in the National Curriculum Framework (NCF) promote critical thinking and environmental action in line with ESD. However, national school syllabi and geography textbooks and syllabi (NCERT and MSCERT) mostly emphasize reproductive pedagogic practice which emphasizes memorization rather than bringing forth the greater transformative potential of ESD principles. A detailed textbook analysis on the topic of "water" demonstrates that teaching content and methods hardly relate to six ESD dimensions, namely social, environmental, and economic sustainability, as well as didactic principles of argumentation, network thinking, and student orientation. Geography syllabi and textbooks display a fragmented, fact-oriented, and definition-oriented approach. Water resources are presented as a fixed commodity, and unequal access to water is not depicted as socially constructed. Differing perspectives on water access in the students' urban and rural environment are not presented. This contradicts ESD principles, which favor a problem-based, integrated, and skill-oriented approach to topics at the human–environment interface. The applied ESD framework offers entry points for integrating ESD in textbooks and syllabi through the selection, sequencing, pacing, and evaluation of knowledge and skills.

Pedagogic practice in classrooms is not taking place in isolation but is being strongly influenced by national and state standards, as presented in curricula, syllabi, and textbooks. In order to examine the instructional discourse and how pedagogic practice is controlled through institutional regulations, this chapter presents the prescribed selection of knowledge and skill development in geographical school education in India. Of specific interest is the question in how far academic frameworks (curricula, syllabi, and textbooks) promote principles of ESD, particularly argumentation. This chapter will start with examining the relevant educational stakeholders and processes of curriculum and textbook development at the national and state levels in India.

S. Leder, *Transformative Pedagogic Practice*, Education for Sustainability,
https://doi.org/10.1007/978-981-13-2369-0_6

Subsequently, ESD principles will be analyzed in educational policies, curricula, and syllabi. The content and didactic analyses of textbooks examine teaching approaches on the topic of "water" both from an ESD perspective and through a Bernsteinian lens. ESD's environmental, social, and economic dimensions as well as network thinking, argumentation, and student orientation (cf. Fig. 3.7) offer didactic insights into teaching contents and methods. A Bernsteinian lens enables an in-depth analysis of the discourse on pedagogic principles and instruction.

6.1 Curriculum and Textbook Development in India

To understand the reasons for the often fragmented approach in textbooks and curricula with a weak didactic grounding in terms of integrating the promotion of critical thinking and skill development, it is important to look into the development process of textbooks and its associated institutional regulations and stakeholders. The wide range of quality within curricula and textbooks reflects the diversity of curriculum designers and textbook authors involved in their development. Textbooks are usually edited after the release of a revised curriculum or syllabus, which is meant to be every five years in India. The last revision of the curriculum was in 2005, which indicates that this process has been delayed by more than a decade, and hence, textbooks have only been sporadically revised for completely new editions.

The process of writing textbooks differs between the states. The most recent National Curriculum Framework 2005, the national NCERT syllabi, and NCERT textbooks function as orienting guidelines for the state textbook development through SCERT. NCERT has its own staff to write textbooks, and the environment education board functions as an advisory body. SCERT either directly selects a committee to develop syllabi and state textbooks or, as it is the case in Maharashtra, SCERT develops guidelines for a separate institution, the textbook bureau Balbharti. At Balbharti, a selected textbook writer committee develops textbooks. The textbook writer committee is appointed by the State Ministry for Education in that invitation letters are sent to principals and schoolteachers to as them to be on the committee. These textbook writers work for the government and are not exposed to competition or quality control, which may influence their motivation and, thus, their performance. In contrast, the state committee for environmental studies, for example, which develops syllabi, is made up of selected individuals with diverse backgrounds from public, private, Zila Parishad, urban and rural schools, as well as NGOs and women. None of the recent members of the textbook writer committee has been involved in the curriculum design in 2004, and hence, a new team was built.

The textbook writing process leads to little continuity within and between topics and the lack of a consistent and comprehensive didactic approach. Textbook development in Maharashtra (and probably similar in other states and at the national level) is a lengthy process structured in several phases[1]: First, a group of diverse educational

[1]The description is based on several focus group discussions with Maharashtran textbook authors.

and scientific experts (NGO staff, government officials, and academics) as well as teachers of diverse backgrounds (from rural and urban schools, male and female teachers) meet twice a month for a couple of days for over one year to decide on topics and approaches to cover in each textbook. In the second year, responsible authors are chosen for each chapter so that each book has a range of authors. A draft of each chapter will be revised by other group members and then finalized. One author of the geography textbooks admitted that he was the only full-time government appointee responsible for writing all textbook chapters, although he still had to integrate the perspectives of others from the geography textbook committee. This indicates a lack of collaboration in the writing processes and limited quality control and competition. Furthermore, the appointed government official used to be a teacher, but has not been teaching for several years. As textbook writers work for their subjects in isolation, subject content is not coordinated between subjects (NGO 8). Especially for crosscutting themes such as environmental education, meetings of textbook authors from different subjects to agree on thematic foci are not conducted (MSCERT 1, MSCERT 3). The coordination between subject content for each class is also not consistent, especially between elementary and secondary school levels (Prof 7). A further problem, which was mentioned by teachers involved in textbook writing, is that the majority of experts at the university level has limited exposure to students and classroom teaching, and therefore, the content load and the methodological approach are not appropriate for the level of students (MSCERT 1). Textbook writer trainings are not mandatory (MSCERT 2). One national textbook writer training on ESD was voluntarily conducted by BVIEER in form of a five-day-long interdisciplinary workshop with seminars, discussions, and group work in January 2013.

A further challenge is the possible loss of content due to the translation from syllabi in Hindi to textbooks in Marathi and English (NCERT 8). MSCERT provides guidelines in Marathi which are based on NCF and syllabi in Hindi. The textbooks are written in Marathi and translated into English and Hindi. Thus, the understanding of textbooks depends on the linguistic capacities of multilingual speakers and content might be changed or simplified due to language challenges (Prof 7). These highlighted issues demonstrate possible problems in the textbook development process causing inconsistency, lacunae, and fragmentation of content in textbooks.

During a session I conducted during an ESD workshop organized by BVIEER in January 2013, textbook authors from different states in India discussed in groups ESD content and ESD competences in NCERT and MSCERT geography textbooks from standard 3–12. The majority of textbook authors clearly agreed that argumentation and network thinking are not sufficiently promoted in textbooks. They rated statements on textbooks in a questionnaire from a scale 1 (I fully agree) to 5 (I fully disagree). Rather negatively rated statements (<2.5) were "Controversies are stated to encourage critical thinking," "Multiple perspectives and different views are stated," "Problem-solving skills are enhanced," "Causes and effects are explained properly," "Discussion and communication between students is encouraged," "Group work is encouraged," and "Students have to interlink knowledge of different topics." Sustainability aspects, particularly social and economic perspectives were rated as not sufficiently addressed (<3.0) (cf. Table in appendix).

Positive evaluation, in contrast, was given to the logic structure of the chapter and its visualization, the amount of environmental information, the use of language considered appropriate for students, and the facilitation of reproduction skills, such as to be able to define terms. This shows that textbook authors are aware that textbooks could better address ESD principles. However, they admitted that they had not been introduced to how to implement these principles in textbooks. In addition to the above-mentioned institutional challenges of textbook and curriculum development, it seems necessary to address ESD in capacity development trainings for textbook authors.

6.2 ESD Principles in Educational Policies and Curricula

This subchapter examines if and how several educational policies and curricula relate to ESD principles (cf. list of documents in the appendix). The educational apex body of India, NCERT, developed the most recent Indian National Curriculum Framework (NCF 2005). The NCF 2005 is an overarching framework for orientation, whose principles are meant to concretely guide the development of syllabi and textbooks. It functions as the current steering, orientation, and legitimization instrument for educational authorities in India as it summarizes the syllabus for each subject for a specific time period. Defined teaching objectives, contents, and methods prescribe official educational standards and represent the currently acknowledged requirements for the next generation to tackle societal challenges. Some ESD principles, such as environmental and social values and student-centered teaching techniques for school education, are mentioned in several policies and curricula. However, concrete means and methodologies for implementing these are not available. While the *National Policy on Education 1986* already enforces the "removal of social barriers" and a consciousness for the "protection of environment" (3.4) and NCF 2000 similarly encourages understanding and preservation of the environment, concrete directives on *how* these challenges can be tackled through school education are not mentioned in these or subsequent documents. The *Curriculum Framework for Teacher Education* 2004 states that the environment needs to be protected, but equally does not provide any explanation how this objective can be achieved through education. Although "child-centered, activity based, play-way and joyful approaches besides oral instruction and demonstration" (NCERT 2004: 27) and role-play, story telling, and creative-thinking strategies are mentioned, the proclaimed social change and the conservation of the environment are not sufficiently grounded through teaching methodology approaches in educational policies and curricula to guide teachers as public individuals to trigger changes in pedagogic practices.

The consensus of the NCF 2005 is "systemic reforms" to strengthen "the processes for democratization of all existing educational institutions at all levels" (NCERT 2009: 13). Similar to the perspective of Bernstein (1975, 1990), NCERT acknowledges that school reflects economic differentiation and reinforces segregation and therefore is the "institution that brings about social transformation" (NCERT 2009:

21). Thus, the educational apex body of the Government of India (GoI) recognizes Indian schools as sites of reproduction, but also of social transformation. The NCF 2005 is in accordance with the values of social justice and equality of the *Constitution of India*, as the following should be passed on through education:

> These include independence of thought and action, sensitivity to others' well-being and feelings, learning to respond to new situations in a flexible and creative manner, predisposition towards participation in democratic processes, and the ability to work towards and contribute to economic processes and social change. (NCERT 2005: vii)

The NCF 2005 is notably influenced by the report *Learning without burden* by the National Advisory Committee (Pal 1993) appointed by the Ministry of Human Resource Development (MHRD). This controversial report identified unsatisfactory pedagogic practices as structural and societal problems of elite development and strongly urged a reduction of the curriculum load:

> The problem of the load on school children does not arise only from over-enthusiastic curriculum designers, or poorly equipped teachers, or school administrators, or book publishers [...]. We continue to value a few elite qualifications far more than real competence for doing useful things in life. (Pal 1993: 24)

The report *Learning without burden* (Pal 1993) and the previous NCF 2000 criticize both the heavy curriculum load and also the prevalent perception of children as receivers of knowledge, instead of as constructers of knowledge. This is manifested through teaching methods that focus on rote learning and memorization as well as examination systems that enforce the learning practice of literally reproducing the textbook contents. Against this backdrop, the latest NCF 2005 states that the amount of information students have to memorize and reproduce has further increased: "Flabby textbooks, and the syllabi they cover, symbolize a systemic failure to address children in a child-centered manner" (NCERT 2005: 3). The NCF 2005 criticizes the inflexibility of the school system and that "learning has become an isolated activity which does not encourage children to link knowledge with their lives [...] schools promote a regime of thought that discourages creative thinking and insights" (NCERT 2005). This indicates that the report *Learning without burden,* which was published more than two decades ago, has not lost anything in its actuality. While the roots of the problem were identified and recommendations were given, the central problem of the curriculum load has continued to increase until today. The immense amount of information to be studied and the associated manner of learning do not leave space for content and skill development relevant for students, as ESD demands.

According to Bernstein's types of curricula (1975), the NCF 2005 can be classified as an *open curriculum*, as it states the overarching aim of school education for all subjects, however, does not provide concrete contents and methods. Thus, the NCF 2005 is weakly framed. An orientation for the selection, sequence, pacing, and evaluation is given and general objectives are stated, whereas concrete measures are missing.

Drawing upon the insights from *Learning without burden*, the NCF 2005 promotes "teaching for construction of knowledge," child-centered and *Critical Pedagogy* with active participation (NCERT 2005). The guiding principles for school education, as cited in the NCF 2005, correspond to principles of an invisible pedagogy (Bernstein 1990):

- connecting knowledge to life outside the school
- ensuring that learning is shifted away from rote methods
- enriching the curriculum to provide for overall development of children rather than remain textbook centric
- making examinations more flexible and integrated into classroom life and
- nurturing an over-riding identity informed by caring concerns within the democratic policy of the country (NCERT 2005: 5).

These guiding principles encourage student orientation and flexibility within the school structure. Hence, the NCF 2005 proposes pedagogy with weak framing and classification (Bernstein 1975, 1990). By distancing itself from performance-based pedagogy, the NCF 2005 suggests the reduction of textbook use in favor of a stronger student focus. Knowledge should be selected, organized, and paced according to students' needs. Thus, the NCF 2005 displayed ambitions to orient school education toward a *competence model of pedagogy*. This model of pedagogy and the proposed pedagogic principles of reasoning, posing questions and critical thinking on social justice and environmental pollution, have relevant overlaps with the educational objectives of ESD and the didactic approach of argumentation. NCF spells out that the principles of ESD and, subsequently, argumentation skill development are of great interest to be embedded within the Indian educational system.

The NCF 2005 encourages participative and discussion-oriented methods and "to inculcate in the child a critical appreciation for conservation and environmental concerns along with developmental issues" (NCERT 2005: 5). For the actual implementation in local classrooms, the translation of these principles into syllabi and textbooks is critical. An example mentioned in the NCF 2005 is the selection of content on water in form of questions relating to gender, caste, and class: "Who controls the village well? Who fetches water?" (NCERT 2005: 53). These questions raise awareness on who governs and who uses water. Particularly in social sciences, students are supposed to be encouraged to critically question social realities and the natural environment, as well as reasoning and arguing on experiences outside of school (NCERT 2005: 52). In the context of a Critical Pedagogy, students should further learn to question critically issues under political, social, economic and moral aspects as well as "received knowledge [...] in a 'biased' textbook" (NCERT 2005: 33, 50).[2] Thus, explicit references to Critical Pedagogy in regard to social and environmental issues and teaching methods are encouraged, as principles of ESD and single examples are given. Separate national syllabi frame values in concrete educational themes and objectives, as will be outlined in the following subchapter.

---

[2]This may directly refer to the controversy on national Hindu identities represented in NCERT textbooks enforced under the National Democratic Alliance (NDA) regime as anti-constitutional approach until 2004.

6.3 ESD Principles in National School Syllabi

As the following chapter will outline, national Indian school syllabi hardly include any ESD principles promoting argumentation skills, network thinking, and student orientation. Although the syllabi were also developed by NCERT, the principles suggested in the NCF 2005 do not translate into syllabi. This suggests either a strong division and insufficient communication within NCERT, or the need to develop individual capacities to integrate a competence approach in the syllabus. By applying Bernstein's concepts of curriculum types (1990), the following analysis will highlight the intention and priorities of the "Syllabus for Secondary and Higher Secondary Classes" (NCERT 2006). The syllabus is more relevant and specific to all textbook authors and teachers within India than NCF, because it prescribes concrete topics and objectives to be covered in lessons, and hence, the syllabus decisively frames pedagogic practice.

Although the guiding principles of the NCF 2005 are mentioned in the syllabus and activities are suggested, the syllabus hardly promotes these: The syllabus includes teaching contents by listing the facts to be studied, while skill development is only exemplary mentioned through activities. In terms of the classification of the syllabi structure, the examined syllabi are *closed curricula,* in contrast to the *open mode* of the NCF 2005 and the principles of ESD (cf. Table 6.1). One exemplary activity in the syllabus is the "estimation of water used by a family in one day, one month, one year" (NCERT 2005: 24). This single example may serve textbook authors and teachers as orientation. However, the integration of this activity in greater critical-thinking tasks that encourage reflecting upon the causes and consequences of their own use as well as reasons of differential water access and use in their environment is not encouraged. This example demonstrates how the syllabus does not provide examples according to the framework and objectives from NCF.

Considering the framing of the geography syllabi, the syllabi for environmental sciences (class 3–5) and geography (class 9–12) represent a *collection* curriculum in contrast to an *integrated curriculum* as per the ESD principles (cf. Table 6.1). The *selection* of themes is not comprehensively integrated throughout the curriculum. Water-related themes are covered in all classes; however, the topics are fragmented and repetitive. For example, in class 9, the topic "drainage" covers major rivers and tributaries, lakes and seas, as well as river pollution and control measures for river pollution, while in class 10, the topic "water resources" covers sources, distribution, utilization, water scarcity, and conservation. Class 11 covers oceans and the hydrological cycle, and class 12 again covers "water resources" concerning the availability, utilization, and conservation of water as in class 9. The lists included in the syllabi show that a fact-oriented and technical coverage of the topics is valued over an integrated and comprehensive perspective in which multiple factors for a particular context are causally interlinked, as demanded by ESD principles, but also NCF.

Unlike the suggestions in the framework of NCF 2005, the *sequence* of topics in the geography syllabi is not progressive and does not build on each other. Topics

Table 6.1 Classification and framing of geography syllabi in India and of ESD principles (own draft)

Bernstein's code	Geography syllabi in India	Education for sustainable development principles
Classification (power through structure)	**Collection curricula** = Strong boundaries between topics:	**Integrated curricula** = Weak boundaries between topics:
	• Regional geographical approach • List of up to 19 chapters per academic year, not interrelated	• Interdisciplinary approach • Few topics for intense study from different perspectives
	→ Strong classification (C+)	**→ Weak classification(C−)**
Framing (control through interaction)	**Closed mode** = Precise objectives, contents, and structure through prescribed method and explicit learning control benchmark: exams with pre-structured answers	**Open mode** = Flexible structure, open question mode, implicit learning control, guidelines
	• Explicit pacing and evaluation rules	• Implicit pacing and evaluation rules
	→ Strong framing (F+)	**→ Weak framing (F−)**

are ordered by continents and countries and are not embedded in greater themes. This regional approach of geography in school has been criticized within NCERT, as knowledge of regional singularities does not promote knowledge that can be transferred (NCERT 2006: 44). The sequence of topics is also contradictory to the neurodidactic principle to study from simple to increasingly complex structures by integrating new knowledge in existing nets of knowledge (Rinschede 2005: 125). As the arranged sequence of topics in the syllabus is arbitrary, new learning units cannot be integrated into existing knowledge nets to enhance growing complexity (Spitzer 2007). To help students categorize geographical content and recognize geographical structures (Rinschede 2005: 130), the coherence of geographical factors, structures, and processes has to be understood through examples, and the transfer of those. The syllabi analysis suggests that students stagnate at the lowest level of cognitive learning goals by the degree of complexity (cf. B.S. Bloom 1972 in Rinschede 2005: 159), in which they only gain the knowledge of facts, but do not acquire skills to apply, analyze, or evaluate this knowledge. Furthermore, instrumental or affective learning goals are not adequately addressed.

The *pacing* of learning is prescribed through lesson periods ranging from 6–12 periods per month, which indicates a short time period in which a great amount of information has to be studied. The limited time available for so much content risks fragmented knowledge and enforces, instead of prevents, children as receivers of knowledge despite the constructivist approach supported by the NCF 2005. It is, however, noteworthy that the syllabi do not have direct influence on *evaluation criteria*. Examination and tests are excluded in the syllabi and outsourced to the educational boards, such as the Central Board of Secondary Education (CBSE) or the Indian Certificate of Secondary Education (ICSE). These so-called board exams are, however, based on the syllabi and further enforce the memorization of facts through multiple-choice tests.

The aspiration of a paradigm shift toward sustainable development and critical thinking as stated in the position paper 1.6 *Habitat and Learning* and the NCF 2005 is barely translated into the syllabus. The syllabus lists relevant topics such as river pollution, scarcity of water and conservation methods such as rainwater harvesting and watershed management. Concrete thematic examples and particular projects or references to sustainable development principles or skill development are, however, barely available. Although according to the Supreme Court order (cf. Sect. 4.1.6) it is mandatory to infuse concepts of environmental education in all subjects, the stated objectives of environmental education do not translate into principles or criteria guiding the choice of content, methods, and evaluation measures in geography syllabi. There is an obvious contradiction between the postulations for environmental education and Critical Pedagogy in the NCF 2005 and its position papers, and the approach in the syllabi, which is more influential on structuring pedagogic practice than the NCF 2005. This indicates a great need for educational research and capacity development on how to introduce evidence-based principles and methods for ESD in geography syllabi. A comprehensive conceptual and methodological framework, which provides systematic pedagogic principles for the selection, sequence, pacing and evaluation of knowledge and skills, needs to be developed to better guide textbook development. The following chapter will demonstrate how the selection, pacing, and evaluation of knowledge in textbooks are closely linked to the principles found in syllabi.

6.4 ESD Principles in NCERT and MSCERT Geography Textbooks

In India, textbooks play a particularly important role for structuring classroom teaching, as the textbook is often the only available teaching medium in schools, particularly government schools. The textbook has three functions: Firstly, it is an instrument of the state to control and structure teaching contents as it presents the pre-structured content of teaching. Textbooks control the selection, sequence, pace, and evaluation of knowledge in pedagogic practice to a great extent. Several studies have shown that textbooks represent norms, values, and political orientations of current ruling powers. Hence, they can be interpreted as a “powerful vehicle for political and cultural control” (Budke 2010; Kumar 1988; cf. Advani 2004; Gottlob 2007; BVIEER 2002). Secondly, textbooks function as an orientation for teachers for their lessons. It is the most commonly used medium to teach, prepare for exams, and thus determine the teaching procedure by the choice of material available for teachers. The teacher reads out each chapter that students have to memorize and reproduce for exams.[3] Thirdly, students use the textbook to work independently on contents. Contents and tasks in textbooks should be motivating, correct, age-specific, and comprehensible for each class. Textbooks can be placed between the priorities of summarized contents to be learnt (e.g., definitions, lists, explanations) and of the functional integration of material to stimulate individual learning processes for students. These three functions of textbooks explain why their analysis is central to better understand pedagogic practice in geography lessons as well as to identify entry points for the implementation of ESD. The presentation and quality of sustainability issues in textbooks can be analyzed as one indicator for the status of integration in teaching. This analysis focuses on how ESD principles, contents, methods, and competences are integrated in textbooks.

The analyzed geography textbooks select and arrange learning contents in a *closed* manner, and thus, these textbooks have strong framing. NCERT and MSCERT textbooks barely present NCF's recommendations of a child-centered and constructivist approach to learning and Critical Pedagogy (cf. Crossley and Murby 1994). Thus, the textbooks scarcely include tasks or texts that fulfill the demand of the NCF 2005. While the framing of the NCF 2005 supports a competence model of pedagogy, contents and methodological approaches in the textbooks of NCERT and the state Maharashtra produce a *performance model of pedagogy*. As the following analysis will demonstrate, textbooks contain copious amounts of fragmented information that needs to be memorized for passing examinations. As the syllabus needs to be covered, there is not enough time for discussions and activities that link learning with students' lives and encourage "creative thinking and insights" (NCERT 2005: 2). Methods other than teacher-centered question–answer conversations are rarely encouraged through tasks.

This subchapter is divided into three sections. The first section will reflect on the overall structure of NCERT and MSCERT textbooks, particularly in regard to the objectives and specifications stated in the national syllabi. The second section presents a content analysis of the topic of water from the three sustainability perspectives (environmental, economic, and social aspects). The third section encompasses a didactic analysis of the water chapters in geography textbooks. The analysis investigates how the three ESD teaching principles: network thinking, argumentation skill development, and student orientation are promoted in geography textbooks. The subchapter will conclude with a comprehensive perspective on ESD in curricula, syllabi, and textbooks.

[3]The important role of the textbook was observed in lessons and is further outlined in Sect. 4.1.4.

6.4.1 Structure of NCERT and MSCERT Textbooks

A comparative analysis of the objectives stated in the NCF 2005 and the syllabi and their realization in textbooks demonstrates inconsistencies in the translation from curriculum to syllabus to textbook. Objectives in syllabi have weaker framing so that they can be reinterpreted for the actual design of textbooks; however, water-related topics in the syllabus and their realizations in geography textbooks show discrepancies (cf. Table 6.2). For example, for water-related topics in grade 10, the syllabus suggests covering water scarcity, the need for conservation and management, rainwater harvesting and watershed management. Six of a total of 50 periods are allotted to cover these topics. However, only one subchapter covers these topics in the textbook. Another example is that one syllabus objective for grade 9 is to understand the role of rivers for humans. In a textbook for grade 9, simplistic statements may even discourage relating freshwater sources to themselves: "Think about life of human beings without fresh water" (N-Geo-9: 23). As the question is broadly framed, the answers are likely to be reduced to answers such as that life without water is not possible, as humans need water to drink, bath, cook, clean, etc. With a more applied or context-bound question on the economic, ecological, and socio-cultural role of a specific river, argumentations on natural resource use could enhance deeper understanding of the need for sustainable water resource use.

The comparison also shows that the postulated focus of human–environment interrelations in geography syllabi was not realized in textbooks, as human and physical geography topics are covered separately and not in an integrated manner.

The structure of the textbook contents is rather fragmented. The analyzed geography textbooks contain between 6 and 22 chapters each, of which only in some textbooks these chapters are categorized into thematic units to enhance interlinked knowledge. The high number of separate chapters indicates a strong classification of knowledge and strong framing of the sequence of content. Chapters usually consist of only two to four pages, of a total of 45–65 pages per textbook. As only a limited time is allotted per chapter, pacing is quick. The chapters are subdivided in a chapter title, texts (sometimes structured with reader-friendly subtitles), exercises, and activities. The difference between exercises and activities is not clear or defined, as activities can also ask for lists as in exercises. There is only limited didactic presentation of contents, specifically according to different class levels. The presentation of a page is rather condensed, even in lower grades where students' curiosity should be raised. The font is approximately size 12 and 1.5 space (slightly bigger in class 3 and 4), which stresses the density of the text, although each page contains visuals in form of drawings, maps, diagrams, or pictures. MSCERT textbooks are written in double columns, while NCERT textbooks are single columned with a wide margin, where visuals and "Do you know" activities as well as "Find out" or "Map Skill" boxes are included. Some NCERT chapters contain more didactic and structuring elements such as specific information boxes, which make the book's appearance less monotonous and encourage the understanding of information (cf. photograph of a double page of NCERT geography textbook for class 8 in appendix). Sources of

Table 6.2 Water-related topics in geography in the national Syllabus for Secondary and Higher Secondary Classes and in NCERT and MSCERT textbooks for class 9–11

Class	Themes suggested in the national syllabi	Water-related topics suggested in the national syllabi	Objectives stated in the national syllabi	Allotted periods in NCF syllabi	Implementation in national textbooks (NCERT)	Implementation in the state textbook of Maharashtra (MSCERT)
9	India—Land and the People	Drainage: Major rivers and tributaries, lakes and seas, role of rivers in the economy, pollution of rivers, measures to control river pollution	"To understand the river systems of the country and explain the role of rivers in the evolution of human society" (p. 82) Activity: Poster on river pollution (p. 83)	10 of 50	Chapter 3: Drainage	Chapter 3: river systems; Chapter 4: ocean and lakes
10	India—Resources and their Development	Water resources: sources, distribution, multi-purpose projects, water scarcity, need for conservation and management, rainwater harvesting and watershed management	"To understand the importance of water as a resource as well as develop awareness towards its judicious use and conservation"(p. 84) Example activity: Poster on pollution of water in the locality (p. 85)	6 of 50	Chapter 3: Water resources	Chapter 3: Water resources
11	Fundamentals of Physical Geography	Water (Oceans) Hydrological Cycle Oceans—submarine relief, distribution of temperature and salinity, movements of ocean water-waves, tides and currents (Unit V, p. 103)	– Familiarization with terms, key concepts, and basic principles of geography – Analyze the interrelationship between physical and human environments and their impact – Utilize geographical knowledge in understanding issues concerning the community such as environmental issues, socio-economic concerns, gender and become responsible and effective member of the community (p. 102)	12 of 88	Unit V: Water (Oceans)	Chapter 4: Drainage System and water resources; chapter 5: Oceans and marine ecosystems

data and a detailed description below diagrams are usually missing, and thus, it is not indicated that information comes from different perspectives, not to mention that teachers explicitly address a reflective dealing with information as they represent only one perspective.

The didactic presentation of the textbook is enhanced through bold printing of important terms to increase the overview of relevant key terms. Some sections are colored, but the use of color in MSCERT textbooks is randomly chosen, so that it does not enhance the methodological structure of the textbook (cf. textbook scans in appendix). Hence, colored thematic sections could improve the textbook structure logically. In general, the quality of public textbooks, particularly of the visuals, needs to be kept at low costs of 18–45 INR (=0.25–0.60 €) per textbook, as students need to buy textbooks for each subject every year, and this can add to the financial burden of many families with low cash inflow.

The difference between NCERT and MSCERT textbooks may be representative for the range of national textbooks and other state textbooks. Despite the opportunity of using national textbooks as templates, state textbooks have an overall lower quality concerning the content in written texts as well as the formulation of exercises. MSCERT textbooks oversimplify information and contain a number of mistakes, and particularly focus on rote learning through a focus on definitions and lists. Water is only considered for its agricultural use, and domestic and industrial purposes are excluded in all MSCERT textbooks. Additional information and detailed chapter structures and thematic units as in NCERT textbooks are not available.

Grammatical constructions used in textbook texts are rather simple because subordinate sentences such as conditional or causal sentences are rarely used:

> Water is an important and valuable resource in India. Monsoon rainfall is the primary source of water resources in India. Monsoon rainfall influences availability and use of water resources. Rivers, glaciers, lakes, springs and wells are secondary sources of water. Rivers and lakes are important for flow, storage and regulation of the currents of rainwater. (MSCERT geography textbook for class 10, scan in appendix)

As language determines the complexity of content, it also encourages the reproduction of simple and only few causal relations by the students. However, difficult vocabulary such as technical or abstract terms is used, but often not sufficiently explained. This is particularly problematic for students in lower classes as they are just learning to use the English language as a medium in school. The following excerpt demonstrates an example of a rather technical use of language in textbooks:

> The Ganga Action Plan (GAP) Phase-II, has been merged with the NRCP. The expanded NRCP now covers 152 towns located along 27 interstate rivers in 16 states. Under this action plan, pollution abatement works are being taken up in 57 towns. A total of 215 schemes of pollution abatement have been sanctioned. So far, 69 schemes have been completed under this action plan. A million litres of swage is targeted to be intercepted, diverted and treated. (NCERT geography textbook for class 9, scan in appendix)

In the opening pages of each textbook, one can find the National Pledge, and in some textbooks also the preamble of the Constitution of India. These are meant to promote national values such as justice, liberty, equality, and fraternity. In Maharashtra state textbooks, one page is dedicated to "life skills education," as promoted by the World Health Organization (2001) with ten life skills, which are mostly congruent with ESD skills (e.g., "problem solving" and "critical thinking"). Through school education, children are expected to acquire creative and critical thinking as well as problem solving. However, as the following chapters will demonstrate, the mentioned skills are barely transferred into the selection of knowledge and methods of the textbooks.

6.4.2 Content Analysis of the Topic "Water" in Textbooks

In the following, textbook chapters on water-related contents are examined for their inclusion of ESD principles (cf. Fig 3.7 in Sect. 3.3). At first, chapters on water are screened for their explicit references to sustainability. Then, the chapter contents are analyzed from environmental, economic, and social sustainability perspectives. Subsequently, water chapters are didactically analyzed on network thinking, argumentation, and student orientation.

6.4.2.1 Water and Sustainability in NCERT and MSCERT Geography Textbooks

Water is an explicit topic in textbooks as almost every textbook has one chapter named "water" or "water resources." The water content in each class level is similar, covering the location of oceans, rivers, dams, and lakes. Water-related topics covered in NCERT and MSCERT textbooks comprise mostly water sources, their use, and water pollution, whereas social, environmental, and cultural linkages and governance issues are sporadic and not integrated into chapters (cf. Table 6.3). Human and physical geography are not integrated but dealt with separately. Especially with the topic of water, an integrated approach could enhance thinking in networks, rather than only deepen fragmented knowledge. Concepts and principles are not sufficiently elucidated, as, for example, the difference between fresh and salt water is not clearly explained (N-Geo-6: 30–32). The selection of knowledge hardly promotes principles of ESD as the perspective of individual students and their possible agency is neglected.

The concept of sustainable development is introduced in an NCERT textbook for class 10. The definition of sustainable development in the Brundtland Report (WCED 1987) is given, but not explained with examples to help students understand its meaning. Furthermore, it is not put in a concrete thematic (water) context for students to understand, but rather elaborated with abstract terms:

> An equitable distribution of resources has become essential for a sustained quality of life and global peace. If the present trend of resource depletion by a few individuals and countries continues, the future of our planet is in danger. (NCERT 2009: 3)

Table 6.3 Water-related topics covered in NCERT and MSCERT textbooks (the numbers indicate the class)

Water-related topics covered	NCERT textbook	MSCERT textbook
Fundamentals: hydrological cycle	7, 8, 11	3
Source: fresh and salt water sources (regional: river systems, oceans)	6, 7, 9 (only rivers), 10, 12	3, 4, 5, 7, 9, 10, 11
Water availability and shortage	7, 8, 10, 12	(5), 12
(Agricultural) use: irrigation and drainage	9, 11	10, 12
Water pollution	7, 8, 9, 10	5, 7, 11, 12
Groundwater depletion	10	–
Cultural references (customs)	5	–

This quote exemplifies the strong rhetoric found in many textbooks. Abstract words such as "future of our planet," "global peace," and "equitable distribution" make it difficult for students to grasp the idea of sustainable development. The textbook chapter "Resources and Development" includes types of resources, resource planning in India, and resource conservation with the example of soils. Without prior thematic introduction on different resources and their depletion, students are asked to "make a project showing consumption and conservation of resources in your locality" (NCERT 2009: 13). This example illustrates how information is fragmented and tasks are abstract in textbooks without concrete methodological suggestions. Environmental and social concepts in textbooks are not clarified with specific cases and linked to the students' own lives, which make it difficult for students to understand these.

Sustainability of water supply and water sources is only implicitly mentioned when the limited water availability and conservation of the resource are addressed (e.g., N-Geo-7: 30-32). Statements and definitions of conservation and pollution are given, but explanations or details to obtain a deeper understanding of the causes and effects of water pollution are not included. Long-term effects of depleting this resource and a reference for the availability of water for future generations (intergenerational justice) are not mentioned. The unequal distribution of water within society (intra-generational justice) is only slightly mentioned in a few textbooks, but not comprehensively covered. Concerning the three dimensions of sustainability, social and economic values are generally not covered, and pollution and environmental effects are addressed, but not explained comprehensively, as will be further outlined in the following chapters.

6.4.2.2 Environmental Aspects on Water in NCERT and MSCERT Geography Textbooks

In most textbooks, water pollution facts are stated, but the human-made causes and the effects on the environment are not explained. In an NCERT textbook for class 7, for example, newspaper headings address "dying rivers," but an explanation regarding the causes and consequences of water pollution is not given in the textbook (N-Geo-7: 33). Newspaper headings in the NCERT geography textbook for class 12 mention "rivers of conflict…but also of piece" or "Climate change? Barmer grapples with floods" (cf. textbook scan in appendix). The related task asks to discuss the highlighted news. In the same textbook, it is mentioned that "poisonous elements […] destroy the bio-system of these waters" (N-Geo-12-IndPpEc: 136), but their origin, the processes, and consequences are not further explained. Furthermore, pollutants are listed in a table, but their effects are not explained. In other textbooks, pollution is elucidated as a hazard for human use, but the tremendous changes within ecosystems, for example through the extinction of species, are not explained (cf. N-Geo-8, N-Geo-9). Therefore, the textbooks follow an anthropocentric approach that considers pollution primarily harmful for human health. However, the long-term consequences of environmental degradation on humans, such as increasing resource conflicts, natural disasters, and displacements, are not reflected on in textbooks. For example, in the NCERT textbook for class 10, a statement reads: "India's rivers, especially the smaller ones, have all turned into toxic streams" (N-Geo-10: 25). In the following sentences, it is stated that river water is hazardous for human use, but wider environmental effects and opportunities for environmental action are not mentioned. In the NCERT textbook for class 9, river pollution is addressed in that it affects water quality and the National River Conservation Plan is explained in an info box, but again, effects on the health of human beings are mentioned but effects on the ecosystem are excluded (N-Geo-9: 23, cf. excerpt of Chap. 3 "Drainage" in NCERT geography textbook for class 9 in annex). The last sentence of the text encourages a debate on "life of human beings without water" and the health effects of polluted river water. This demonstrates how a relevant environmental and social problem is not debated, but rather only personal and one-dimensional questions are asked for. This stands in stark contrast to the abstract information provided in difficult language in the text.

While in the MSCERT textbook for class 12, causes for water pollution in Maharashtra such as untreated industrial sewage dumped in rivers as well as the use of oxygen and chlorination for treatment are mentioned, impacts on both human health and ecosystem degradation are not referred to (M-Geo-12: 12). In the NCERT textbook for class 8, the value of ecosystems itself is not considered; instead, it is only claimed that water "overuse and pollution make it unfit for use" (N-Geo-8: 16). This demonstrates the exclusion of the harm water pollution causes to species and ecosystems. Furthermore, a long-term perspective of water access and use as well as the need for sustainable water management is not adequately given. In the MSCERT textbook for class 4, a definition of water conservation is stated, but examples or ideas for student projects are not given. Other textbooks do not relate water to sustainable measures at all. In the NCERT class 6 textbook, the statement "We face a shortage of water!!" (N-Geo-6: 33) is explained with the small percentage of available fresh water sources in comparison to water stored in ice sheets, groundwater, and the salt water of oceans.[4] This is not complemented with further information on man-made causes and effects of water shortage. Thus, water shortage is considered as natural and not further reflected as human-induced and the human–environment interrelation is neglected. While in the NCERT class 8 textbook, sustainable development is defined and resource conservation is explained in dialogues between children, and exercises in this chapter do not deepen this understanding. Instead of encouraging thinking about the fundamental resources of water, a task "for fun" (N-Geo-8: 7) asks to think of the wind, a stone, and a leaf as important resource.

These examples show that water pollution is covered in most MSCERT and NCERT textbooks, but coherent explanations, underlying causes, and environmental effects are not comprehensively given or encouraged to think about. In contrast to public textbooks, private textbooks follow a more comprehensive approach with more detailed, outlined principles. For example, the private ICSE geography textbook for class 9 explains complex environmental effects of water pollution on vegetation, marine life, and animals and thus includes effects other than human health (ICSE-Geo-9: 171–173). Eutrophication and biomagnification are explained, and thus, the processes of environmental pollution on different scales are outlined. This indicates the difference in textual quality of public and private textbook providers.

Water-related topics in textbooks cover environmental aspects, while the social and economic value of natural resources such as water is not explicitly stated. A reason for this may be the infusion of environmental education into school curricula by the Supreme Court since 2002 (cf. Sect. 4.1.6), through which it was mandatory to cover environmental aspects in each subject. For the subject geography, an environmental approach to water is partially infused, which is marked by factual statements considering human resource use. A comprehensive approach with detailed environmental consequences and value-oriented environmental education is not strongly developed.

6.4.2.3 Economic Aspects on Water in NCERT and MSCERT Geography Textbooks

Economic aspects of water supply are briefly mentioned in NCERT textbooks, but barely mentioned or indicated in MSCERT textbooks. Concerning the use of water, agricultural purposes, but not industrial or domestic uses, are explained in MSCERT textbooks. Whereas in the MSCERT textbook for class 10, a number of irrigation projects in India are extensively listed, issues around irrigation, as for example its involved costs and the conditions and consequences for farmers, are barely explained. In NCERT textbooks for classes 8, 9, and 10, the use of water for hydroelectric power

[4]Such a natural shortage of water was also observed in geography lessons in which a teacher-related water shortages solely to globally available water resources.

is stated, but barely emphasizing the economic value and use of water power, as well the controversies around this (such as resettling villages and impacting ecosystems). In class 11, the death of livestock as a consequence of draught is mentioned, but economic or social consequences for farmers are not explained. In the NCERT textbook for class 12, the demand of water for agricultural irrigation is a separate topic, but the social and economic impacts, e.g., through the loss of harvest and livestock in droughts, are again not explained (N-Geo-12-IndPpEc: 64). This shows that the far-reaching economic impacts of water access or its lack are barely outlined.

A positive example of the integration of the economic value of water is in the NCERT Environmental Studies textbook for class 5, in which a water bill is printed and students are encouraged to check their water bill at home (cf. textbook scan in appendix). In the NCERT geography textbook of class 8, the water market is shortly addressed in one sentence: "Amreli city purchasing water from nearby talukas" (N-Geo-8: 16, cf. textbook scan in appendix). These two examples encourage students to realize that water is paid for and hence has economic value, but tasks to critically reflect on this are missing. The high economic value of water for agriculture, industries, and domestic use is also not explained. Discussions on the promotion of diesel subsidies for agricultural water use, but also the exploitation of groundwater for irrigation, are relevant societal topics omnipresent in newspapers, science, and government policies. However, such controversial topics are not included in textbooks. The reason for the insufficient representation of water as a valuable good may be based on the common assumption that water is a free good, which is not paid for (Prof 7). Especially in the context of sustainable development in India, students should be sensitized to the fact that water is a scarce resource and that access to it is conflictual and highly contested.

6.4.2.4 Social Aspects on Water in NCERT and MSCERT Textbooks

Differential access to water for people within India is stated in many textbooks, whereas causes and consequences are not further considered. In MSCERT textbooks, the only mentioned aspect of water for human use is irrigation, and some irrigation and drainage techniques are explained in detail. However, exceptions are the textbook for class 3 in the Mumbai district, in which the sources of drinking water supply are mentioned, as well as the one page covering the different sources of water in a textbook for class 5 (N-EVS-5: 57). The page states: "This is how we get water," and drawings depict a jal board water tanker, a well, a hand pump, a canal, a borewell, a jal board pipeline pump, as well as taps. Despite presenting the diversity of water sources, social disparities that determine access to water, as well as the consequences in terms of health effects and time burden, are not included. In MSCERT textbooks, only few references are drawn to the scarcity of water to urban and rural societies.

In a Maharashtra geography textbook for class 3, remedies against pollution are listed and can possibly have counterproductive effects as over-generalizations such as "not allow the slums to grow" (M-Geo-3M: 44) encourage students to think that stopping slum development can solve social problems. This promotes a simplified understanding of social problems related to water, or could even enforce students to think toward excluding or resettling slum dwellers from cities—a highly politicized and contested discussion.

The importance and shortage of water are addressed, whereas local specificities or the difference of urban and rural water use are not explained, which would help students to grasp water as a social problem. In class 8, the issue of having no taps at home and making the effort to fetch water from a nearby well is mentioned: "Mamba fetches water in Tanzania" (N-Geo-8: 9). Relocating "fetching water" to another continent may discourage students to realize that the same circumstances of water access are also a widespread reality in India. The reasons for displacement of daily water realities in India to an African country are not explained and are not made clear.

Thus, inequalities of water access are not explicitly discussed as a relevant local or regional problem. In the NCERT textbook for class 10, "unequal access to water among different social groups" (N-Geo-10: 24) is stated, but not further explained. In class 12, the task "the depleting water resources may lead to social conflicts and disputes. Elaborate it with suitable examples" (N-Geo-12: 71) is not backed up with further information in the text on which a dispute in class could be based to sensitize students for the multiple causes and consequences of differential access to water.

In contrast to geography textbooks, the environmental studies textbook "Looking Around" for class 5 covers social aspects of water to a greater extent. Unequal access to water is mentioned and the differential methods of access to water are demonstrated, such as continuous supply through taps, a motor pump with regular electricity shortages, hand pumps, tankers, wells, bottle buying, and fetching water from canals (N-EVS-5: 57). Another chapter describes a situation, where (urban) children play on a water sled while a woman from a village angrily states:

> Our wells have no water. We get our water only when the tanker comes once a week. Today, even that has not come. And here, there is so much water everywhere- just for you all to play and enjoy. Tell me, what should we do? (N-EVS-4: 148)

Thus, the daily challenges of water shortages from the perspective of a woman from rural India are described. As predominantly found in textbooks, water scarcity is presented as a rural and not an urban phenomenon. Nevertheless, students are asked whether a shortage of water has occurred at their home and how they dealt with it. Playing with water is also addressed in the exercises, asking students to list reasons why one should not play with water (N-EVS-4: 149). Water pollution and its social effects, e.g., for not being able to offer guests a drink, health effects, and waiting long hours for drinking water in a queue, are also mentioned (N-Geo-8: 9). As a task, students are encouraged to collect newspaper articles on water that may help them realize that water shortages and water pollution are daily phenomena in the media. Furthermore, Indian customs related to water are also addressed, such as festivals for newly rain-filled lakes or brides worshipping the water taps in their new homes (N-EVS-5: 54). This indicates cultural practices related to water and offers a different perspective to the utility approach to water prevalent in most textbooks. In the EVS textbook for class 3, students are encouraged to sing songs and poems related to water, to "bring them closer to their community and arts of their region" (N-EVS-3: 19). They are sensitized to the amounts of water for different uses (bathing, drinking, cleaning, fields, kneading dough) and water sources and storage around their own house. Furthermore, a discussion on the fact that "certain people are not allowed to take water from the common source" is encouraged so that "children are sensitive towards issues like discrimination" (N-EVS-3: 21). Discrimination in general, and particularly in regard to water access, is a relevant issue not mentioned in geography textbooks. Interestingly, environmental science textbooks integrate this issue. However, motives for discrimination are not mentioned nor critically reflected upon, which might hinder students in understanding discrimination and its negative effects. This rather factual approach to such a sensitive topic could be challenged through clear statements stressing values of equality, as stated in the preamble of the Constitution of India or article 24 of the Children's Bill of Rights in the opening pages of the textbooks, and giving explanations and examples for these.

6.4.3 *Didactic Analysis of Water Chapters in Geography Textbooks*

In the following, textbook chapters on water-related contents are examined for their inclusion of ESD principles. This subchapter examines how far network thinking, argumentation, and student orientation are promoted.

6.4.3.1 *Network Thinking on Water-Related Issues*

Information and tasks given in geography textbooks are fact-based and fragmented and barely facilitate network thinking by interlinking ecological, economic, and social aspects. The analyzed geography textbooks indicate a factual approach to the topic of water as definitions and monocausal statements are listed rather than interlinked. As stated in the prior content analysis (Sect. 6.4.2), multiple causal relations with regard to the topic of water are explained only one-dimensionally, and hence, the issue is not approached comprehensively in MSCERT and NCERT textbooks. For example, in an NCERT textbook for class 12, the following statement is made without further explanation: "Resources are unevenly distributed" (N-Geo-12: 30). Thus, causes and effects of unequal access to resources are not mentioned and complex issues are reduced and not explained in an intelligible manner. Even if the accurate answer to a question in the exercise section would involve a text interlinking several factors and its interrelationships such as "Explain the major causes for water scarcity in Maharashtra" (M-Geo-12: 12.13), students are not encouraged to explain complex interrelations as causes are given in the form of lists in the text:

> Water scarcity originates due to two causes. They are:
>
> A) Natural water scarcity: This arises due to a) Inadequate rain b) More evaporation and c) Lack of retention of soil water.
>
> B) Man-made water scarcity: 1. Poor water management 2. Urban-rural conflict 3. Canals are not repaired 4. Absence of harvest water. (M-Geo-12: 12.9)

Thus, reproduction of information for the exercise questions is encouraged; however, tasks could be presented more learner-friendly if information given in the text would be elaborated upon and had to be transferred rather than reproduced. From an ESD perspective, textbooks should promote thinking in networks on complex topics such as linking water scarcity to consequences for women and girls, e.g., in terms of time spent, as it is usually their role in the household to fetch water.

Water scarcity is not approached from an integrative perspective, as a result of human–environment relation. In the NCERT textbook for class 10, causes, uses, and effects of water scarcity are addressed, and the effects of industrialization and urbanization on water scarcity are mentioned (N-Geo-10: 25). In contrast, in the NCERT textbook for class 8, the text includes a short paragraph describing that water is scarce in many "countries located in climatic zones most susceptible to droughts," mentioning "variation in seasonal or annual precipitation" or "overexploitation and contamination" (N-Geo-8: 16). Maps or diagrams could explain regional variations in precipitation and the stated human-induced scarcity could be exemplified, so students link water scarcity not only to the natural environment, but also to human interventions.

The discrimination between natural and human–environment is also strengthened through separate textbooks on human and physical geography for different classes. Class 9 and 11 deal with physical geography, and class 10 and 12 deal with human geography. Therefore, an integrated approach to both fields is not encouraged at the school level. This becomes also obvious in the chapter "rain" (M-Geo-6: 24 f.) in which types of rain and the distribution of rainfall are explained from a purely physical geography perspective. However, relevant environmental problems such as acid rain or human-induced climate change as well as human dependence on rain are not addressed. Although the first chapter of the NCERT textbook for class 7 introduces the natural and human–environment, an integrated approach in the textbook is not followed. The features of the natural habitat are dealt with, while consequences of human behavior on the natural habitat are not critically addressed (N-Geo-7).

Water conflicts prevalent in urban and rural areas of India are not dealt with, although students may know about them from their own experience, as, for example, they have to walk further to fetch water or wait longer hours due to water scarcity. In the MCERT textbook for class 4, the reason for a low groundwater table is related to the rocks and not further explained, while the prevalent problem of falling groundwater tables in India due to water extraction for irrigation purposes is not mentioned. Thus, textbooks ignore the causes and effects of human–environment relations on water scarcity.

In the MSCERT textbook for class 5, the question: "Why is it necessary to make proper use of our water resources?" (M-Geo-5: 29) is answered in the text: "We should use water carefully and stop wastage and pollution of water, maintain groundwater level and thus make proper use of our water resources" (M-Geo-5: 28). This example gives general duties to students such as not to waste and pollute water without further explanation as to *how* and *why* students could prevent water pollution. That falling groundwater tables could be human-induced is not explained and how people can maintain the groundwater level is not comprehensively interlinked and thus does not promote network thinking for students.

Textbooks are written with a regional and thus primarily descriptive approach to geography. Topics are repetitive in title and content and hinder cross-linkages over different class levels. The focus of geography textbooks grows in scale, from district level in class 3, via countries to continents from class 6 onwards, which have to be learned by heart. In class 4 of MSCERT textbooks, students learn about the climate, agriculture, forests and wild life, water and marine resources of their state Maharashtra (M-Geo-4). In class 5, the same topics are repeated for India. In class 6, the world is introduced and physical features and resources for selected countries of Asia are addressed as well as for selected additional countries (Israel, Saudi Arabia, Malaysia, Japan, and Sri Lanka). In class 7 and 8, other continents and countries are introduced. Although it seems logical to follow from small scale to larger scale throughout the school years, it is ignored that students may understand the concept of a country before the concept of a state. Further, students do not learn to transfer concepts to different countries but rather learn facts about countries.

The tasks in each chapter barely enhance network thinking. Multiple-choice questions postulate single-factor understanding and do not encourage the critical understanding that there might be several answers to one question. For example, the task "Tick one: Water is: abiotic, biotic, cyclic, non-renewable" (N-Geo-12: 70) is scientifically not sound, as both abiotic and cyclic answer this question correctly. As students are not used to more than one right answer, students, as well as the teachers, most likely will not notice the mistake in the task. Tasks to "Fill in the blanks" hinder network thinking. Although open questions are posed, the space to answer these is limited to a specific number of words to explain a controversial or complex topic. Activities such as "Look […] up" or "Think of a slogan" encourage creative thinking, but do not enhance network thinking. Thus, facts are merely memorized and reproduced, which stresses a *performance model of pedagogy*.

6.4.3.2 *Argumentation on Water-Related Issues*

The textbook tasks are pre-structured and give little space for students to promote critical thinking and to develop their own opinion on controversial topics. Although stated in and promoted by the NCF 2005, controversial topics such as dam building or falling groundwater water tables are not introduced from multiple perspectives. In general, different perspectives on irrigation projects (cf. M-Geo-10) and domestic water supply are not postulated, although topics like these are frequently critically discussed in Indian and international media (NCERT 2005). In textbooks, causes of water pollution are presented as given and presented uncritically. For example, water pollution and scarcity effects are not differentiated according to the impact on different socio-economic groups of society: "Explain the effects of pollution on living things" (M-Geo-4: 59).

The analysis of geography textbooks shows that tasks in textbooks are often multiple-choice questions or ask for definitions and lists given in texts, which encourage rote learning. Even if questions are not multiple-choice questions, a limited format of the answer is expected in 30 words: "Discuss factors responsible for the depletion of water resources" (N-Geo-12: 71). These tasks only provide space for two or three sentences, in which students can simply list the facts written in the textbook itself. This type of question requires a low textual level and merely asks to reproduce information given in the text.

Approaches to topics do not focus on a topic or a conflict, but are presented as a short collection of information without interlinking reasons, consequences, and controversies. For example, in the MSCERT textbook for class 12, the task "Explain the causes of water pollution in Maharashtra" (M-Geo-12: 12.13) and in the NCERT textbook for class 10, the task: "What is water scarcity and what are its main causes?" (N-GEO-10: 33) are supposed to be both answered in 30 words, which does not leave enough room for more than three or four sentences that only list factors. Even tasks that should be covered in 150 words, such as "Discuss the availability of water resources in our country and factors that determine its spatial distribution" (cf. textbook scan in appendix), could be covered in a term paper of several pages. Thus, the space for students to communicate their opinions in tasks is limited.

Furthermore, the meaning of the verbs used in the textbooks to pose the task (so-called *operators*) is not clear or can be often wrongly interpreted by the students and the teachers. For example, the task "Discuss how rainwater harvesting in semi-arid regions is carried out." (N-Geo-10: 33) requires as the correct answer a description, not a discussion, of rainwater harvesting, as the interrogative particle "how" points out. Therefore, the operator "discuss" is misleading for this task, as the task does not cover controversial aspects of water harvesting, which should be covered in discussions. Thus, the verb used in the task should have been described: "Describe how rainwater harvesting in semi-arid regions is carried out."

One positive example of a discussion generating activity is given in the NCERT textbook for class 10: "Enact with your class mates a scene of water dispute in your locality" (N-Geo-10: 33). However, to ensure whether this task is fulfilled in a relevant and realistic context, more steps how to proceed are not given. Conversely, for this task, background information on water disputes or information on how to prepare a dispute are not covered in the text. Another example raises awareness on the diverse socio-economic backgrounds in a task in the NCERT textbook for class 12 on human development:

> Talk to the vegetable vendor in your neighbourhood and find out if she has gone to school. Did she drop out of school? Why? What does she tell you about her choices and the freedom she has? Note how her opportunities were limited because of her gender, caste and income. (N-Geo-12 FHumGeo: 25)

A comparable task on water promoting communication on resource access within the community was not found in any textbook. The NCERT textbook for class 12 includes an example of how the water topic could be approached in a discursive way as discussion tasks are included in the text, and are not separated. For example, students are encouraged to discuss newspaper headlines on river conflicts, floods and insufficient water supply (N-Geo-12 FHumGeo: 25). Although these problems are not further elucidated in the text, this task may still encourage discussions on various conflicts related to water. In the same textbook, discussions on recycling and reusing water (N-Geo-12 FHumGeo: 25) and the effects of salinity in the soil and the depletion of groundwater for irrigation (N-Geo-12 FHumGeo: 65) are encouraged. Despite the limited information given to enrich discussions, these indicate relevant disputes on water.

In conclusion, although ten life skills by the WHO (2001) are listed on the front page of MSCERT textbooks, tasks in textbooks do not promote these skills. Interestingly, NCERT textbooks include some discussion tasks, while MSCERT tasks are restricted to reproducing information. Despite the diversity of the most common task types (cf. Table 6.4), most of them require answers that simply repeat the information as given in the text. Even if formulated differently, the interpretation of task types clarifies that most tasks ask for a reproduction of the wording found in textbooks. Thus, the development of argumentation skills to express own opinions or to differentiate opinions is not promoted.

6.4.3.3 *Student Orientation on Water-Related Issues*

Students' environmental awareness and actions are barely encouraged in textbooks through exercises or activities addressing local specificities. Texts and tasks in geography textbooks do not cover the wide diversity of students' socio-cultural backgrounds. Despite the vast socio-cultural diversity within India, state and national textbooks often homogenize students' backgrounds by only presenting and focusing on urban than rural living environments and thus addressing primarily urban middle class students. This lack in student orientation due to the neglect of urban lower class and rural environments and the diverse socio-economic backgrounds in textbooks has an impact on the motivation of learning, as relevance to students' lives is not sufficiently given. Thus, the different needs of rural and urban audiences of textbooks in India do not differentiate sufficiently the social and economic backgrounds of the students. However, there is some focus on traditional knowledge systems such as rainwater harvesting: "Find out other rainwater harvesting systems existing in and around your locality" (N-Geo-10: 31). Participation in community actions is not encouraged, but students are encouraged to write a proposal on how to save water (N-Geo-10: 25). Thus, opportunities to promote student initiatives, solidarity, and participation in community projects could be enhanced by further integrating activities outside of classrooms in students' environment.

Field visits with observations, guided tours or interviews with experts to help students realize water problems in their own city are encouraged in class 5: "Visit

Table 6.4 Sample task types from MSCERT and NCERT geography textbooks and an interpretation with example

Common task type in textbooks	Task interpretation	Example
Answer in one word	Practice technical vocabulary; reproduce a definition from the textbook	What do you call water naturally stored under the earth's surface? (M-Geo-5: 29)
Answer in thirty words	Summarize textbook content; reproduce 2–3 sentences from textbooks	Discuss the factors responsible for depletion of water resources? (N-Geo-12: 71)
Answer in about 150 words	Formulate a text with information given in textbooks	Discuss the availability of water resources in the country and factors that determine its spatial distribution? (N-Geo-12: 71)
Answer in one sentence each	Summarize/reproduce a definition from the textbook	(1) What is a rill? (2) What is a creek? (3) How are lakes formed? (M-Geo-3P: 19)
Answer the following questions in one sentence	Summarize/reproduce information from textbook	What gets polluted by the use of chemical fertilizers? (M-Geo-5: 59)
Choose the right words from the bracket to fill in the blanks	Reproduce a definition from the textbook	Water gets polluted by … (smoke from vehicles, sewage, loud noise) (M-Geo-5: 59)
Choose the right answer from the four alternatives given below	Multiple choice on information found in textbooks	Which one of the following lakes is a salt water lake? (a) Sambhar (b) Dal (c) Wular (d) Mahanadi (N-Geo-9: 24)
Explain the terms	Reproduce a definition from the textbook	Explain the terms: (1) Oceanic sediment (2) Trench (3) Continental slope (M-Geo-7: 18)
Discuss	(Mostly understood as) describe and compare with information from the textbook	Discuss the significant difference between the Himalayan and the Peninsular rivers. (N-Geo-9: 24)

the water supply centers in your village/town/city. Find out ways of purifying water at home, write them down in your notebook" (M-Geo-5: 59). Although the task is very specific, a field trip would encourage multi-dimensional aspects of water supply, which could be given more space in the textbook to prepare students for a field trip. In the MSCERT textbook for class 11, a field study to identify water pollution (M-Geo-11: 230) is recommended, but detailed objectives and information on the methodology is missing. As scientific results are included in the textbook, the motivation and curiosity to conduct such a field study with an expected high environmental outcome may not be encouraged sufficiently.

Causes and measures appropriate for students to reduce pollution or waste are not clarified. In the MSCERT textbook for class 3 it is stated: "We should use water very carefully. We should take care that it does not get polluted or wasted" (M-Geo-3M: 22). General remedies against pollution are mentioned in Mumbai and Pune district textbooks for class 3, but these are not meaningfully explained and related to the students' local environments in Pune nor to their cognitive development at that level: "We must reduce pollution by using organic manures and pesticides" (M-Geo-3P: 44). Remedies such as "Banning the use of plastic. Building as many public toilets as possible" (M-Geo-3M: 44) may be even counterproductive, as students can feel helpless in their position from where they cannot initiate these. This may lead to an attitude blaming the government for any pollution.

Textbook contents are not put into the local contexts of students to achieve skill development. The request to become active and the need to conserve natural resources is framed in a vague statement in a geography textbook for class 8 of the state Maharashtra: "It is the duty of man [humans] to protect all components of the environment and make judicial use of natural resources" (M-Geo-8: 32). This generalization neither encourages students to take concrete action nor does it promote values encouraged by the WHO (2001), and also ESD. Although activities are given, they are not meaningful since they do not have an impact on the students' awareness of their own action. For example, tasks that state make a poster with a slogan such as "Water on earth is for one and all" (N-EVS-5: 59) are unspecific and use a strong rhetoric instead of convincing arguments to stimulate environmental discussion among students or in students' homes or even motivate for action.

To strengthen the ESD principle on student orientation in textbooks, relevant exercises practicing the transfer of knowledge to their own lives should be promoted, rather than fragmented reproduction, as this very broad question demands: "What is sustainable development?" (N-Geo-8: 6). Thus, neither the WHO (2001) life skill of "self-awareness" nor the ESD principle of student orientation are encouraged through detailed contents, material or exercises.

6.5 Summary on ESD Principles in Curricula, Syllabi and Textbooks

The remarkable difference between the weak framing of the *open* curriculum NCF 2005 and the strong framing of the *closed* syllabi and textbooks shows that two contrasting types of pedagogy are promoted at different institutional levels. While the NCF 2005, with a strong influence on instructional discourse, promotes a *competence model of pedagogy*, a comprehensive conceptual and methodological framework that offers systematic guidelines for the selection, sequence, pacing and evaluation of knowledge and skills is not available. To implement pedagogic principles of Critical Pedagogy and environmental education in existing educational structures to facilitate change in current pedagogic practice, context-specific challenges need to be explicitly addressed. Hence, the strong gap between the educational objectives stated in the NCF 2005 and the framing of the syllabi and textbooks needs to be overcome as the latter structures pedagogic practice. Despite the promotion of weak framing and ESD principles in the NCF 2005, textbooks represent strong framing. The translation of the pedagogic principles mentioned in the NCF 2005 to textbooks' texts and tasks needs to be facilitated through a methodological framework. Besides stating educational objectives, NCF barely offers concrete guidance for the knowledge organization and presentation to syllabi designers and textbook writers in their respective contexts with specific restrictions.

In addition to educational objectives, curricula also need to state the concrete manner for choosing criteria of selection, sequence, pacing and evaluation of knowledge and skills. Thus, guidance needs to be given to syllabi designers and textbook writers to recontextualize and reinterpret the NCF 2005. Through strengthening the link of curricula, syllabi and textbooks, educational objectives could work toward the aspired transformation of pedagogic practice. A practical option to better formulate and translate educational policies into pedagogic practice would be if curriculum designers were to work with teachers and in teacher trainings to include their learning experiences in the future NCF. In this way, a critical review of the appropriate accordance of pedagogic practice, textbooks, and curricula has to be reassessed and revised.

A methodological concept oriented on ESD principles and the promotion of argumentation skills could improve the textual and didactic depiction of the topic of water in NCERT and MSCERT textbooks by facilitating a critical awareness of water conflicts and its effects, or help problem solving through the promotion of network thinking, argumentation skills, and student orientation. As the textbook analysis showed, most of the present exercises and activities do not encourage communication and critical-thinking skills. Controversies are not adequately handled and only a single-factor understanding of causal relationships is advanced. At the task level, there is a strong need for tasks with a greater range of clearly defined operators. Moreover, working with different sources and perspectives also needs to be promoted. Thus, a concept promoting a precise application of teaching methods in textbooks is either not provided or is lacking.

Particularly an argumentative approach to water conflicts comprising environmental, social, and ecological causes and consequences of water use, water pollution or water availability could promote ESD in geography education. In textbooks, an anthropocentric approach prevails, e.g., when effects on health and agriculture of water pollution are stressed, but limited reference is drawn to the destabilization of ecosystems or the extinction or endangering of species. Causes and consequences for water-related issues are neither sufficiently stated nor analyzed, and information is rather repetitive and not focused on developing a progressive understanding of principles. To integrate a more critical and student-oriented approach, principles need to be applied to a specific example and then need to be reinterpreted in different contexts afterward. Furthermore, textbook tasks need to be adjusted for different state, urban, rural, and socio-cultural contexts (cf. UNESD objectives) because of the great heterogeneity in India.

The regional approach to the selection and sequence of knowledge hinders the learning of concepts and principles. Although the content in geography textbooks at different class levels is ordered according to the scales of district, state, national and international level, principles do not gradually develop in complexity. Focusing on principles does not only infer a change of methods, and it also implies superseding the regional approach to geography education. A strategy to change the conceptualization of textbooks would also entail different principles of the learning progression at different class levels. To structure these principles for textbook design, however, is not only the task of textbook authors, but also of curricula and syllabi designers, whose guidelines textbook authors follow. This indicates the reciprocal interdependence of curricula, syllabi, and textbooks, which has been neglected and needs to be considered, especially when introducing transnational educational policies for pedagogic practice.

References

Advani, S. (2004). Pedagogy and politics: The case of English textbooks. In A. Vaugier-Chatterjee (Ed.), *Education and Democracy in India*. New Delhi: Manohar Publishers.

Bernstein, B. (1975). *Class, codes and control. Towards a theory of educational transmission* (Vol. III). London: Routledge.

Bernstein, B. (1990). *Class, codes and control. The structuring of pedagogic discourse* (Vol. IV). London: Routledge.

Bharati Vidyapeeth Institute of Environment Education and Research. (2002). *Study of status of infusion of environmental concepts in school curricula and the effectiveness of its delivery*.

Budke, A. (2010). *Und der Zukunft abgewandt - Ideologische Erziehung im Geographieunterricht der DDR*. Göttingen: V&R unipress.

Crossley, M., & Murby, M. (1994). Textbook provision and the quality of the school curriculum in developing countries: Issues and policy options. *Comparative Education, 30*(2), 99–114. http://www.jstor.org/stable/3099059.

Gottlob, M. (2007). Changing concepts of identity in the Indian textbook controversy. *Internationale Schulbuchforschung—International Textbook Research, 29*(4), 341–353.

Kumar, K. (1988). Origins of India's "Textbook culture". *Comparative Education Review, 32*(4), 452–464. http://www.jstor.org/stable/1188251.

National Council of Educational Research and Training. (2004). *Curriculum framework for teacher education*. New Delhi: NCERT.

National Council of Educational Research and Training. (2005). *National curriculum framework 2005*. New Delhi: NCERT.

National Council of Educational Research and Training. (2006). *National curriculum framework 2005. Syllabus for classes at the elementary level* (Vol. I). New Delhi: NCERT.

National Council of Educational Research and Training. (2009). *National curriculum framework 2005. Position papers on national focus groups on systemic reform* (Vol. II). New Delhi: NCERT.

Pal, Y. (1993). *Learning without burden*. New Delhi: Ministry of Human Ressource Development, Government of India.

Rinschede, G. (2005). *Geographiedidaktik*. Paderborn: Schöningh.

Spitzer, M. (2007). *Lernen. Gehirnforschung und die Schule des Lebens*. Berlin: Spektrum Akademischer Verlag.

World Health Organization (2001). *Skills for health*. http://www.who.int/school_youth_health/media/en/sch_skills4health_03.pdf Accessed Mar 10,2014.

Chapter 7
Pedagogic Practice, ESD Principles and the Perspectives of Students, Teachers, and Educational Stakeholders

Abstract In this chapter, I analyze how power relations and cultural values in pedagogic practice observed in geography lessons at five English-medium secondary schools in Pune relate to the principles of Education for Sustainable Development (ESD). The analysis of geography teaching contents and methods is structured according to the selection, sequence, pacing, and evaluation criteria of knowledge and skills by Bernstein (1975). Pedagogic practice in the observed lessons is strongly framed by the regulations formulated by educational institutions at the state and national level. Teachers' agency is bound by an overloaded, fact-oriented syllabus that prescribes a strongly framed teacher–student communication with limited space for students to develop critical-thinking and argumentation skills. The analyses shed light on the role of the textbook in determining the authoritative position of the teacher, as the textbook governs the method of communicative interaction and learning content. Students, teachers, and educational stakeholders reveal similar perspectives on geography education and the role of ESD. These findings suggest that there is a need for strengthening teacher's agency and a more open framing of textbooks for transformative pedagogic practice as inspired by ESD.

In classroom teaching, the interplay of a variety of factors influences the development of students' knowledge and skills. Apart from syllabi and textbooks, the use of classroom space and resources and, most importantly, the agency of the teacher shape pedagogic practice. The analysis of videotaped classroom lessons at English-medium schools in Pune exemplifies the structure and ordering of teacher–student interaction and displays how knowledge and skill development are communicated and controlled. The analyses of geography teaching contents and methods shed light on the steering role of the textbook and the authoritative position of the teacher. As the following sections will show, the textbook governs the method of communicative interaction and learning content transmission. Teachers are transmitters of a strongly framed syllabus and thus have limited control over content presentation and learning. Strongly framed classroom communication and learning leave limited space for the participation of students in classroom discussions as demanded by ESD. The differential use of classroom space by teachers and students and the limited availability

S. Leder, *Transformative Pedagogic Practice*, Education for Sustainability,
https://doi.org/10.1007/978-981-13-2369-0_7

of teaching resources show how the physical setting promotes explicit boundaries between the teachers and the students rather than implicit boundaries as promoted by ESD. These findings are also prevalent in students, teachers, and educational stakeholders' interviews, which articulate at the same time aspirations and doubts toward a successful implementation of ESD.

In this chapter, I analyze pedagogic practice in geography lessons on water from an ESD perspective at five English-medium secondary schools in Pune, which I observed between January 2012 and February 2013 (cf. Table in annex). For an in-depth analysis of power relations and cultural values relevant in pedagogic practice, I structure the analysis according to the selection, sequence, pacing, and evaluation criteria of knowledge and skills by Bernstein (1975). At the start of the chapter, I reflect on the use of classroom space and resources, as this is particularly relevant in under-resourced educational contexts. In the last subchapters, I present the perspectives of students, teachers, and various educational stakeholders on geography education and the role of ESD (cf. Table in annex).

7.1 Pedagogic Practice in Geography Education

7.1.1 Classroom Space and Resources

The socio-cultural stratification at urban English-medium schools in Pune was not directly visible during classroom observations. The students were seated according to their academic abilities; i.e., academically better students sit in the front rows. Different than in observed rural classroom interaction, girls also sit in the front and participate in classroom communication in equal measure to boys. The school uniforms and the same amount of space at their desks in the classroom make students appear as a rather homogeneous group.

The physical arrangements of classrooms demonstrate the differential use of space by the teacher and the students. In the observed lessons, the teacher stands continuously facing the classroom and the students, while between fifty and sixty students sit in pairs on benches, which are placed in rows oriented toward the teacher. The students' desks cannot be shifted or rearranged as they are screwed to the floor; thus, the physical arrangement discourages group work. Students only stand up if the teacher asks them to answer a question, while the teacher has about one-fourth of the classroom's space in which she or he stands or walks (cf. photographs in appendix). Because of the teacher's standing posture and therefore higher position, the communication with the students is explicitly centered on the teacher. In the dramaturgical terms of Goffman (1971), the teacher is the actor who performs on stage and faces the audience, while the students are the spectators who watch and react to the performance. This classroom arrangement did not change during my observations. The posture of the teacher and the students expresses the institutional expectations of their social position: The teacher is the transmitter of textbook information, while the students are the recipients and reproducers of it. The physical arrangements of the classrooms and the standing position of the teacher visualize the explicit boundaries between the authoritative role of the teacher and the sitting, information-receiving students. This illuminates the hierarchical relation between the teacher and the students as these kinds of arrangements stress the respected position of the teacher.

Teaching resources are confined to textbooks, notebooks, blackboards, a cupboard and occasionally posters on the wall. The textbook is the central, and often the only medium as other teaching tools and media are rarely used. Each classroom has a blackboard, which is not used for explanatory drawings, but for sporadically writing down key terms. Students have their textbooks in front of them or only listen, while the notebooks are only used to complete tasks from the textbooks. Some classrooms had beamers and whiteboards, which were used to screen short videos supplementary to textbooks. This functional and formalistic use of classroom space and resources present a challenge for the implementation of ESD principles. Particularly the benches screwed to the floor symbolize the deeply entrenched unidirectional communication between teachers and students. The classroom setting underlines the authoritative and central role of the teachers, which poses a challenge to the claims for multi-directional classroom communication and dialogue, which promotes debates and argumentation skills.

7.1.2 Selection of Knowledge and Skills

Teaching contents and methods in the observed geography lessons substantially correspond to the contents and methods prescribed in the textbooks. As knowledge in geography lessons is selected according to the texts and tasks in the respective chapters, the pedagogic principles of the textbooks (cf. Sect. 6.4) are transferred to geography lessons. The strong framing of knowledge in textbooks is thus translated in pedagogic practice. Knowledge prescribed in the geography syllabi and textbooks is transmitted through the teacher to the students, and only occasionally enriched with additional information from the teacher's own experience or knowledge, or interrupted by teacher's questions on textbook contents to the students. The information in textbooks is implemented, or "delivered," in classrooms and reproduced by both teachers and students through word-by-word repetition. Thus, the textbook, rather than the teacher, governs classroom communication and interaction.

The teacher's speech is marked by declarative and factual knowledge rather than procedural and integrated knowledge and follows the strongly framed selection of knowledge according to the textbook. Knowledge is structured regionally, and concepts are not cohesively taught. Thus, knowledge in pedagogic practice is strongly classified, as it is in textbooks. Chapters on water are read out or are reproduced by the teacher word by word. For example, reasons and effects of water pollution are not explained, and the contested nature of water resources is not addressed in an argumentative manner (cf. textbook analysis Chap. 6.4).

Classroom observations show how teaching and learning are structured through the textbooks. At the beginning of the lessons, the teacher calls out the name of the chapter's topic. In the observed geography lessons, teachers neither posed introductory questions connecting the new topic to students' perceptions and existing knowledge nor introduced a stimulating and motivating "hook" through visuals or objects. In some lessons, teachers give lectures or read out of the textbook, and in other lessons, teachers lead an inquiry-response cycle and ask for definitions, facts, or lists that students memorize and reproduce unanimously. Students often do not answer in full sentences, but say words or phrases, to which the teacher has little or no time to react, which stresses a performance-oriented model of pedagogy. There is no space for individual answers, and correct answers as literally given in textbooks are expected. An observation in a geography lesson of grade 7 illustrates an inquiry-response cycle between teachers and students in which students practiced for exams based on what they have learned from the textbook:

Excerpt of a geography lesson at school 4 in grade 7 on March 16, 2012

Teacher	"Okay now, let us discuss about the physiography of the place (pointing to a big hanging map of North America). Now we are aware of the physiography of the place. Is that clear? Now we are talking not about America … (pointing to North American continent on the map). So how many countries do we have? Canada, United States of America, Mexico."
All students	"Canada, United States of America, Mexico." (repeating unanimously with the teacher)
Teacher	"Three countries are there."
All students	(nodding in agreement)
Teacher	"And Alaska is the part of United Sates of America. Is that clear? Then we come to United States of America. It's written here. So its short form we generally use is USA. Now we come to very important thing. I will explain you all the gulfs. Now Gulf of California (pointing on the map on the gulf). When we talk about California, what comes to your mind, you have to express it?"
Student m1	"Important fruits and orchids."
Student f 1	"Beaches."
Teacher	"Okay. Beautiful beaches."
Student f2	"Cement industry. Hardware."
Teacher	"Yes. Cement industry. Okay."
Student m2	"Black canning."
Teacher	"Black canning will come to it. We will come to the other parts. Let us divide. Now what's the famous range here?" (pointing to the Appalachian range on the map).

All students	"Appalachian."
Teacher	"We will come to it after that. I will remove this map and we will go to the second map. Is that clear? When we will talk about the physiography. Now I have this bay with you called Hudson Bay (pointing to the Hudson Bay on the map). Now what's the definition of a bay? We discussed. Last time we talked about Bay of Bengal in India. What is a bay?"
Student m3	"Um, large portion of water which is surrounded by three sides by land."
Teacher	"Yes. How many sides of land is [sic] there?"
All students	"Three sides."
Teacher	"One side the outlet is there for the water to flow. Is that clear? Now we will come to very important thing. I did talk to you about Greenland."

This lesson excerpt shows a strong regional approach of geography teaching, focusing on declarative and not procedural knowledge. Students learn the names of places, but these are not linked to scientifically, politically, or socially relevant issues such as resource conflicts (except when listed in the textbook), and thus, students rarely understand why these are important to learn. Further, students are not encouraged to think about problems and develop solutions. Instead, the teacher gives facts prescribed by the textbook, leaving students little space to control the selection of knowledge, stressing performance-based teaching methods. As students hardly speak full sentences and repeat answers simultaneously, they cannot contribute to the selection of knowledge through their own experiences and knowledge, and they also cannot develop their own communication or even argumentation skills.

Teachers occasionally have mental leaps or switch the type of questions (e.g., location questions, definition questions, association questions), which make it difficult for students to follow. At the same time, students are encouraged to answer. As multiple questions are asked at the same time, the transmission of declarative knowledge often leads to fragmented knowledge. The following excerpt demonstrates that students are given little time to think about a question posed by the teacher:

Excerpt of a geography lesson at school 4 in grade 9 on March 16, 2012

Teacher	"What is this river famous for? What is the multi-purpose project? Okay, which are the famous rivers flowing here? Come on. Let's make an account. All the important rivers. Come on. I want everyone to answer. I am not going to ask you. Come on."

One lesson of grade 6 (cf. box below) on water pollution demonstrates how the teacher is unable to organize students' verbal contribution in a comprehensive answer.

At the same time, the teacher is continuously distracted by various students' inattention. When the teacher asks the students to revise how water gets polluted, only half of the students raise their hand to participate, while the others are talking and creating some kind of turmoil. The students' answers on how water gets polluted range from water-saving methods to domestic and industrial wastewater; however, this diversity of answers is not reorganized by the teacher. This indicates that students become confused and do not learn to give structured answers to questions, but rather list everything they remember on the topic. As the following excerpt demonstrates, the teacher focuses on maintaining discipline and calming the class down throughout the lesson, rather than reacting to and paraphrasing students' answers.

Excerpt of a geography lesson at school 4 in class 6 on March 16, 2012

Student m1	"Rainwater harvesting"
Teacher	"Rainwater harvesting, yes. One of the solutions is rainwater harvesting"
Student f1	"Madam, factories lead harmful chemicals into rivers and if it is possible to purify the dissemination of water, the harmful chemicals should be converted into harmless!"
Teacher	"Anushka said something. Janaki, were you attentive? Yes or no?
Student f2	"No"
Teacher	"Saurabh?"
Student m2	"No"
Teacher	"See, students, you should be listening to each other. If you want to answer, you should respect others also. Yes, Sakshi?"
Student f3	"We should tell people to stop washing utensils and clothes in the river"
Teacher	"Yes, rivers should not get polluted. They should stop washing"

Especially, the last comment implies that both the teacher and the students blame "people" for polluting rivers by washing clothes. The reasons leading people to wash their clothes in rivers are not further explicated. Network thinking on interlinking environmental, social, and economic aspects of washing clothes in the river is not promoted, and thus, students have little space to enrich information in the textbooks. Thus, the selection of knowledge in geography lessons is standardized according to the respective textbooks.

These examples demonstrate that strong framing and strong classification as prescribed in the textbooks mark the selection of knowledge in observed geography lessons. In such a setting, it highly depends on the competence of the teacher to make the information in textbooks coherent to the students.

7.1.3 Sequence and Pacing of Knowledge

Similarly to the selection criteria, the sequence and pacing of knowledge in geography lessons are strongly framed through the academic frameworks of the syllabi and textbooks. The contents for each class level are determined through the syllabus, and the lessons per year are covered as suggested in the syllabus. The teacher controls that the sequence of knowledge rigidly follows the chapter order of the geography textbooks. Neither the teacher nor the students reorganize the arrangement of information to be learned through discussions or in activities. Hence, the sequence is linear and static as in textbooks, because teachers do not prepare individual scaffolding of contents according to the needs of the learning group. The teacher functions as a timekeeper and is in charge of the strict obedience of the curriculum following the sequence of the chapter. These expectations are expressed by principals, as well as parents who hope their students will do well in competitive exams. Students have to fulfill activities and tasks at the end of the chapter as homework; these are usually not integrated into the lesson.

In a geography lesson of grade 9, the teacher reads out paragraph by paragraph of the textbook and interrupts herself to repeat the definition given in textbooks. She asks students for confirmation without looking up and she does not wait for any response from the students. Although short videotapes are included in the lesson according to the sequence of textbook paragraphs, these are not embedded for reflection or changing the strong framing of the lesson:

Excerpt of a geography lesson at school 2 in class 7, on January 15th, 2013

Teacher (reading out of the textbook) "[...] from the above conversation, you understand that the place, people, things and nature that surround living organism is the environment. You understand what exactly the environment is? The environment is the place, things, people, and nature that surround the living organism. Who is living organism? It is anything like it is animal, human being or birds. Fine? It is a combination of natural and human-made phenomena. While the natural environment refers to both abiotic and biotic condition existing on the earth, human environment talks about the activities and creation and interaction amongst human beings. What is the human environment? It is the activities and creation and interaction amongst human beings... Got it? We will now take the first topic which is natural environment."

Teacher shows a short video on the natural environment on the smartboard.

Teacher "You can see the physical environment?"
All students "Yes."

Teacher clicks the video forward on different short chapters, such as landscape, air, water bodies and weather.

Teacher "Why is it such a short period of time?" (referring to the definition of weather) Students do not react.
Teacher "Because it is changing."

Teacher continues the video.

Teacher "Children, when you go on holidays, you want to know the climate, whether you have to wear a sweater or a kurti. So you want to know the climate, which is the weather pattern over 30 years. Everyone understands?"
All students "Yes."
Teacher "Anyone wants to ask a question?"

The students do not answer.

Teacher "Everyone understood?"
All students "Yes!"
Teacher "Anyone has any doubts?"
All students "No, ma'am."
Teacher "That is such a good class. We move on to the next class...."

While the teacher tries to explain the terms of "weather" and "climate" with going on vacation, she mixes them up so their definition does not become clear. The teacher insures herself that everyone listens by asking them whether they have understood. As the students agree or do not react, she continues the class on another topic. This shows that lessons stay strongly framed, even though additional teaching tools, such as a smart board, are introduced. "What is...?" questions are posed, to which single sentences as answers are easily and quickly evaluated, unlike group work, for example. Thus, there is limited space for students to intervene in the teacher's speech. The teacher maintains control over the sequence of knowledge and strongly frames the order of contents. Apart from this, a "good class" reflects the view that unquestionable acceptance of what the teacher says is considered as appropriate.

One teacher offered me the chance to observe her lesson and was caught by surprise that the students had not received their new textbooks yet. While she reproduced information she remembered from the first chapter in the textbook, she sometimes stumbled while trying to recall the exact word-by-word information used in the text. The following excerpt exemplifies that teachers strongly depend on the textbooks to conduct a lesson:

Excerpt from a geography lesson at school 2 in class 9 on September 11th, 2012

Teacher	"[...] the tropic of cancer is passing through our country, through five states of our country and the meridian is of 82 degrees 30 min nor..., uhm, east, which considers, ...uhm, defines the local time of our country. You understood until here?"

Students do not react.

Teacher	"When are we going to get the books?"
Students	(mumbling) "In a week."
Teacher	"So next week you probably will have books with you, so bring them so I can give you work."

The *pacing* of knowledge is mediated through the teacher in lessons. The teacher steers the transition of topics and activities. However, these are not necessarily coherent. Sometimes, spontaneous activities are integrated; in the geography lesson without books, for example, every student in the class had to name five natural and five man-made things without repeating anything previously stated. Students react to questions of the teacher, but do not individually participate through comprehension questions or own comments. This leads to a high speaking time for the teacher while steering the classroom interaction. Additionally, students are not given any time to work individually, in pairs or in groups and mainly repeat textbook information to perform well in exams.

The pacing of pedagogic practice is also determined by a specific number of lessons per chapter (cf. Table in appendix). For example, in class 12, 6 out of 85 periods are allotted to a unit of water resources. Due to the high pacing in teaching, students who miss some lessons do not have the opportunity to learn contents at a later point, as information is not repeated and included in textbooks in an integrative manner. The teachers' pacing in leading through chapters is high as they jump from one to the next piece of information without linking them. Thus, the use of time is strictly governed through syllabi that prescribe the number of periods spent on which topic. This is in contrast to ESD principles in which student-oriented methods are meant to encourage that students themselves decide the sequence and pacing of knowledge and skills to be learned.

7.1.4 Evaluation of Knowledge

The acquired knowledge in classes is evaluated through oral questioning or in written exams. In both exam types, the question's format is that of closed questions: definitions, facts, single-word responses, single or fragmented sentences, or lists make up the answers. Examinations are mostly standardized multiple-choice tests at the end of each school year, whereas the board exams in the twelfth standard are the most important ones, as they lead to the final degree from secondary schools.

Although "Continuous and Comprehensive Assessment" has been introduced as part of the Right to Education Act at the national level, this formative evaluation added additional tests for which students have to study for, and teachers find it difficult to evaluate other forms of performances apart from tests, such as oral or social performance. This is because they are not sufficiently trained in applying more qualitative forms of evaluation (T11_S4). Thus, the evaluation of knowledge remains strongly framed, as it only covers textbooks' contents.

In oral questioning, teachers neither show individualized reactions to students' comments nor point out why the students' answers are incorrect. They give feedback in form of "correct" or "no," or repeat the question. Thus, the evaluation is rather depersonalized as the incorrect answers are ignored until the correct answer is mentioned. Often, students answer in unison. This demonstrates the idea of collective (in contrast to individualized) learning, in which all students learn the same information. If students do not react, teachers give the answer to the class. Often, answers are already suggested in the question. The answers are repeated several times. When a teacher asked an open question, students did not respond. This demonstrates that students are rather used to fixed answers to questions.

The following sequence of a classroom situation in class 9 illustrates the central role of the teacher and the strong framing and control of evaluating teaching content:

Excerpt from a geography lesson at school 2 in class 7 on January 17, 2013
(A short video sequence on spheres of the earth is shown)

Teacher "Can you see where the lithosphere is located?"
Students "Yes."

Another short video sequence is shown.

Teacher "Did you understand?"
Students "Yes." (some are nodding)
Teacher "Has anyone any doubt? Do you have any questions?"
Students "No, madam."

To check whether students understood the learning content, the teacher asked a polar question to which students unanimously agreed. Without asking the students open questions, for which they would have to develop their own answers and transfer knowledge, it is difficult to tell whether students really understood. As students were not actually stimulated to ask original questions, they did not interrupt the teacher's speech and instead reacted with a nod. This shows a strong framing through the teacher, as the amount the teacher speaks is very high, whereas students communicate little and cannot influence the content or the order of learning. A performance model of instruction with strong framing, strong classification, and visible principles of instruction mark the observed lessons. As the repetition of information was expected in examinations, knowledge is understood as transferable.

In a lesson to prepare for exams, the fragmentation of knowledge becomes even more detached from contexts; the teacher jumps quickly from one topic to the next (in this order): from the temperature in the arctic ocean to the trees found in the frigid zone, to the countries of the North American continent, the location and names of the five big lakes of America, the four points of the compass, the important rivers of the USA and the definition of a slope (classroom observation in school 2 in class 6 on March, 16, 2012). This demonstrates that the evaluation of knowledge is fragmented and thus marked through strong classification of knowledge, rather than integrated knowledge. Students study for multiple-choice questions, which contradicts ESD principles in which students should interlink knowledge through argumentations.

7.1.5 Relating ESD Principles to Observed Pedagogic Practice

Unlike the requirements given in following ESD for a competence model of pedagogy, the pedagogic practice in the observed geography lessons reflects a performance model of pedagogy. None of the observed lessons included elements of the ESD principles: student orientation, critical and network thinking, as well as argumentation skills. The teacher functioned as a transmitter of textbook contents, which the students had to reproduce word by word. The role of the teacher is also that of timekeeping, as the tight syllabus allots only a few periods per chapter. The exact sequence of chapters in the textbook is followed, which indicates that neither the students nor the teacher has control over the selection, sequence and pacing of knowledge and skills in class. Because of the steering role of the syllabus and the textbook, learning contents are hardly linked to the learners' interests. Since the teachers recite the textbook, they have the most speaking time and the students have little space to take control over the teaching processes. Students are expected to be obedient and affirmative to the teacher, and they are taught that only one answer is correct: The one given in the textbooks. Students often have to reply to the teacher in unison, which reflects the notion of collective, rather than individualized learning. This reflects the notion of correct, standardized and collective knowledge, which students have to learn by heart.

ESD principles, in contrast, demand critical thinking as well as network thinking and the promotion of argumentation skills, as illustrated in this book. These aspects are hardly promoted in the observed geography lessons. While the geography textbook contents follow a regional approach, the information to be learned is factual, such as definitions and lists, which feed into declarative and not processual knowledge. Teachers mostly use closed questions, with fixed one-word or even yes/no answers. This implies that it is very difficult for students to relate and integrate new knowledge to existing knowledge, and that knowledge is not interlinked

Table 7.1 Pedagogic principles observed in geography lessons and the objectives of Education for Sustainable Development (ESD) in comparison through Bernstein's categories

Categories of pedagogic principles	Observations in geography lessons	Transformative objectives of ESD
Selection of knowledge	Strong classification (C+)	Weak classification (C−)
Sequence of knowledge	Strong framing (F+)	Weak framing (F−)
Pacing of knowledge	Strong framing (F+)	Weak framing (F−)
Evaluation criteria and processes	Explicit	Implicit
Space and resources of interaction	Limited	Extensive

as promoted by ESD. Furthermore, topics such as (water) resources are approached through facts, such as exact definitions of rivers, climate, and multi-perspectives and resulting conflicts on water resources are not addressed.

The pedagogic principles observed in geography lessons can be contrasted with the transformative objectives of ESD through Bernstein's codings of weak and strong classification and framing (Table 7.1).

7.2 Students' Perspectives on Water and Geography Education

While academic frameworks strongly influence pedagogic practice, it is important to understand how students perceive pedagogic practice in geography lessons. Students' perceptions and knowledge on geography education and the topic of water demonstrate what they praise and criticize in class, and whether they demonstrate a critical awareness of current teaching methods. This would be encouraging for integrating ESD principles in pedagogic practice. A total of 52 students from grade 9 at the selected five English-medium schools in Pune filled out questionnaires on teaching methods in the subject of geography and on their perceptions on the topic of water (cf. questionnaire in appendix). These answers were reflected upon and validated in student interviews and focused on in group discussions. They will be introduced in the following sections.

7.2.1 Students' Knowledge and Perspectives on the Topic of Water

Students' knowledge about water demonstrates that they have some awareness on the causes of water scarcity and water pollution, as well as effects on different sections of society. ESD postulates topics relevant to students' environments, and the topic of water is highly relevant in their everyday lives. It even seems that their everyday experiences seem to be more relevant than their textbook knowledge. Students either experience water shortage themselves or have heard of it in the media (newspaper, Internet, or TV). Many students stated that they do not always have sufficient water, but as they have access to water tankers or storage containers, they did not face serious problems. 37 out of 52 students claim they have not been sick due to waterborne diseases, whereas 15 students stated that they had had stomach infections or typhoid. Students reasoned the cause was drinking water in villages or hotels and not at home. Concerning the quantity of water, students underestimated the amount of water used daily at their households. Most of them answered 10–80 L, and only some students mention that they know that the water tank of 1000 L is refilled every day. A principal who looked at the questionnaire commented on this question with "how are they supposed to know? I don't even know" (T5_S3). This shows that water amounts are used unconsciously despite water scarcity being a recurring topic every summer.

The majority of students stated that they drink filtered water, and they know that the Pune Municipal Corporation provides water from the Khadakwasla dam through pipelines to tanks. However, most students mixed up the question of the source and provider of domestic water, which was clarified in interviews. When students were asked whether they had visited a dam, waterworks, or treatment plants, most students answered that they had been to dams for recreational activities, such as picnics, but not for learning purposes. Single students have observed water pollution there and state to have been impressed by the amount of water. One student stated "Yes, I have visited dam. I cleaned the place. And I learned that if we keep the area clean around dam, do not contaminate it, we can get pure water." (St20_S4). One student stated that he saw people throwing polythene bags and "told them not to throw" (St21_S2). These answers demonstrate students' intrinsic motivation to learn about water scarcity and pollution, but also show some generalizing and abstract conceptions they learned from textbooks.

All students agree that it is very important to learn about water. Some students gave a reason for their opinions, either related to the importance of water for humankind, or because it is scarce, or because it is "nature's gift" (St18_S2). Some mentioned that they need to learn how to conserve water and how it becomes polluted. The topics that some students are interested in are the importance of waterborne diseases, their treatments, and water transportation. One student stated that she would like to find out about "the actual position of the water crisis" (St19_S4). Almost all students mentioned that they would like to learn how to save and not to pollute water.

When questioned as to what they had learned about water in school, many students answer normatively that it is a precious and important resource, and that they should save water as it is "God's gift" (St_26_S1). Students repeated slogans from their textbooks, such as "It is an important resource because man cannot live without water," "Close the tap after use" (e.g. St26_S4). A great number of students listed the topics covered in the textbooks, such as the water cycle, irrigation, rivers, pollution, and rainwater harvesting. Interestingly, most students did not answer this question or simply wrote "save water." Students remember fragmented facts given in textbooks, which they do not link to an argument to point out the relevance of the learnt content, e.g., water pollution or water scarcity. For example, one student wrote a technical answer "water is a limited resource on earth, 98% of water in seas and oceans, only 2% of freshwater, of which only 1% is in the form of glacier" (St17_S3).

When students try to bring in their knowledge from school, there are uncertainties in using their knowledge as explanations. For example, some students confused water saving with water pollution, as they answered how to purify, and not how to save water. These findings are coherent with the results of the study by Caroline Saam (2013), which found that students' knowledge on water pollution consists of fragmented pieces of knowledge according to the theory of diSessa (2008). The qualitative analysis of 32 problem-centered interviews with students on their perceptions of water pollution at the same[1] English-medium schools in Pune shows that the three dimensions of sustainability are not interlinked. Particularly, environmental aspects are mentioned in monocausal and not coherent argumentations. Students also tend to combine all their knowledge on environmental problems into one subjective theory, e.g., when linking water pollution and global warming (Saam 2013: 34). This demonstrates that students take over the fragmented and factual content and knowledge from the textbooks as transmitted by their teachers; as expected, this does not correspond to the ESD principles of network thinking and argumentation skills.

As one of the ESD's objectives is to create environmental awareness on water conflicts, students were openly asked whether there are problems with water in Pune. A number of students mentioned water pollution and water shortage because of both less rain and wastage. Some stated consequences of pollution like diseases and that "aquatic life gets disturbed" (St22_S2). When asked for differential water access, students mentioned the differences between rich versus poor, rural versus urban areas, as well as dry areas, and differentiated between states such as Rajasthan and other states in the northern part of India. Some related water scarcity to the growing population and water usage for agriculture, and one student reasoned "in slums, they have to fight for water or steal water from the other" (St2_S2). Students further pointed out that the water quality they receive in developed housing societies is better, as it is purified, than that of poor people, who live in slums. One student elaborated the effects of differential access to water for domestic purposes:

> In my area, there are less [sic] problems, but other areas face very severe problem [sic] regarding water. They don't get enough water to drink. Also there is scarcity of water for agricultural purpose and daily use. Many times people get water but water is highly polluted and cause harm to their health. Condition of water is so bad that they cannot even use this water to wash clothes or take bath. Those people who pay for water can get sufficient water but those who cannot pay for it can never get good supply of water. (St8_S2)

One student perceived water scarcity as a distribution problem and was of the opinion that authorities are bribed for better water access (St9_S5). Another student mentioned that the water provided by the Pune Municipal Corporation is filtered, but

[1]Caroline Saam conducted the study on student perceptions on water pollution for her state examination thesis at the University of Cologne at the same English-medium schools where my study took place.

that "villagers and slum people drink contaminated water from the river" (St23_S2). Another student stated that "some people do not know where good water comes from" (St10_S5), relating the disparities in access to awareness. One student blamed an indifferent government for water scarcity: "There is not enough drinking water for everyone as I think the government is not doing anything that will get the water to the villages" and the "chief minister is not paying attention" (St6_S5). These answers show that student knows about the differential access to water through diverse sections of society, but that they can only sparsely explain the coherence. This shows that there is a great gap between the expectations of ESD of network thinking and being able to state and defend their opinions with the use of arguments.

Concerning possible solutions or changes in current environmental concerns, a number of students answered that awareness and rules like "not take bath, wash clothes, utensils and stop throwing industrial waste directly into rivers" (St11_S2) would help to solve water conflicts. A number of students suggested that water should be conserved and not polluted and that "people should not throw polythene bags and garbage in the water" (S14-S2).

Concerning multi-perspectives on water conflicts as required by ESD, students attributed the responsibility for water pollution and scarcity to either the government or the poor in slums or rural areas. Pollution through industries and urban wastewaters was not mentioned, and awareness on collective contribution to water pollution was not raised. One student suggested that the "government provides slum areas with well organized water supply connections" (St12_S2). Another student even proposed a "cooperation of people and government" (St13_S1). With respect to water pollution, some students stated that:

> Village people wash clothes or clean their animals in the river, water gets dirty, and after the river water through the pipes comes into their own house and they drink the dirty water and their health get many diseases even the small children in their house. (St7_S1)

This quote, which was also similarly stated in lessons by teachers and students, demonstrates that water pollution is associated with "the poor" and farmers rather than with urban lifestyles and domestic wastewater. One student mentioned "teaching uneducated farmers" (St9_S5). This puts the blame on others—supposedly less educated—for water pollution. Interestingly, 10 students from school 3 admitted that they did not know how to solve the problem of differential water access.

In regard to critical and solution-oriented thinking, as required in ESD, a few students answered the question whether they could save water with water harvesting and wastage prevention; however, they did not mention specific measures *how* they could or actually do prevent water wastage. One student suggested reducing drinking water (St24_S2). The main reason given as to why water should be saved was because of water shortage in societies. Students stated that it is important to save water as it is a basic need and it must be saved "for the next generation" (St25_S1).

One student contrasted India to other countries and said that in India, the "government does not provide water" (St15_S3), while other countries have sufficient water distributed by the government. The perception of water problems in other countries ranged from similar problems to no problems with respect to water. The

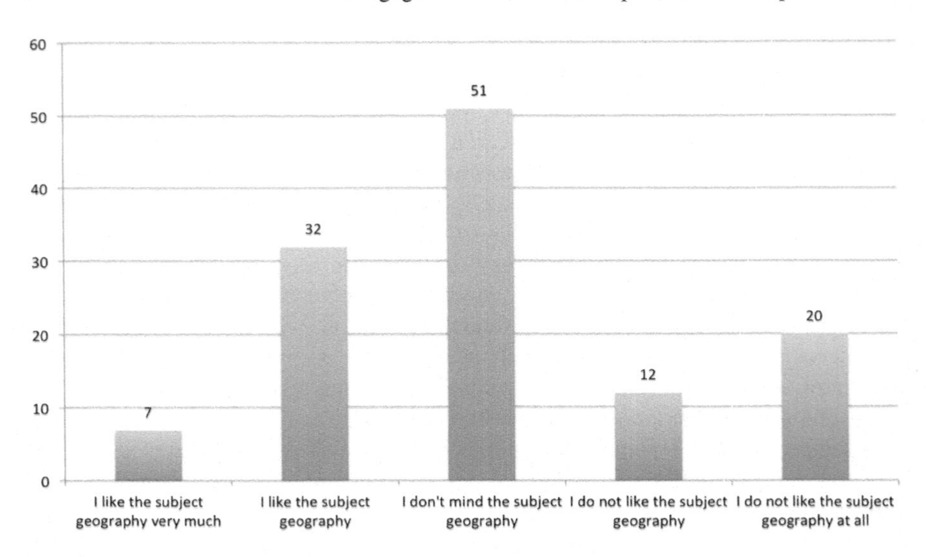

Fig. 7.1 Students' overall rating of the subject geography

reasons are that pollution is everywhere or that "they are not wasting or making it dirty" (St3_S2). Several students gave the example of China, which they might have learned in school, and stated "they could have sufficient water to drink because rules are strict there and people are not allowed to throw garbage around" (St16_S4).

7.2.2 *Students' Perspectives Geography Teaching*

To understand how students perceive geography teaching, they were asked to rate the subject in general and to answer open questions to understand the reasoning for their study experience. Students related the subject of geography to the memorization to a lot of facts. On a five-level Likert scale from 1 (very much) to 5 (not at all), 122 students rated the subject of geography at a mean of 2.35 (Fig. 7.1). Almost half of the students (52) were indifferent to the subject, whereas one-sixth of the students (20) admitted that they do not like geography at all. Although overall 83 students rated geography 2 or 3, the rating demonstrates that only one-third of the students enjoy geography.

The open question on what they like about geography included answers on learning about countries, cities, environment, fauna, physical features, biodiversity, solar system, resources, and maps. These answers indicated a general interest in geographical topics. Concerning the open question on what they do not like about geography, students named memorizing longitudes and latitudes, capitals, rivers, currents, as well as "answers for test questions." This demonstrates that the majority of students do not like detached rote learning of regional facts in geography lessons, which ESD also aims at overcoming. Some students also mentioned that they do not like to "learn negative things like pollution" and "big answers", which shows that students do not necessarily associate positive messages with geographical topics. Moreover, terms in textbooks that are too general and abstract indicate that the presentation of textbook contents does not motivate children. The phrasing "big answers" indicates that students know they do not have to find their own answers to question or even pose questions, but that they have to learn given answers in textbooks, which makes learning rather boring. This shows that student-oriented and action-oriented principles, as desired by ESD, are also demanded from the students' side. One student stated her opinion on geography class in an interview:

> Geography is good, when it comes to natural vegetation, what is in a place, I like it. But when it comes to remembering numbers, like this much space is this landmass, I don't like it. (St5_S6)

With reference to the teaching style in geography lessons, some students stated in interviews that they like geography class because they get "good grades" and it is interactive; however, others answered they would like it to be "more practical," as the lessons are "boring" and the "teacher explains each paragraph of the textbook." Some say it is "easy to remember," whereas others stated it is "difficult to remember." Some said that locations are not shown on maps, whereas others mentioned that "maps help." Students further mentioned that they would enjoy field trips and 3D animations for an innovative change of teaching methods. One student said: "we have to mug up each and every question and answer in geography" (St1_S1). These diverse answers indicate that students do not enjoy the strong framing through textbooks. These various perceptions of the subject of geography indicate that the answers are highly dependent on the teachers and their teaching style in class.

To understand how students perceive student participation in class, they were asked several questions regarding *discussions* in geography lessons. Students had various perceptions of what a discussion is. Some answered that they discuss physical features of states and countries, which shows that they might have understood discussions as students' general classroom participation, rather than as a debate, in which they can develop and present their own opinions. The confusion about discussion and general oral communication became obvious with a student's statement: "We discuss about water crisis, noise, global warming and we are also taught to conserve our natural resources" (St2_S2). In observed classroom lessons, these topics were, however, approached through the delivery of facts, rather than discussions. Some students said they "discuss" once in a while on exam preparations and organization; hence, administrative procedures are addressed in class, rather than geographical contents. Others mentioned that they discuss how to save water and how not to pollute it, which indicates lively discussions on water use. In contrast, one student answered: "the teacher comes, explains, and goes," which illustrates that students do not have any space to participate in geography lessons. These answers indicate that students' experiences of discussions in class varied from none to some. One explanation is that in textbooks, "discuss" is used interchangeably with "describe" (cf. Sect. 6.4), and hence, students also do not understand what a discussion is. This demonstrates that there is a need to integrate a more discursive teaching style in geography education, as postulated by ESD.

Being asked whether they (would) like discussions in geography class, students gave generic answers such as that they like to become familiar with other opinions, to share their feelings, to express and to exchange their views and ideas and to solve problems. One student stated that it is easier to remember discussions as they are more interesting and interactive:

> "I like discussions because […] when we discuss with each other we get to know the things which our friends know more than us. I think discussion is the best way to study because we can remember it more easily than reading or writing." (St3_S2)

Some even say they learn through debates more that concepts become clearer, and their awareness is raised on particular topics. A student stated the importance of discussions "to become responsible citizen" and "discussing issues makes us aware of many things which is very important, as we are the future citizens of our country" (St4_S2). This demonstrates that some students are aware of the relevance of discussions in lessons for their own lives and their motivation in class, and that they are interested in presenting their own and listening to other students' perspectives. These answers demonstrate students' interests in discussions on controversial and relevant topics concerning natural resources, which correspond to the central concerns of ESD. The critiques voiced by students in regard to a teaching methodology centered around rote learning of textbook facts in geography education express the objectives of ESD: to make education more student-oriented, to interconnect various perspectives, and to critically argue on topics from social, economic, and ecological sustainability perspectives.

7.3 Teachers and Educational Stakeholders' Perspectives

While the students are "receivers" within the pedagogic device, the teachers as "transmitters" structure pedagogic practice at different levels. Hence, the most central and direct role in pedagogic practice is that of the teachers, which makes their perspectives central for implementing ESD. Other educational stakeholders' perspectives at the state and national level are transmitted through curricula, syllabi design, and textbook writing, which convey their perspectives on teaching contents and methodology in pedagogic practice. In the following, the different perspectives of teachers, MSCERT, NCERT, NGOs, and academic institutions are summarized to demonstrate how they define their role and impact in concern to pedagogic practice and ESD, and how their role is perceived through other institutions. Interviews with teachers and educational stakeholders from NCERT, MSCERT, or educational non-governmental organizations that were conducted over a period of three years (Sept. 2011 to Sept. 2014) provide evidence.

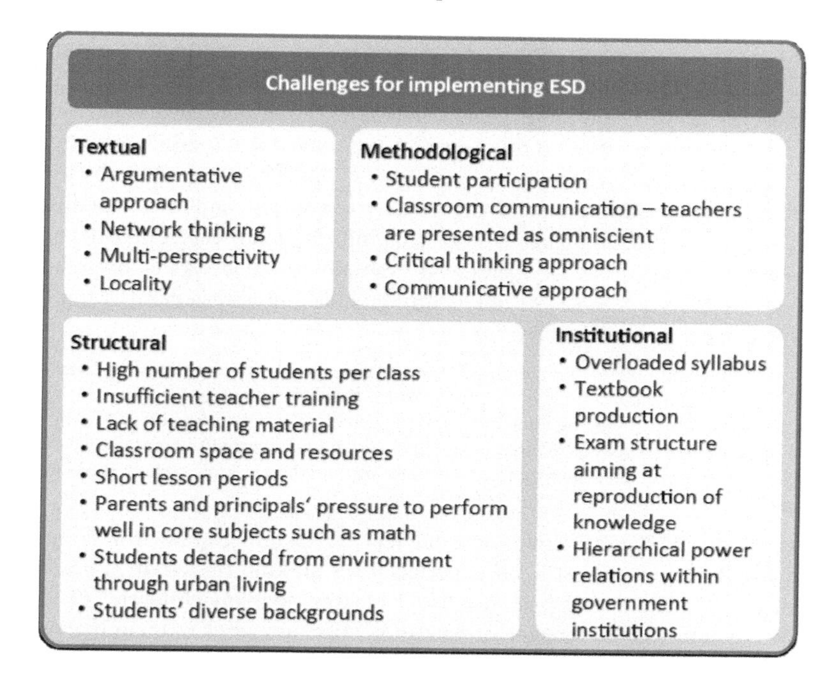

Fig. 7.2 Summary of the challenges for implementing ESD according to teachers and educational stakeholders

The challenges for implementing ESD as perceived by teachers and educational stakeholders can be divided into the perception of ESD having high-level textual and methodological expectations, which are difficult to realize under existing structural and institutional conditions (cf. Fig. 7.2).

7.3.1 Teachers' Perspectives on Barriers and Opportunities of ESD

Teachers feel they are bound to the syllabus and hence do not have time for teaching approaches as postulated by ESD. As teachers revealed in interviews, completing the textbook is their most important task in geography lessons. As one teacher explained, "our aim is to make sure that our students do well in competitive exams" (P2_S2). As a limited number of periods are appointed to specific chapters, teachers have to complete the topics in a prescribed time schedule. As lessons last only 35 min, merely a quick review of a topic can be given. Other activities at school, such as religious festivals and competitions, take a lot of time and cause them to often be behind with the syllabus. Some committed teachers say they would like to spend more time on environmental topics and projects; however, parents do not value these: The "parental attitude is that less importance is given to the subject of geography or environmental sciences, math is more important" (T1_S2).

Further barriers to the possible introduction of ESD that were mentioned by teachers are that students are used to rote learning and expect the "perfect answer of the teacher and only copy from books for exams" (T1_S2). This observation by teachers demonstrates that students conform within the given system and do not challenge it, possibly because reproducing and memorizing information is easier than developing independent thoughts. Teachers perceived examples in textbooks as not related to "day-to-day life" as textbooks are not locally specific and only general examples are used. One teacher critically stated:

> when you are testing them, you are putting it in a very water tight compartment. Then it is restricted only to the definitions, then give reasons, then questions and you are restricting them to answer only to the point and only to that manner. When there is a discussion in the class, there are no restrictions. Children can think beyond that particular point. So we don't believe in testing each and everything. It has to go beyond the marking system. It has to go beyond the grades. (T10_S4)

However, during the same interview, the teacher acknowledged, "for us, it's very difficult to give children first hand experience for all kind of activities because of the sheer number of students per class. Huge classes, so many children." (T10_S4). Students may either know through observation, "because some of them live in areas where they may have slums very close by. So they see ladies fetching water from a common tap. […] In other cases, teachers tell them" (T10_S4). A perceived characteristic of students in urban environments is that

> [urban students] are too cut off from the actual process of agriculture or farming. They are absolutely cut off. They don't know what is happening. They know the food comes and their idea of food coming onto the table for most of them is a mall or a super shop. Their understanding is that you get foods in packets. (T10_S4)

Student culture is perceived as such that their attitude toward the environment is "casual and not interested" and that students are relieved whenever they have to study less: "Thank God we have to study one subject less" (T1_S2). Further reasons for students' low commitment to environmental issues are that "urban children do not introspect about environmental problems as they are not exposed to natural things" (T1_S2). Hence, this teacher implied that environmental topics are more difficult to link to urban students' environment. Furthermore, it is difficult for teachers to multi-task: To explain and maintain discipline, while observing the class, as "few students who can express themselves will overpower quieter students" (T4_S1).

Teachers see opportunities for ESD in activities and discussions on the conservation of water as these raise students' environmental awareness. Interrelating ESD topics such as rainfall, wastage, pollution, increased population, and water resource availability promotes the integration of factual thinking and network thinking (T4_S1). According to one teacher, students learn to understand and accept the view of other students and ESD could "change the mind of a child, our teaching would reach their heart" (T4_S1). These statements signify that teachers acknowledge the benefit of changing teaching methods toward ESD, and that learning is ensured through interaction and critical thinking.

7.3.2 *Educational Stakeholders' Perspectives on the Role of Teachers*

State and national educational stakeholders see teachers as the transmitter of knowledge. Expectations toward their agency vary, as it is recognized that rural teachers have limited access to information and material, while urban teachers, particularly those at English-medium schools, should take the full responsibility in classroom teaching (Prof_7). This indicates that teachers are not viewed as a homogeneous group but are rather associated with the binary concepts of "urban/rural" or "English/vernacular" groups. Even if teachers are interested and they may have access to teaching material, they are thought of as not being used to "organizing and planning a lesson" (NCERT_7, 41: 54).

The potential to change habits of teaching practice is rather seen in young teachers, as changes are slow and will take time (MSCERT_2, 15:53). According to the perception of one NCERT professor for geography, teachers simply lack time, knowledge, and detailed teaching materials on how to implement ESD in classes:

> At least give them the ready materials that you can do this way and then they can go and implement that. But nothing is available. First they do not have the time for planning, then other thing is that they do not have the resources to do it. The thing is that they do not know how to do it. So they cannot do it. (NCERT_7, 1.42.33)

In the same interview, the professor, however, stresses the teachers' agency despite a lack of resources and the centrality of the teacher's willingness to teach in under-resourced contexts:

> How will the teaching learning process happen based on the textbooks if they do not do the activities. Again many reasons they will give, they will say we do not have this resources [...] they have all these excuses, the problem is with the whole mindset; you have to change the mindset of the teachers also, they really need to start doing the activities. (NCERT_7, 1.45.28)

On the one hand, this draws attention to the role of skill development through teacher trainings and appropriate teaching material necessary for the implementation of ESD. On the other hand, underlying reasons for teacher's motivation, such as cultural appropriateness and exposure to learning new activities for teaching, need to be understood and addressed. Structural reasons to not implement ESD are large numbers of students per class and an overloaded, fact-oriented curriculum that already constrained teachers' agency. The head of a textbook writing committee at MSCERT explains the difficulty of implementing ESD and focuses on teachers as the transmitters of contents, and that their training is very relevant:

> There will be some committee, some authors will sit with our syllabus and prepare the textbook. It's easy. Because all this will be the expert doing that work. Then there are pupils in the class. In between is a teacher. You have to train the teacher and if we fail in training the teacher, our entire program will collapse and we are sure about it. All is going to be sent through the medium of the teacher. (MSCERT_2, 17:02)

He further admits that teachers do not have the sole responsibility, but rather that there is a general problem of changing attitudes and behavioral patterns:

> There is always a kind of inertia to resist the change in the minds of Indian, any change. He will enjoy whatever he was doing all these years. If we ask him to change, he will always [demonstrate a] little bit of resistance which we have to overcome. And that's the challenge this committee will face. (MSCERT_2, 17:47)

A geography textbook author noted the challenge of a "deep clash" of ESD and "traditional methods" (MSCERT_3, 18:46) in teaching and stated that change in pedagogic practice will be slow because of the great diversity within the country. He supported this statement by pointing out that educational reforms are difficult to link to the various social backgrounds, differences of urban, rural, and tribal cultures in different regions. The head of the MSCERT curriculum design group for environmental studies stated the increasing role of the curriculum for the teacher next to the textbook:

> Curriculum in the past was only a book which was not used by the teacher. It was never seen by the teacher who is working in the classroom. But we are now giving the scope for the classroom activity in curriculum. (MSCERT_2, 16:23)

Hence, it is acknowledged that teachers need, besides textbooks, further access to lesson planning material and academic frameworks. Teachers' capacities can only be strengthened if they have more access to teaching resources. As the classroom analysis has shown, teachers have limited control over the selection, sequencing, pacing, and evaluation of pedagogic practice. While the role of the teachers is acknowledged by educational stakeholders, facilitating their agency is partially neglected because of the perception that teachers need to be guided through their duties. However, teacher's agency can be increased through teacher workshops and trainings, more methodological guidance and support, and less strict content prescriptions through syllabi and textbooks.

7.3.3 *Educational Stakeholders' Perspectives on Curriculum Change*

While NCERT develops national guidelines for school education, MSCERT is responsible for developing the state curriculum framework (SCF), syllabi, and textbooks for the schools of the Maharashtra state board. The task is to discuss and develop contents for the syllabi of each respective subject. Curriculum and subsequent syllabi and textbook changes occur every decade, and to introduce these changes to teachers, the District Institutes of Education and Training (DIET) conduct teacher trainings before the schools start. However, there is a discrepancy between different stakeholders' perception as to what extent NCF is relevant to the state: NCF is a framework and not compulsory; hence, SCERTs are flexible to change any sections of the national curriculum. The "central government prepares guidelines and [the] state government takes the liberty of re-telling." (MSCERT 21: 54).

NCERT views SCERTs as the state authority to develop syllabi and textbooks used in the respective states. NCERT sees itself as a different entity than the SCERTs and transfers the responsibility in the federal system to the states as education is on the concurrent list. There is no direct cooperation between NCERT and SCERT to ensure that the central guidelines are recontextualized at the state level. Hence, a strong classification can be attributed to the categories of curriculum, pedagogy, and evaluation. That is, according to Bernstein (1975), a strong hierarchy within and between institutions. This indicates a great need for an integrated view, exchange of messages, recontextualization to make educational reforms effective. An NCERT professor admitted that there is limited interaction between curriculum design and research (NCERT_5), indicating that NCERT could improve its impact by linking syllabi and textbooks to educational research.

Within NCERT and MSCERT, internal hierarchies indicate a strong classification in which only a few committed people can actually participate in decision-making processes. Due to a rigid system of limited responsibilities and low educational ownership, periodically shifting positions and difficulties in implementing ideas against a strong hierarchical internal system, inflexible decision-making processes often hinder innovative approaches (NCERT_3). Because of internal hierarchies, external forces are highly influential, for example, the government passed the "Right to Education Act" as well as the "Right to Information Act" that put institutions under pressure to rethink and publish their practices (MSCERT_1, 01:53). One curriculum designer stated that "the government policy has decided to decrease the content load. Now we are doing our own work [...] according to their instructions" (MSCERT_1, 04:32). This indicates that curriculum designers feel obliged to transmit government policies into their curriculum design without necessarily considering it appropriate or useful.

These reasons indicate a gap between institutional authorities. The perceptions of educational stakeholders in national and state government institutions, non-governmental organizations and academic institutions give insights into their strongly classified voice, identity, and the specialized rules of internal relations. This leads to a rather fragmented discourse in which responsibilities are often associated with other institutions than their own. National stakeholders transfer responsibilities to state stakeholders and the other way around. The high degree of "insulation between categories of discourse, agents, practices, contexts" (Bernstein 1990: 214) hinders the flow of messages within and between national and state stakeholders as well as academic and/or government institutions, as the carrier of the message is more important than the message itself. In this way, some kind of "othering" (Spivak, 1985) within the education system takes place: "teachers," "government," the "state," "NCERT," "MHRD," "MoEF" are associated with a certain degree of power and thus the responsibility to introduce ESD and student-oriented pedagogic practice. Stakeholders within educational authorities feel they cannot take the full responsibility as they have limited space to act. As a result, none of the mentioned authorities takes the responsibility of implementing educational reforms in pedagogic practice. In Bourdieu's terms, the "distinction" (Bourdieu 1977), especially the binary concept of the "teacher" and his or her duty and agency opposed to those of the "government" amplifies the tendency to pass responsibility to others.

7.3.4 Educational Stakeholders' Perspectives on the Role of ESD

When asked about the role of ESD for curricula, syllabi, and textbooks, a representative of the Maharashtra state board answered "it is implied, it is not exclusively mentioned. But it is implied in all the subjects we have thought of." (MSCERT_2, 07:21). A textbook author stated "constructivism, then self-learning, self-awareness, gender equity—these are the major spirit used for our syllabus" (MSCERT_3, 09:54). In contrast, an NGO director expressed what a majority of teachers wonder: "Eighty children and one teacher. How is it ideally possible to follow constructivism?" (NGO_10).

One objective of ESD is the integration of social aspects in environmental sciences. One MSCERT syllabi developer hopes that this integration means "to decrease the content load" (MSCERT_1). One textbook author thought that the concept of sustainable development is rather for the university level, as it is too abstract for school students and thus not integrated in the syllabi (MSCERT_2). The term "development" is associated with physical development such as roads, buildings, factories, and cars, another author stressed how the concept of development is linked to "Western" lifestyles:

> Adopting Western culture is called development so far in India also and in our syllabus also. Adopting Western culture, adopting those things in our home, adopting TV in our home was developmental indicator in past. (MSCERT_3)

> We do not directly talk about sustainability. But we generally talk how human being is interfering in the natural aspect and how things are deteriorating. How we are Western. How the globalization and marketization […] are coming up to us. How the media is coming and interfering very much in our lives. (MSCERT_2)

However, one state stakeholder stated that the focus of the term sustainable development is the reverse; instead of attracting villagers to cities, attitudes should be changed to respect home, parents, and culture through the curriculum:

> When the whole village is developed then no one wants to go to city. Now the development pushes the villager to the city. How should we stop the migration of people which is very big problem in cities and in villages also? (MSCERT_1).

The coexistence between humans and nature is stressed; nature, instead of humans, is discussed in the center. This attitude developed "not from information; we give them that kind of feelings from our attitude." (MSCERT_1). The topic of the village, for example, should be approached in such a way that children know about social and environmental aspects, for example, animals' relationship with them (MSCERT_1).

Syllabi designers would like to present living realities as they are and include conflicts over resources from different angles. The equal distribution of water and education is discussed as "rights given by the law, but the reality is something different. So that also we want to club two things. There is ground-level reality which is something different." (MSCERT_1). A number of textbook authors perceive changes in their own syllabi making, as now they want to develop knowledge-based constructivism, and not solely give information (MSCERT_1, 05:05). The focus in syllabi and exams is on skill development and not content. Concrete changes in the approach are that students are encouraged to explore:

> in this syllabus I will not give information for students. But through information I want to give in; I will ask some questions to students. What you know about that subject? Let them tell some. Because they also know many things about our subject and the Indian environment. So first the questions, and after the information. (MSCERT_1)

> Our exam system was again also information based. [...] Now the new focus is; it's like exploration. The content is secondary. But the child has to develop the capacity to critical thinking, inquisitiveness, asking questions, building confidence, gathering information, curiosity. All these things are focused in the current syllabus. So the evaluation system also doesn't focus now on content information but it focuses more on developing abilities of inquisitiveness, curiosity or making a project or maybe interviewing technique. [...] And this is a major shift in focus happening in India, almost all India. (MSCERT_2)

These interview excerpts show that educational stakeholders perceive a shift toward skill development that ESD, and also the NCF 2005, wants to encourage. In a group interview, one committee member of the environmental studies committee agreed that within the committee, relevant discussions take place, for example whether castes should be included in textbooks to explain different communities.

Concerning the organization of the Indian education system, one member of the committee confessed: "we get confused, because there are many bodies" (MSCERT_1, 1:03:20). This indicates that for a transnational policy, as ESD, clear responsibilities and directives need to be transferred to institutions to be recontextualized at the national, state, and district levels.

7.3.5 *The Role and Perspectives of Extracurricular Stakeholders*

The rigid organization of the classroom setting is in contrast with extracurricular environmental projects detached from the curriculum, which promote a weaker framing, and thus, modify the teacher–student relation. These projects are not linked to or built on syllabi and textbooks, but give a cause for thought for daily pedagogic practice. A number of NGOs who conceptualize, organize, and implement environmental projects in schools or explicitly work toward ESD follow different approaches to promote environmental action. Most prominent are the so-called green school projects of different NGOs that develop manuals and activities, which can function as best-practice examples. The *Center for Science and Environment (CSE)*, for example, encourages students to measure their schools' environmental practices by conducting audits of the water, energy, land, air, and waste of their schools. Yearly, the greenest school is nominated for an award. The NGO *Greenline*, which is active in Mumbai under Don Bosco, offers workshops with innovative projects such as "cut the flush," for which students convince their neighborhood to put a water bottle filled with sand in the toilets' water tank to minimize the water use with each flush. The NGO encourages students to develop their own ideas to green their schools. Participating schools are visited by a jury annually to evaluate the participation of students, initiatives of schools, the creativity of ideas, the impact on the school environment and neighborhood, as well as the sustainability of their activities. Urban gardening projects, composting, and rainwater harvesting are some ambitious projects that depend on a committed teacher and the support of the school management. In many other NGOs, slogan and drawing competitions are encouraged. Slogans, such as "every drop counts," or "save water, save life," indicate a strong rhetoric that honors the earth; however, concrete courses of are often not suggested. Similarly, drawings romanticize nature and are often related to goddesses and festive events such as Ganesh Chaturthi. Audits strengthen measuring and analytical skills; however, environmental-friendly practices and a grounded discussion among students are not sufficiently conducted to strengthen argumentation skills.

The MoEF initiated the program National Green Corps (NGC) in 2001 that funds eco-clubs at schools to promote environmental conservation. By 2007, there were 91,000 eco-clubs in the country, which made the program the largest and most far-reaching program of its kind. The *Center for Environment Education (CEE)* is supported by the MoEF and executes environmental projects of the government. Paryavaran Mitra is a program that explicitly focuses on promoting ESD. In Pune, for example, CEE offers a water kit to school classes, with which students can measure substances, such as fluoride, in the water. However, a staff member criticized that network thinking is not encouraged and background information on the causes and impacts of fluoride is not given (NGO_11).

A critique by the educational stakeholders of the formal educational system is that there is not enough time and resources to include ESD in lessons (NGO_6). Extracurricular projects are seen as the only opportunity to implement pro-environmental activities in schools. As these educational stakeholders do not find environmental topics adequately covered in textbooks, they develop their own material and publications.

The mentioned NGOs aim at filling the gap of a lack of environmental projects in formal education, as they accuse state boards for "not taking ESD seriously" (NGO_9). The success of educational reforms is dependent on dedicated individuals within governmental institutions. However, multiplier effects as well as both activity and reflection (Freire 1990) can only be achieved extensively through the formal education system. Therefore, linking extracurricular projects to the curriculum would promote an integrated approach for promoting environmental knowledge, awareness, and action.

7.4 Summary of ESD Principles in Pedagogic Practice and Triangulation of Perspectives

Pedagogic practice in the observed schools is strongly framed and closely bound to the regulations formulated by educational institutions at the state and national level. Teachers' agency is bound by an overloaded, fact-oriented syllabus that prescribes a strongly framed teacher–student communication with limited space for students to develop critical-thinking and argumentation skills. While the textbooks hold a steering role for both teachers and students, teachers transmit the contents to the students, who reproduce and memorize the contents to be a successful student. Since a "good class" is considered to be one that demonstrates obedient and affirmative behavior toward the teacher, students are given little space and control over the selection, sequence, pacing, and evaluation of knowledge. Interestingly, the teachers' agency is similarly restricted. The teachers' main task is to keep pace with the syllabus that is to follow the sequence and pacing of the textbook chapters' content, for which a particular number of lessons is allocated. Since the syllabi are overloaded and some textbooks have up to 22 chapters, these regulations strongly structure teaching. This also demonstrates that contents are separated in small pieces, which lead to a fragmentation of knowledge rather than an integrated knowledge, as demanded by ESD. In the terminology of Bernstein, the observations of pedagogic practice in geography lessons in English-medium schools in Pune indicate a performance model of pedagogy in which the reproduction of contents is intended. Unlike the demands for a weaker framing and classification of pedagogic practice, the observed geography lessons display a visible pedagogy that stands in stark contrast to ESD principles of invisible pedagogy.

The strong framing and classification of pedagogic practice are critically perceived by both teachers and students. Students stated that they do not enjoy memorizing facts for reproduction. They demand teaching contents to which they can relate to and want to enjoy debates in which they can learn from other students and express themselves. This shows that students' criticism of current pedagogic practice in geography education and their demand for changes correspond to those of ESD. While teachers criticize their limited influence because of the overloaded curriculum and expectations from students, parents, and principals that teachers "deliver" textbook chapters on time, they acknowledge that ESD could restructure contents in a more integrative manner. Teachers feel that students expect the teachers to be omniscient, which demonstrates how students have internalized and accepted the authoritative position of the teacher. These results demonstrate that teachers lack agency within classroom teaching to structure the selection, sequence, pacing, and evaluation of knowledge according to their own criteria.

The general consensus is that educational reforms such as ESD highlight the necessary change of pedagogic practice. However, to transform practices is perceived to be very slow unless stakeholders and institutions acknowledge their ownership in the process of transforming pedagogic practice. Educational stakeholders state that they want to change current pedagogic practice as stated above and promote ESD principles such as critical thinking on water conflicts. However, their influence is limited by internal institutional hierarchies and bureaucracies and insufficient capacity development opportunities for textbook authors, curriculum designers and other educational stakeholders who could act as multipliers for ESD. There is a large gap between the postulations for ESD principles promoted by educational stakeholders and the observed everyday pedagogic practice in classrooms. Negotiations on what and how change in pedagogic practice is possible and desirable are ongoing but yet, new approaches have to be successfully implemented into practice. Although stakeholders are aware of the struggles in diverse and, most importantly, under-resourced educational contexts, conceptual frameworks and methodologies of how to develop capacities of teachers and textbook writers toward ESD principles in their particular socio-cultural context have to be directed toward actual changes in pedagogic practice. While some stakeholders admitted that teachers lack time, knowledge and teaching materials to organize and plan lessons themselves, for rather structural and institutional reasons, another stakeholder stated a certain resistance to change in behavioral patterns of both educational stakeholders and teachers. This specifically applies to the postulations to ESD objectives, as they are both understood in an aspirational and vague manner. The spirit of ESD is particularly picked up by NGOs who develop extracurricular environmental activities that motivate and educate students on water-related issues in their local environment. The advantage of NGOs is that they are more flexible and dynamic in developing and implementing extracurricular educational projects and teaching material than the government bodies of the formal education system. The ideas of non-formal educational organizations could be stronger linked to or even integrated in the formal school system.

Following the insights of current pedagogic practice as well as the perspectives of students, teachers, and educational stakeholders on the opportunities and challenges of ESD, the next chapter will describe and analyze the development and implementation of three ESD teaching modules aiming at triggering changes in pedagogic practice and teacher–student communication. It further reflects upon the concrete challenges and opportunities of introducing new teaching methodologies at the classroom level and the form in which new teaching methodologies can be implemented.

References

Bernstein, B. (1975). *Class, codes and control. Towards a theory of educational transmission* (Vol. III). London: Routledge.

Bernstein, B. (1990). *Class, codes and control. The structuring of pedagogic discourse*. London: Routledge.

Bourdieu, P. (1977). *Outline of a theory of practice*. Cambridge: Cambridge University Press.

diSessa, A. A. (2008). A Bird's eye view of the "Piexes" vs. "Coherence" controversy (from the "Piexey" Side of the Fence). In S. Vasniadou (Ed.), *International handbook of research on conceptual change* (pp. 35–60). New York: Routledge.

Freire, P. (1996). *Pedagogy of the oppressed*. London: Penguin Books Ltd.

Goffman, E. (1971). *Relations in public. Microstudies of the public order*. New York: Harper and Row.

Saam, C. (2013). Schülervorstellungen zur Wasserverschmutzung an English Medium Schools in Pune/India. University of Cologne.

Spivak, G. (1985). The Rani of Sirmur: An essay in reading the archives. *History and Theory, 24*(3), 247–272.

Chapter 8
Opportunities for Interpreting ESD Principles Through Argumentation in Pedagogic Practice

Abstract This chapter examines the transformative potential of Education for Sustainable Development (ESD) for pedagogic practice in Indian geography education. I analyze how the implementation of a democratizing teaching approach of ESD challenges power relations and cultural values reproduced in pedagogic practice in five diverse English-medium schools in Pune. To examine how ESD principles can be translated into pedagogic practice, I develop three ESD teaching modules "Visual Network", "Position Bar", and "Rainbow Discussion" in collaboration with teachers. I operationalize ESD objectives by designing teaching material and a teacher training which promotes students' argumentation skills on water conflicts in Pune, instead of memorization. I analyzing how teachers perceive, develop, and implement these three ESD teaching modules and how students cope and perceive shifts toward an argumentative framing in classroom interaction. The activities cause a change in the use of classroom space and teaching resources beyond the textbook. Students actively participate in multi-directional classroom communication. However, underlying principles of pedagogic practice persist during the implementation, as the focus on presentation, sequence, and formal teacher–student interaction remains intact. The latter continues to shape how teachers and students bring ESD teaching modules into practice. This exercise reveals how ESD and the promotion of argumentation skills only partly intervene in prevalent principles of pedagogic practice. This underscores the need for contextualizing ESD through continuous collaboration with teachers, textbook authors, and syllabi designers.

In order to analyze the transformative potential of pedagogic practice with ESD principles through argumentation, action research was conducted with local teachers in Pune. Changes in pedagogic practice through modified teaching contents and methods are analyzed in form of an intervention study in geography lessons. Analyzing the classroom observations, curricula, syllabi, and textbooks determined a strong discrepancy between institutional objectives, structural conditions, and pedagogic practices. The qualitative interviews showed students and teachers' great interest in teaching methods, which promote argumentation and network thinking. Educational stakeholders at the state and national level acknowledge the key role of teachers in

S. Leder, *Transformative Pedagogic Practice*, Education for Sustainability,
https://doi.org/10.1007/978-981-13-2369-0_8

pedagogic practice. However, their space for agency is perceived to be limited due to binding academic frameworks and insufficient methodological teacher training. The following intervention study is framed by the concepts of Bernstein (1990). The strong classification and framing of the selection, sequence, pacing, and evaluation in classrooms (cf. Bernstein 1990) in Chap. 7 indicate that alternative teaching methods should be introduced in geography lessons. These relate to the transformative objectives of ESD:

1. The *selection* of knowledge should be relevant to the students and hence related to the students' environments. Multiple perspectives on urban water conflicts should be presented and argumentations should be interconnected and not fragmented.
2. The *sequence* of knowledge should be arranged consecutively and integrated, coherent in content and with an increase in content and argumentation complexity and skill development level.
3. The *pacing* of knowledge is assigned by the teacher. However, within a greater timeframe, students can individually pace several steps of their argumentation development.
4. The *evaluation* criteria are not explicitly defined; the process of achieving and explaining argumentation is more important than giving a simple answer.
5. The *use of space and resources* is arranged according to the needs of stimulating and engaging group work.

In contrast to observed teaching methods in geography lessons, ESD represents a competence model of instruction, marked by weak framing and classification with invisible principles of instruction. Weak framing, for example, refers to a balanced teacher–student interaction, in which the selection, sequence and pacing and evaluation criteria are flexibly oriented to the learner group. Weak classification refers to weak boundaries between contents and integrated instead of sectorial approaches to topics. Like the NCF 2005, ESD promotes weak framing a constructivist understanding of learning with learner-centered teaching methodologies, flat hierarchies, and the teacher in the role as facilitator. In this chapter, I analyze in how far these ESD principles can be introduced in geography lessons at English-medium schools in Pune. In order to do so, I conceptualize three levels of intervention:

1. Three ESD teaching modules were developed of which one was pretested in 14 group works at seven schools.
2. The ESD modules and the results of the group work were discussed and modified in a teacher workshop.
3. The teachers of five cooperation schools implemented one ESD module of their choice in their geography lessons.

With regard to possible changes in classroom interaction through changed teaching methods, these interventions will demonstrate how existing power relations and control mechanisms are revealed and challenged in the teacher–student relationship. However, it is beyond the scope of this intervention study to shed light on the sustainability of classroom changes and the impact on everyday school routines, as further sources of data and long-term intervention are needed. Methodical constraints and their consequences and restrictions for interpretation are pointed out in Chap. 5. The intervention study analyzes the following three questions depicting the teacher, student, and interactionist perspectives:

1. How do the teachers transmit and perceive ESD teaching methods in their class?
2. How do the students cope with and perceive ESD teaching methods?
3. How do the boundaries and communication patterns of the teachers and the students change through the ESD teaching methods?

In the following, I describe the conceptualization of the three teaching methods and their modification in a teacher workshop. Then I will analyze teacher and student interaction during the method application in group work and in classroom teaching.

8.1 Development of ESD Teaching Modules

As curriculum pre-structures contents and limits available time, teachers have a strongly restricted space of agency. I decided to tackle this by infusing weakly framed teaching modules that challenge teacher and textbook-centric methods, but yet fit into the format of a lesson and the topic of a textbook chapter. Hierarchical communication patterns led by the teacher are disrupted and discussions in group work are encouraged. The aim of visual instead of textual input is to weaken the strong classification and framing of classroom interaction. The expectations are that the teaching module affects the hierarchical relationship between teachers and students and hence allows the promotion of student argumentations on urban water conflicts in Pune. To achieve this, the teaching modules "*Visual Network,*" "*Position Bar,*" and "*Rainbow Discussion*" (Fig. 8.1) promote weak framing and classification. They fulfill the following didactic ESD principles:

1. Network thinking: Social, economic, and environmental aspects are interlinked from a sustainability perspective so that network thinking is promoted.
2. Argumentation: The students' productive and receptive argumentation skills are promoted.
3. Student orientation: Through visual inputs, reproduction is avoided and students can relate to their environment. Group work facilitates active student participation.

8.1.1 Teaching Module 1: Visual Network

To reduce oral reproduction and to disengage students' minds from textbooks, visual material is the input for this teaching method. The method of "Visual Network" was chosen as it is both a learning technique for students to interlink knowledge and a

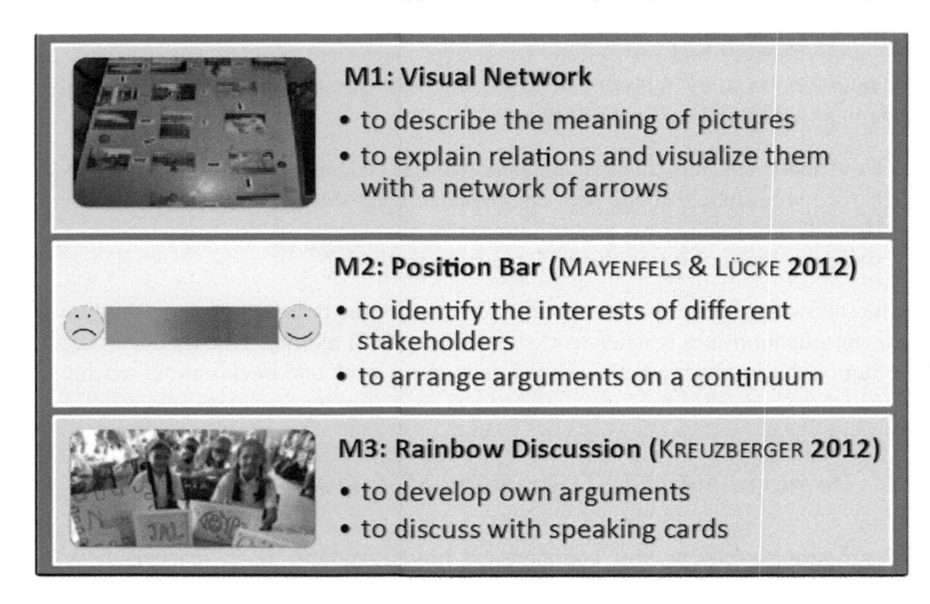

Fig. 8.1 Overview of the three ESD teaching modules

research method to analyze students' perceptions, knowledge, and creative and analytical skills. According to Vester (2002: 9), visual perceptions are assimilated and interconnected. To visualize causal relations, a structure-laying technique ("Strukturlegetechnik") by Groeben et al. (1988) is used to reconstruct subjective theories. This teaching method is used to explore individual argumentative approaches to own knowledge and attitudes (Geise 2006) and to promote systemic thinking (Bette 2013). Visual techniques, instead of solely verbal communication, have evolved in developmental research to empower local people and to facilitate access to their perceptions. As a method of Participatory Rural Appraisals (PRA), participatory linkage diagramming was developed to "discover in this way the richness not just of the knowledge of villagers, but of their creative and analytical abilities" (Chambers 1994: 963).

The central objective of a "Visual Network" is to identify argumentative coherences and explain linkages on water through pictures of students' own environment. The students receive the following task: "Arrange the pictures of Pune in a network on urban water conflicts. Use arrows to visualize relations between pictures." The learning objectives of the "Visual Network" concerning contents and skills are:

1. Students should be able to describe and arrange pictures in a network which reflects interconnected factors leading to water supply conflicts in their own city.
2. Students should be able to explain interlinkages between two pictures in form of an argumentative statement.
3. Students should be able to explain different causes (water pollution, precipitation, etc.) and effects (unequal access to water, diseases, etc.) of urban water conflicts in their own city.

4. Students should be able to write a well-structured essay explaining their opinion on urban water conflicts and their interlinkages in Pune with their own arguments.

The developed teaching module "Visual Network" consists of 12 laminated pictures taken in Pune that present different situations, for example a dead fish in the river, a woman fetching water with a bucket from a river and greenery in a gated community. These pictures can be arranged with arrows in causal relations. This activity facilitates network thinking because social, economic, and ecological dimensions are flexibly interlinked. To help students think through the picture arranging process and to follow their thoughts, two worksheets are given to the groups, one to write titles for each picture and one to explain the link between two pictures visualized by arrows. The reproduction of knowledge is avoided as only visual, and no textual input is given and students can note down their own experiences and associations from their urban environment. Detailed tasks for different lesson phases (cf. teaching module in appendix) guide teachers through a lesson introducing the "Visual Network."

8.1.2 Teaching Module 2: Position Bar

As textbook analysis and classroom observations demonstrated, students are accustomed to "right" and "wrong" answers. They expect correct answers from the teacher and seldom question statements in textbooks. Students are rarely exposed to multiple perspectives in geography lessons. However, statements sometimes cannot be classified into contrasting binaries as they express differentiated and balanced opinions. To challenge thinking in opposing categories, students are encouraged to identify the interests of different urban water stakeholders and arrange arguments on a continuum of opinions ranging from "pro" to "contra" on 24/7 water supply. This method was suggested by Mayenfels and Lücke (2012) to help students position and visualize their own and other people's points of view in a conflict. In group work, students have to discuss where to position a statement along the bar.

The learning objectives of the application of the "Position Bar" in a geography lesson are:

1. Students should be able to arrange arguments on the sustainability of 24/7 water supply in their own city on a "Position Bar" between pro and contra, so they realize that there are not only extreme positions but also balanced ones.
2. Students should be able to identify various stakeholders who could have made each argument.
3. Students should be able to take over the role of stakeholders and give arguments for their position.
4. Students should be able to express their opinion in writing and ground it firmly with pro and contra arguments.

The students are given a "Position Bar" and 15 cards with arguments of different stakeholders stating their opinion on 24/7 water supply in Pune. Additionally, a worksheet with starting sentences for an argumentative essay on the introduction of 24/7 water supply in Pune supports the formulation of their own opinion (cf. teaching module in appendix). This teaching method considers social, economic, and environmental arguments for and against sustainable urban 24/7 water supply in India. Students' receptive argumentation skills and critical thinking of multiple perspectives are challenged as they have to position statements of multiple interests on a continuum. The argumentative essay promotes argumentative writing skills on an issue relevant to the students' environment. The strength of this method is to visualize a gray zone of opinions. Nevertheless, a linear continuum is not able to visualize different levels of abstraction or norms underlying opinions.

8.1.3 Teaching Module 3: Rainbow Discussion

To react to communication partners in debates, receptive, productive, and interactive argumentation skills need to be promoted. In contrast to the method "Position Bar," in which students arrange other people's arguments, this method supports the development of students' own arguments. The method "Rainbow Discussion" is based on Kreuzberger (2012) who suggests a method of discussion and reflection in the argumentation process in which dominant and rather silent students are encouraged to participate in discussions. Heterogeneous student groups are supported through a specific number of colorful speech cards that need to be placed after a student stated her or his opinion. This way, an equal amount of speaking opportunities is given within groups. This seemed to be necessary as the pretest of the method "Visual Network" shows that single students take a dominant role in completing the task on their own, while other students are only observant.

The topic to be discussed is: "Shall we save water?" The positions are those of a politician, a common man, an environmentalist, and a technician. The learning objectives through the method "Rainbow Discussion" are:

1. Students should be able to come up with a list of pro and contra arguments on sustainable water supply in Pune with the help of information material.
2. Students should be able to present arguments for both positions, pro and contra.
3. Students should be able to back up their opinion in a discussion with relevant, objective, and well-grounded arguments.
4. Students should be able to argue against and refer to counterarguments.
5. Students should be able to evaluate arguments on whether they are relevant, objective, and well-grounded.

Materials needed are twenty colorful speech cards (in red, blue, yellow, and green), role cards assigned to each group, a stakeholder per group for whom they have to develop arguments, and worksheets with material (text, diagrams, critical comics) to support arguments for economic, environmental, and social sustainability. Additionally, on a worksheet, students write arguments in six categories (cf. ESD teaching module in appendix). One member per group gets an observation worksheet in which the process of the argumentation can be documented.

This method is the most demanding for the teacher, as it requires extensive content preparation and a fixed lesson course. Students learn to prepare, conduct, reflect, and evaluate various viewpoints of a discussion. Furthermore, the discussion on the use of sustainable resources promotes value education, which is central to ESD: Students reflect the underlying social, economic, and environmental values of their own attitudes and behavior and define for themselves whether their behavior contributes to sustainable resource use.

8.1.4 Modification of Teaching Modules in a Teacher Workshop

In geography lessons, textbook contents are recontextualized and translated through the teacher. When textbooks are replaced through teaching modules aimed at promoting argumentation skills, patterns of pedagogic practices in geography lessons are challenged. A workshop on ESD was conducted to discuss principles and methods for student-centered, argumentative geography education in general and these three ESD teaching modules. The objectives of the teacher workshop were:

1. To communicatively validate preliminary findings from classroom and group observations, interviews and textbook analyses
2. To evaluate and modify three conceptualized ESD teaching methods
3. To examine teachers' perspectives on the barriers and opportunities of ESD principles and methods in their own lessons based on concrete methods
4. To encourage teachers to introduce one method in their own lessons.

During the workshop, the participating teachers were highly engaged and motivated, which was demonstrated in their high oral participation and focused attention. Although teachers feel they have some access to supportive teaching material besides textbooks, they perceive the introduced teaching modules as opportunity for students not to study too many facts and details, but to practice group work, which is usually not integrated into classroom teaching. As the teachers are used to the role of a lecturer rather than a facilitator, they are worried about multitasking as they have to explain, maintain discipline, keep students' concentration, and observe the class at the same time during group work. However, they felt their school management and the students' parents will be supportive to focus on the topic of water from an argumentative perspective and not a fact-oriented approach as it has been the practice for examination preparations. The fact that geography, as part of the social sciences, is not a core subject made it easier to experiment with and create some space apart from textbook teaching. Besides insights into teachers' perspectives on ESD, their comments and suggestions for the modification of the teaching methods aimed at simplifying and structuring the method for implementation. Teachers' suggestions and concerns included the following:

M1: Visual Network

1. The method is applicable from lower to upper-class level and to a number of topics.
2. Headlines and papers from magazines and newspapers can be used, which make the module cheaper for teachers and students.
3. A question, rather than a task should be posed.

M2: Position Bar

1. Multiple stakeholders with multiple statements at the same time are confusing, especially since some statements could be from different stakeholders. It would be easier if stakeholders had only one statement.
2. The argumentative essay sheet should be separate for different stakeholders.
3. Students may think of stakeholders as stereotypes.

M3: Rainbow Discussion

1. More textual information should be provided prior the introduction of the method.
2. The teacher should define groups and roles based on the students' capabilities, i.e., the observer should not be a biased person.
3. One student should write the minutes of the discussion.

The suggestions for M1 were incorporated into the teaching modules' application. The first suggestion of M2 was discussed and finally not incorporated, as one stakeholder can have different opinions and statements, an aspect which students should be exposed to and deal with. Suggestions 2 and 3 for M2 were meant to be commented on and discussed by the instruction teacher to the class. For M3, additional textual information was added to the method. Suggestion 2 was not considered, as every student should have the same opportunity to take up different roles required in the group work, and groups should also be adjusted in exceptional cases if they do not work well together. The third suggestion was incorporated, and one student will be assigned to write the minutes of the discussion.

8.2 Analysis of ESD Teaching Module Implementation

8.2.1 *Changing Space and Resource Use in Pedagogic Practice*

During regular classroom lessons, the teacher stands in front of the students, but for all ESD teaching modules, the teacher asked students to sit in groups, and they changed their seats and moved together. When classroom space is seen as materialized effect of societal power relation, the ESD teaching modules initiated a differential use of

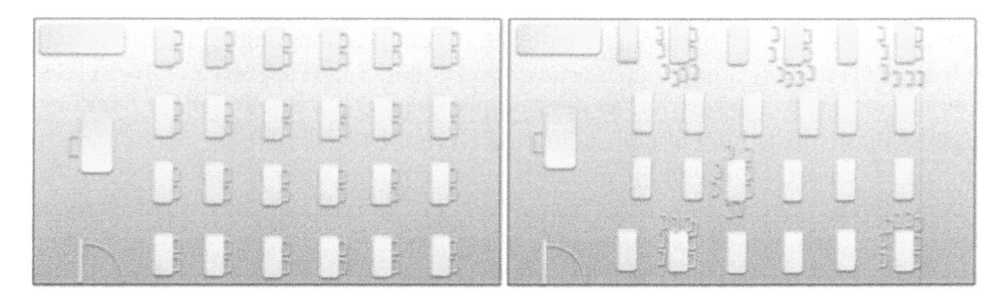

Fig. 8.2 Classroom arrangement during regular geography lessons (left) and during the ESD teaching module 2 (right) in school 2

space and resources as well as a visible change of power relations. As Massey (1994) describes the relational space concept, in which she conceptualizes "space as constructed out of interrelations" (Massey 1994: 264), the new seating arrangements represent an open and flexible use of space that does not reproduce dominant structures, but rearranges existing structures in that no hierarchically higher position is visible. In contrast to unidirectional communication patterns in regular classroom lessons, circular arrangements offer multi-directional interaction. However, as the tables and benches are screwed to the floor in four of the five schools, they could not arrange circles to sit. Instead, groups of around eight students were standing, sitting, or leaning around one table, whereas the teacher was still sitting in the front (cf. Fig. 8.2).

Because the desks are screwed to the floor and not movable in classrooms, the teacher in school 3 took students to a separate room where they could sit in groups on the floor. Similarly, the teacher in school 4 decided to move the lesson with 60 students in the courtyard, where students could sit in six circles on the ground. This shows that teachers felt that they could not implement group work in regular classroom settings and took the effort of an irregular logistic rearrangement, which created a setting different from usual classroom teaching. The desks were only movable in school 5, so that the students could move two tables into a block and students could sit around it. This visualizes how in normal classroom sessions, communication is unidirectional, while in a circle it is multi-directional (cf. photographs in appendix). The former reinforces the central position of the teacher as transmitter of the curriculum, while the latter can present a democratic, participative teaching approach, if it facilitated as such by the teacher.

The ESD teaching modules require additional efforts from teachers and students: They need to collect information and search for resources in the Internet and in newspapers, and they need to prepare material, such as to photocopy worksheets and to cut individual comments. ESD changes lesson planning and preparation for teachers and also students, since teachers and students do not have to prepare additional resources when using regular textbook-centric lessons. The amount and diversity of resources teachers used for the implementation of the teaching modules varied from no additional material except the teaching module worksheets and materials provided at the teacher workshop (school 2), to teachers undertaking the effort with their students to collect newspaper articles and conduct Internet research for several days to prepare the lesson and to develop posters (school 1 and school 5). Some teachers implemented the lesson as introduced in the teacher workshop (school 2), while other teachers adapted the method to the current chapter in class so that it fits in the syllabus of the week, e.g., a "Visual Network" module on the topic of garbage (school 3). Teachers were encouraged to reinterpret the teaching methods according to their needs. However, this led one teacher to choose pictures with different types of waste, which made it difficult for students to logically arrange the photographs with arrows in a cause and effect network. This demonstrates the need to train teachers in choosing resources and using material that are adequate to the contents and methods they would like to apply in their teaching. Another teacher told the students to collect newspaper articles on either pollution or (over-) population prior to the lesson, which they arranged to draw arrows and relations with chalk on the ground of the courtyard (school 4). The arrows explained changes in the population and environmental and social consequences of overpopulation and thus were a successful adaptation of the method to a different topic. This demonstrates that the method of implementation in the five schools varied according to the teachers' choices on the setting and spatial arrangement, and the ability to use additional material and the amount of time invested to prepare the lesson.

One teacher made an exceptionally great effort to apply one teaching module in her teaching as she took some extra lessons to prepare the teaching module. The weekend before the arranged classroom visit, she called me and desperately divulged that her principal did not want to allot additional lessons for the preparation of the ESD teaching module. This demonstrates that teacher's agency to implement new teaching methodologies in pedagogic practice was constrained by the existing school structures that are marked by strong classification in the selection, sequence, and pace of educational contents.

The group dynamics varied between the schools as well. In school 3, students were arranged to sit according to their gender, because the teacher felt they would work more comfortably and hence more efficiently. Particularly in this lesson, however, students seemed not as interested and engaged as in other schools where students were in gender-mixed groups. While younger female and male students of standard 7 at school 4 were engaged and vividly discussed how to arrange the pictures, in observed gender-mixed group discussions, several girls and boys spoke predominantly, others did not say anything and some boys fought about who could speak. In all schools, observations were that particularly articulate students were participating while other students remained silent and hardly participated. It also happened that single motivated students took over the group work by themselves and hence dominated the outcome of the method.

8.2.2 Interlinking Thoughts: The "Visual Network"

The ESD teaching method "Visual Network" offered students space to discuss and creatively develop and reflect their own thoughts with their peers. This also highlighted how teaching contents and methods need to be better linked to students' perceptions to improve their argumentation skills. The "Visual Network" was introduced in schools 3, 4, and 5 and stands in contrast to the teaching methods M2 and M3 introduced in school 1 and 2, as the teacher had a minor role in instructing and steering the class, which resulted in comparably more space for student interaction. As group work is central in this method, teachers are in the background and give students space to develop their own ideas. In contrast to the other methods, in which a single student presented group results in front of the class, presentations were done in groups standing in front of the class. Although mostly one student spoke to explain their "Visual Network," other students supported with additional comments. Even when students were stumbling, the teacher did not interrupt the students, as their presentation visualized their understanding of the topic, and hence, there were no obvious corrections necessary in content. However, this also led to irrelevant statements not being questioned, e.g., one student gave additional information he learned from the textbook that "70% of the human body is made of water." The student tried to include as much information as he knew about water, without selecting the information on its relevance to the topic, which could have been a pointed out by the teacher to help students focus on the task.

While some groups arranged the "Visual Network" with many interlinkages (cf. photographs in appendix), other students used a linear approach, showing the flow of water from the dam via the pipes to the people, and eventually leading to pollution. One group presented a network in which the flow of water from dams, through pipes to people in bottles or hand pumps is provided. They added that "poor people do not get water." Another group of students developed two separate "Visual Networks": one "positive," in which water is conserved and sufficiently available to people, and one "negative", in which water is polluted, which leads to water scarcity. Mostly, students give monocausal relationships while arranging and explaining their network with the phrase: "This leads to this…, this leads to that…" this indicates that students had difficulties to interlink several factors by other means than consecutive sentences.

All students focused on water pollution as an environmentally relevant factor in the "Visual Network." Students perceived the pictures to be strongly related to their environment and their everyday experiences of water and environmental pollution, as they are exposed to it and see it in the city. One picture with garbage at the riverside was labeled in a generalizing manner: "Dirt all around." This indicates that the topic has a negative connotation, which might lead to discouragement in regard to environmental action. However, students stated that the method met their interests and motivated them.

Concerning argumentative content in students' explanations, students' answers varied in complexity. On the one hand, simple coherences indicate a false understanding of water scarcity and water sources, such as "Less water is there in river, so people are actually fighting for water." (M1:P14, 3:37). On the other hand, some students could relate and integrate environmental, social, and economic aspects, for example:

> Because of pollution, these fishes, as they are not good for health, are sold for less price in the market, and poor people take these fishes for eating. And their health is deteriorating and also they have to go miles for water, too, for they don't have water for drinking. From that we can conclude that conserve water. (M1, 4:59)

Most students stated that garbage and pollution in slums lead to water scarcity, which leads to effects they explained by pointing to the picture of a woman fetching water with a bucket. Another group explained that "dirty water from a slum leads to river pollution and polluted river banks" (St88_S5, 1:30). This group further explained water harvesting and added a drawing of domestic rainwater harvesting, answering "how?" and "why?" this should be done.

Many students assume that water in rivers is used as drinking water and polluted river water impacts people's access to drinking water. As, however, river water is used for bathing, washing the dishes and clothes, it could impact people's health. For example, one student's comment shows insights of the effects of water pollution, however, with the false assumption of river water as source for drinking water:

> Because there is garbage on the river banks, the river water is getting dirty and because of the river water is getting dirty, the river water is getting polluted, and because the river water is getting polluted, there is no water to drink. So the ladies and the people have to go miles and miles away to bring water. (M1:P12)

The assumption that washing cattle with river water crucially contributes to river pollution was observed in a primary school in urban Pune, in which a poster painted by the teacher illustrated the causes of water pollution in rivers: activities such as bathing, washing clothes, and cleaning the dishes, and also buffalos, takes place in the river water, which is also used for drinking, demonstrated in a woman carrying water on her head for drinking purposes (cf. photograph of teacher's drawing in annex). On the one hand, this shows how misconceptions on the source of drinking water and relevant pollutants are transferred to the students through textbooks and teachers. On the other hand, this leads to students not reflecting on their own contribution to water problems.

Students often conclude water problems as problems of "the poor," in that they generate the problem and bear the burden of the consequences. Repetitive statements such as "village people pollute the water because they wash their buffalos in the river" (M1:P8, 6:34) indicate that "dirty water" or water pollution is associated with others. Interestingly, some students did not differentiate between rural areas and slums, as they misinterpreted the picture of a slum with rural areas. Although students realize the differential availability and access to water due to socio-economic disparities, they express this in generalizations on "poor farmers" with "poor slum dwellers." Blaming the people living in villages or in slums indicates what (2007; 1985) calls "Othering": The perception of others polluting visibly the water, while the own garbage is properly disposed. This may also mirror the public discourse in which state representatives' governing practices are strongly marked by "Othering," as Zimmer (2011) demonstrated on waste waterscapes in informal settlements in Delhi.

Interestingly, some students mention the importance of education for environmental awareness, and the spatial inequities concerning access to education. This, however, also leads to one student again blaming river water pollution to the "uneducated," those in slums and villages: "People living in the slums are not highly educated, so they don't know more about water pollution. They like just dump garbage into the water." (M1:P8, 6:34).

Several students mentioned industrialization, along with the waste of people living at the banks as source of pollution, affecting again the poor as they "drink" the water:

> Industrialization happening in Pune, many of these industries do throw out their chemical wastes in the river itself and along with this, people who live by the banks, they are adding to it and the river just gets dirtier and dirtier [...] And many people living down there also suffer from many problems caused by these chemicals because they use this water for drinking. (M1:P9)

Other students mention corruption and the priority of economic interests as reason for environmental indifference, "This is all the world of corruption, all the people [...] run to get to money, and they are carelessness about water." (M1:P4, 7:29). The responsibility of tackling urban water problems is accredited to the government, as they should regulate industrial wastewaters:

> I think that as the Pune citizens are facing water problems, they should complain to the government as they are having water shortages. [...] As the industrial waste waters are often in rivers which is harmful for aquatic animals, government should take action on it. (M3:P30)

Students also present strong normative statements which reflect their moral perspective, or rather what they have been taught in school and society. Abstract appeals, similarly as found in textbooks (cf. Chap. 6), were stated in a group presentation in school 5, in which students cite together a text they wrote: "We need water. If we lose water, we lose hope. Hopes of life, hopes of living. So save water" (St85_S5, 0:20). One student concludes that water should not be polluted and rather should be conserved. Another group wrote, among other phrases, on their poster: "If our mother earth provides us with what we need, why can't we provide her with one thing she needs?" This mirrors an environmentalist attitude, which supports environmental action and critical perspectives on the causes and consequences of water pollution. This indicates the need to strengthen students' argumentation and also action competence on the differential access and use of water leading to water conflicts in their own environment.

8.2.3 Reinterpreting Strong Classification and Framing: The "Position Bar"

The teaching modules developed according to ESD principles challenge the power relations and control mechanisms in pedagogic practice between teacher and students. The teaching module "Position Bar" visualizes the difficulty to change from strong classification to weak classification in pedagogic practice: When boundaries between right and wrong answers and categories of thoughts are clearly distinguished, placing an answer along a continuum is challenging for both, teachers and students. Similarly to students, teachers are not used to giving answers on a continuum with multiple possible perspectives. Teachers and students are used to a strong classification, which is sustained during the teaching module implementation in class by reinventing categories (pro/contra/in between) for the "Position Bar," instead of using it as a continuum. Similarly, a "discussion" is interpreted as "presentation," and the strong framing of pedagogic practice is maintained. In the following, the teachers' and students' interpretation of and their interaction patterns during the introduction of the ESD teaching module "Position Bar" are analyzed based on classroom observations in school 2 on Feb 7, 2013 and its video analysis.

During the implementation of the "Position Bar" in a geography lesson, the teachers adapted the module to simplify it for their students. Instead of explaining the continuum, one teacher told the students to place the answers on the question of whether 24/7 water supply is needed in Pune according to three categories:

> T2_S2: This is your 'Position Bar'. On one end, the face is smiling. On one end, it is sad. In the center, you can draw a face, which is neither happy nor sad. It is the center one. […] Write your group number on the sheet. Secondly, number the arguments. The third step is to read each comment carefully. Decide whether it is positive, negative or in between. This has to be positioned on the 'Position Bar'. So if you get 24/7 hours of water, and it is positive according to you, this number you put here. […] And also kindly think who could have said these comments. Who are the stakeholders? If you have any questions, kindly ask now itself. According to you, these dialogues are spoken by which people? Who are the stakeholders? Read them, label them and put them on the 'Position Bar'.

Hence, the teachers structure the teaching module according to their well-known categories of agreement and disagreement and further complement these categories by a third category in between the poles. This contradicts the initially introduced method's purpose of arranging statements on a continuum with gradual progression toward two poles with a positive and negative answer. Furthermore, teachers provided copies of the statements but did not ask the students to cut each one, but told students to write the numbers of the respective statement on the "Position Bar" in one of the three categories. This had the effect that students could not flexibly move the comments on a continuum throughout the group discussion. Although the thought-provoking method of a "Position Bar" was not established according to the initial idea, the teachers transformed the method in an easier approach for their students. The pre-structure of two opposing perspectives "pro" and "contra" was sustained, and teachers invented a new third category "in between." The purpose of arranging multiple opinions along the "Position Bar" apparently needs a scaffolding approach. Yet, the teachers "translated" this method for their students in a way that allowed the students to learn that there is an opinion "in between."

In the following, the process of the lesson is described in detail to understand teacher and students' agencies in interpreting the new ESD teaching method "Position Bar" in form and content.

Excerpt of a geography lesson in which the teacher introduces the "Position Bar" method in school 2 at class 9

During group work, one student reads the comment and suggests where to place the comment on the "Position Bar." Another student disagrees and suggests another position. All students seem interested; however, only single students actively participate, whereas the other students observe the process. Meanwhile, the teacher repeats her instructions to ensure the students understand and walks around to ensure the students are working. Even though the groups have not finished their work, the teacher tries to increase the pacing of the lesson and asks the students to complete the next task, which is identifying and matching each statement to a stakeholder [common man, politician, environmentalist, technician] they assume would advocate such an opinion. [...]
After this, single students are instructed to present their arguments for or against 24/7 water supply in front of the class taking the position of the teacher. The first female student presenting the "common man" argues that a certain amount of water should be freely available for everyone, so that only the overuse of water is paid for. She further argues that if every common man was educated how to use and conserve water, everyone should have proper access to water and contribute to equitable use. When the student asks her peers: "What do you think?", the students have hardly time to respond as the teacher adds: "Do you agree with her statements?". When the students agree by a unison "yes," the teacher explains: "You should not answer out of friendship. This is a project, so kindly cooperate with her. Just out of grudge, don't say yes or no." The teacher asks again: "So do you agree with the opinion with the common man or not?" One student repeats the argument of the presenting student, which shows the students are used to reproduce the exact words usually spoken by the teacher. The teacher does not further challenge the student to express their own opinion and instructs the next student to present the arguments in front of the class.
The next presenter, a female student states as an "environmentalist" that 24/7 water supply should not be promoted as there is not enough water in the area and water supply is expensive. The teacher interferes, "we already receive ample supply of water," and the student repeats the sentence. The teacher further adds to her statements that the environmentalist claims that everyone has to conserve water.
After the other two male students finished speaking, the teacher summarizes their arguments. [...]
Only after all four students presented their arguments on the specific stakeholders, did the teacher ask the students for their opinion. One student got up from her seat to explain that she would like people to pay for excess use. The teacher interrupts her to paraphrase her statement and adds that appropriate use and water harvesting systems should come up for future practice. [...] (S2:M2)

The lesson with the method "Position Bar" shows that the teacher strongly controls classroom communication during the ESD modules, which attempt to introduce a weakly framed classroom discussion. The method's purpose of a discussion is interpreted by the teacher in form of student presentations. Four students take the teacher's position and present the statements of a respective stakeholder. This changes the method's idea of an interactive, argumentative discussion between students into four separate student presentations that are not related to each other's arguments. A unidirectional communication from the teacher to the student rather than a multi-directional discussion among different students is practiced. This demonstrates the difficulty for students to express their own opinions and to respond to prior given statements as well as the difficulty for teachers to recognize this and facilitate a discussion.

The classroom space is rearranged for group work at the "Position Bar," in that each group gathers around one table. For the discussion, the regular classroom arrangement of tables facing toward the front is kept with the only change of a student, and not the teacher, standing and speaking in front of the class. As the student takes the position of the teacher, unidirectional communication patterns remain. The teacher only completes the student's arguments, while the other students remain passive. Statements are not criticized and contested, and students are not sufficiently encouraged to add their perspectives and feedback. Although the teacher is aware of the students' uncritical agreement to opinions of their peers, she hardly gives the students time to critically think about the arguments presented or to state their own opinions and feedback to each other. Instead, the teacher remains in full control of the communication and strongly frames it by adding further facts to complete students' statements. In the meanwhile, the students seem more excited about the new classroom arrangements such as forming groups or single students presenting from the front of the classroom, as if being in the position of the teacher, rather than listening and reacting to the content of what was said.

The difficulties to formulate own opinions or the opinion of a stakeholder, as well as to react to arguments, shows that a gradual approach to this new teaching method would be helpful. This needs more time, and both teachers and students require methodological guidance on how to promote their argumentation skills.

8.2.4 Focusing on Sequence, Order and Form: The "Rainbow Discussion"

Explicit power relations and control mechanisms in prior observed classroom teaching are partly changed to implicit through the implementation of ESD teaching module 3. The geography teacher of school 1 chose to introduce a discussion on the topic: "Should we save water?" to a geography lesson in class 9, which was filmed. In the following, the lesson is shortly summarized and analyzed.

The lesson was not a normal lesson in which a teacher infused a new methodology, but rather an exceptional and formal situation. During the lesson, the principal and the senior teacher sat in the back to observe the lesson. The students were from two different classes, class 9 and class 10. The teacher chose the best students and arranged them to sit three to a bench that normally sat two students. Some students in the back row were assigned to hold up posters during the course of the lesson. Other students held a nameplate of their group in the four Hindi meanings of water: "jal," "pani,", "neer," and "toya," and designed posters with newspaper articles, and drawings. At the beginning of the lesson, the teacher greeted the students very formally and introduced me. She told the students that I will observe their performance and see whether the students follow the teaching module appropriately. My presence was interpreted as evaluating students' performance on stage. Students worked in groups to discuss and to prepare pro or contra arguments for a class discussion which gave them space to share their thoughts with their peers. However, after the group discussion, the students returned to their seats in the classroom and listened to student presentations of final statements without a discussion taking place.

The observation of the lesson on the "Rainbow Discussion" at school 1 demonstrates a similar interpretation of a discussion through the teacher as during the introduction of the "Position Bar" in school 2. The lesson consisted of presentations in which the teacher did not facilitate student interaction. The students presented statements, which they had almost learned by heart, in an order of speakers prior determined by the teacher. Hence, the idea of a discussion could not take place as the students were not interacting, but only presenting prepared statements. As there was a clear course of speakers arranged, the rest of the students did not pay attention to the content, as they were not asked to give their opinion. The idea of the speaking cards, which should be used spontaneously in group discussions and placed in the middle, was transferred into speaking cards which were raised to show that someone was the next to speak, according to the order determined by the teacher. This shows that the intention of speaking cards was not transferred into use.

Before the group work started, the teacher chose a particularly strong student to be the "group leader" to present a final statement after the group discussion. Through this, a clear hierarchical position within the students was established. Other students could participate in the group discussion, but knew that they would not have to present the results in class. Thus, only a small number of students, who were selected as group leaders by the teacher, could participate in the discussion. The selected group leaders were strong, articulate students, while the other students were deliberately excluded from the discussion. Hence, the teacher enforced and maintained hierarchical relationships within students.

Students' speeches were descriptive and not argumentative, as they usually consisted of statements followed by the appeal not to overuse or pollute water. The statements and the conclusion, however, were not clearly causally linked. Some students said that they would like to save water; however, they did not explain the causes, means, and consequences of saving water. One student argued for water harvesting stating that water could be conserved and reused from the rooftops and pipes. The content of the speeches was well-prepared and contained detailed information. However, this was not formed into an argumentation as it was not causally interlinked, e.g., on the historical development of water supply in Pune or of water sample results taken by the PMC and the varying water level of the Khadakwasla Dam based on information from a DNA newspaper article. One student cited the PMC stating that leakages are responsible for 25% of the water loss. Some statements were political and criticized the government's capacities: "PMC does not have a groundwater map, so they are unable to identify leakages" (St92_S1, 27:47). Most comments, however, did not refer to each other and therefore did not represent a reaction to other students' statements. After each statement, students said "thank you" which stressed the formal ending of their speech and obedient behavior which leads to a lack of interaction.

After each group presented their statements, the teacher asked each group leader to read their conclusions. The following sequence shows how social hierarchies were reaffirmed in the classroom and how the teacher keeps the central agency:

Excerpt of a geography lesson in which the teacher introduces the "Rainbow Discussion" method in school 1 at class 9

Teacher "Ok, so now, groupwise, the leaders will tell their conclusion. Group leaders, first give the name of your group. Yes, Neer."

Student 1 "Neer. 1. We have 70% of water and only 29% of land. 2. Citizens should report PMC as they face a problem. 3. The shortage of water because of misuse by some strong political personalities. 4. As more dams are built, much more water can be saved. 5. We should use another technique to generate electricity rather than using water. 6. People increase day by day the importance of rivers."

Teacher "Next group"

Student 2 "After listening to this, we have come to the conclusion, that the example of Pune is not economic sustainable. Pune wastes sources, which should be for conversation. Water excess is strong. We suggest water conservation."

Teacher "Next"

Student 3 "We would like to give the message that people who are aware of water, so go to government, will do and take some action. We should save our environment and help our government. Save life, save water" (S1:M3, 3:48)

This example demonstrates how the teacher steers through a predetermined teaching structure without opening up the discussion. Although the students are given time to speak, the teacher strongly frames the pacing of the concluding remarks by only saying "next." Students do not receive any feedback or questions from the teacher, and the statements are not interlinked. This leads to a series of disconnected statements rather than an interactive, argumentative process, which puts the order and form, not

Table 8.1 Students' evaluation of the teaching modules (scale: 1 = I fully agree, 2 = I agree, 3 = I partly agree, 4 = I disagree, 5 = I fully disagree)

Statements	M1 ($n = 79$)	M2 ($n = 44$)	M3 ($n = 25$)
I did not understand the different arguments	3.93	3.75	4.19
I felt that I learnt about different perspectives	1.86	1.77	1.50
I could relate the topic to myself or my environment	2.06	1.77	1.85
I feel now that I can judge the topic better than before	1.79	2.05	1.81

the content in the focus, as the statements are fragmented and do not correlate to the other groups' statements. Thus, the statements are not tested on their argumentative content. Consequently, new evaluation criteria for the selection of knowledge, which stress the importance of argumentation, need to be explicitly addressed when ESD principles are to be implemented in pedagogic practice.

8.2.5 *Relating to Student Perceptions: Questionnaire and Interview Results*

Students' perceptions of the teaching modules were evaluated through student questionnaires after the lessons in which the teaching modules were introduced. Students stated to have understood and learnt about different perspectives in all three modules. Students felt they could relate to the topic and weigh different opinions on urban water conflicts (cf. Table 8.1).

The visualization of the urban water supply situation and its related conflicts through pictures, comics, and newspaper articles promoted network thinking on the causes and effects of differential access to water as well as students' capabilities to feel empathy with the poor. One student stated:

> It's better to visualize than to read, because reading doesn't show us the condition of people. I have read it many times in newspapers like that people suffer due to shortage of water [...] But here I saw it live and I am really touched by it. I also think now that something should be done for this. (M1:P23, 2:34)

Many students stated that the module raised their interest in water conservation. One student explained, "due to photos, we understood that how [the water] cycle is actually formed, from the rainwater to [...] how it gets polluted and we need to conserve it, we got it into our mind [...]. For us conservation is very much necessary." (M1:P18, 4:52). Students were enthusiastic about the pictures of their own city and could relate them to their own environment. Interlinking pictures from their own city helped students to obtain a better understanding of urban water conflicts in Pune and realize that their own behavior is relevant in their society. Other students recognized the need to communicate on environmental and social issues:

> the lesson was interesting because it rises awareness in students for saving the environment. I feel that we should have interactive sessions with pictorial information and we must exchange our views to one another. (M1:P30, 0:40)

Especially, more silent students who do not speak much in class are encouraged to participate in discussions: "usually I talk very less but due to this discussion I got to talk in front of my friends" (St16_S4). This shows that participative teaching approaches help students to open up, as they only speak in front of their friends and not the whole class. However, the students' enthusiastic feedback on the methods must be viewed carefully in regard to their excitement about the interaction with the researcher, similar to the positive feedback on the teaching methods in the questionnaires.

The evaluation of the three ESD teaching modules (cf. Table 8.2) accentuates the three ESD principles of network thinking, argumentation, and student orientation differently (Fig. 3.7). The "Visual Network" mirrored students' (mis-)conceptions on water pollution and their often culturally shaped argumentations. The "Visual Network" facilitated the greatest space for group interaction and creativity. The new focus of expressing their own opinions differs from studying ready-to-learn scientific concepts, which usually characterizes geography lessons. The "Position Bar" challenged both teachers and students, as categories of thought were not prescribed. The "Rainbow Discussion" demonstrated that students find it challenging to formulate argumentations and that teachers do not facilitate that students relate to others' statements in debates. In conclusion, every method aimed to weaken the strong classification and framing of pedagogic practice. However, their interpretation was strongly dependent on the selection, sequence, pacing and evaluation of knowledge and skills in regular pedagogic practice, which resulted in shaping the outcome of the geography lessons in which new ESD teaching modules were introduced.

8.3 Summary of the Intervention Study on Opportunities of ESD Principles in Pedagogic Practice

This chapter depicted the development, modification, and analysis of the three ESD teaching methods "Visual Network," "Position Bar," and "Rainbow Discussion" in five English-medium schools in Pune. The aim of these teaching modules was to weaken the strong classification and framing of pedagogic practice observed in geography lessons. This served to demonstrate in how far ESD principles of network thinking, argumentation and student orientation could be introduced in geography lessons despite the structuring institution regulations. The results shed some positive light on the transformative potential of pedagogic practice through ESD.

Due to the openness of the teachers and students toward changing teaching contents and methods, it was possible to introduce ESD on water conflicts in Pune,

Table 8.2 Evaluation of the teaching modules on the three didactic ESD principles

ESD modules/ESD principle	M1: Visual Network	M2: Position Bar	M3: Rainbow Discussion
Network thinking	Access to student perceptions, which mainly do not correspond to scientific concepts	Continuum was unusual for students because there was not a clearly defined answer	Water pollution and effects are explained, however, to present interlinkages are challenging
Argumentation	Interrelations are often religiously or culturally coined	To arrange arguments challenged the binary nature of answers usually taught in class (correct or false)	Discussion is reduced to separate presentations, despite extensive research, students stay on descriptive level
Student orientation	Group interaction, new perspective on own city, suggestions how to act	Students could find compromises on continuum	Strong group interaction also outside of the classroom, strong local community focus

which aimed at transforming an authoritative into a participative teaching style. This led to a change in the use of space and resources in the classroom, as the setting was rearranged and group work encourage student participation. This encouraged multi-directional instead of unidirectional classroom communication and a flexible and open use of space. The circular classroom arrangements of space weakened the hierarchically higher and central position of the teacher who usually reinforced unidirectional communication.

The weakening of classification and framing through the ESD teaching modules change the role of teachers and students and their communication in the classroom. Teaching methodologies promoting argumentation and network thinking relate strongly to the students' environment, e.g., local water conflicts, and increase students' interests and motivation. The amount of students speaking increases, but also the role of the teacher changes into a facilitator and observer, rather than a presenter. Teachers have to apply different teaching skills, as they need to elaborate or correct students' comments and relate them to each other. This results into facilitating a communication flow different from the prescribed order in textbooks. As a result, the ESD modules require that teachers have to integrate longer lesson planning processes to prepare their teaching. This process further changes teachers' roles, as they need to plan and develop new selection and evaluation criteria of knowledge, in an ideal situation with the support of the syllabus, textbooks, teacher trainings, and additional teaching resources. Although longer teaching processes are necessary to promote students' skills to transform descriptive statements to argumentative ones, the results' analysis showed a great potential for teachers to link to students' perceptions to facilitate their argumentation skills.

The ESD teaching module "Network Thinking" demonstrated a wide range of students' answers interlinking social, environmental, and economic aspects of water conflicts. To prevent oral reproduction, visual inputs encouraged productive well as receptive argumentation skills in group discussions. Students' argumentations highlighted generalizing assumptions, e.g., that the poor contribute to water pollution, and that the government is to be blamed for not taking responsibility. This "Othering" (Spivak 1985, 2007) shows that the understanding of different perspectives as well as a critical reflection on one's own perspective should be promoted in classroom teaching. The "Position Bar" demonstrated the difficulty of weakening classification in classrooms. While a continuum was introduced, categorical thinking of binary concepts of "pro" and "contra" was sustained, while the teacher added a third category "in between." Similarly, the strong focus on sequence, order and form were also sustained in classroom teaching. The speaking cards in the teaching module "Rainbow Discussion" were meant to include weaker students in debates, but their intended use was not realized as the teacher pre-arranged the sequence of statements selected students had to present. The presence of head teachers enforced a formal situation and promoted the students formal and obedient behavior, rather than a spontaneous discussion. All three ESD teaching methods demonstrated how declarative regional geographical knowledge of the topic of water could be turned into processual knowledge on water conflicts through participatory teaching methods. The implementation of the teaching modules also showed that ESD principles are translated differently as intended by teachers and students, and that power relations and cultural values are sustained to some degree in weakly framed teaching methods. This indicates that the introduction of ESD teaching modules requires a comprehensive teacher training approach, in which cultural constructs and power relations are addressed, so that teachers become aware of these and decide how to integrate new teaching methods into their teaching approaches.

References

Bette, J. (2013). Erdölinduzierte Urbanisierung – Das Beispiel Dubai. Ein Beispiel zur Förderung der Systemkompetenz mittels Struktur-Lege-Technik. *Praxis Geographie, 43*(11), 16–21.

Chambers, R. (1994). The origins and practice of participatory rural appraisal. *World Development, 22*(7), 953–969.

Geise, W. (2006). Zur Anwendung der Struktur-Lege-Technik bei der Rekonstruktion subjektiver Impulskauftheorien. In E. Bahrs, S. von Cramon-Taubadel, A. Spiller, L. Theuvsen, & M. Zeller (Eds.), *Unternehmen im Agrarbereich vor neuen Herausforderungen. Schriften der Gesellschaft für Wirtschafts- und Sozialwissenschaften des Landbaues e.V., Bd. 41* (pp. 121–131). Münster: Hiltrup.

Groeben, N., Wahl, D., Schlee, J., & Scheele, B. (1988). *Das Forschungsprogramm Subjektive Theorien. Eine Einführung in die Psychologie des reflexiven Subjekts*. Tübingen: Francke.

Kreuzberger, C. (2012). Regenbogen-Vierer - Diskussion mit Redekarten In A. Budke (Ed.), *Kommunikation und Argumentation*. Braunschweig: Westermann.

Massey, D. (1994). Space, place and gender.

Mayenfels, J., & Lücke, C. (2012). Einen Standpunkt "verorten" - der Meinungsstrahl als Argumentationshilfe. In A. Budke (Ed.), *Kommunikation und Argumentation* (pp. 64–68). Braunschweig: Westermann.

Spivak, G. (1985). The Rani of Sirmur: An essay in reading the archives. *History and Theory, 24*(3), 247–272.

Spivak, G. (2007). Can the subaltern speak?

Vester, F. (2002). *Unsere Welt - ein vernetztes System*. München: Deutscher Taschenbuchverlag.

Zimmer, A. (2011). *Everyday governance of the waste waterscapes: A Foucauldian analysis in Delhi's informal settlements*. Bonn: ULB.

Chapter 9
The Transformative Potential of Pedagogic Practice

Abstract This chapter introduces the concept of *transformative pedagogic practice* as an intermediate, transitional form of pedagogy which gradually links the conflicting priorities of reproductive and transformative pedagogic practice. It illustrates how educational policies such as Education for Sustainable Development (ESD) pose an opportunity to rethink the structuring elements of pedagogic practice. ESD can be situated in the attempt to change existing patterns and forms of teaching contents and methods toward empowering students to become critical, responsible citizens. Based on the empirical findings (Chaps. 6, 7, and 8), the three central research questions of this book are re-visited: (1) How do academic frameworks in geography education in India relate to ESD principles? (2) How do power relations and cultural values structure pedagogic practice in Indian geography education, and how are these linked to ESD principles? (3) How can ESD principles be interpreted and applied to pedagogic practice in Indian geography education? I reflect upon the practicability of the didactic framework and the transferability of the theoretical and methodological approaches to other contexts. Further, I discuss strategies and perspectives for implementing ESD in Indian geography teaching.

Within the setting of a changing educational landscape in India, this book originated from the question as to how pedagogic practice in geography education relates to the principles of the transnational educational policy Education for Sustainable Development (ESD). The results of this study demonstrate how the perspective of ESD provides a normative lens for the interpretation of processes and factors shaping pedagogic practice. The principles of ESD, realized through the didactic approach of argumentation (Budke 2012a; Budke et al. 2010a, b), help to identify challenges and opportunities for promoting a critical environmental awareness in the formal educational system in India. The descriptive–analytical concepts of Bernstein (1975a) depict how the implementation of ESD results in a transformation of strong classification and framing principles of observed pedagogic practice into weaker classification and framing. The study further demonstrated how cultural values of learning and teaching guide educational stakeholder negotiations on geography curricula, syllabi, textbook development, and educational reforms. The understanding of how peda-

S. Leder, *Transformative Pedagogic Practice*, Education for Sustainability,
https://doi.org/10.1007/978-981-13-2369-0_9

gogic practice shapes and reproduces power relations is necessary in order to better link educational reforms to social reality.

The empirical analysis found a gap between the transnational educational policy ESD and its possible implementation in classroom realities. It also identified structural and institutional conditions constraining educational reforms of teaching methodologies to be implemented in pedagogic practice. Currently, there is not enough space for teachers' agency to allow the re-contextualization of ESD principles, e.g., to adapt natural resource use and management topics to students' perspectives and to include an argumentative approach to local water conflicts. In this situation, the implementation of teaching methodologies needs to build on existing capacities and institutions, and a reform of the educational system from within. These include the centralized apex body NCERT, and state authorities responsible for teacher education and textbook development bureaus. Government agencies are crucial for strengthening teachers' agency through teacher training and the provision of textbooks and innovative teaching materials to introduce environmental education into classroom teaching alongside the curriculum. Furthermore, influential think tanks and NGOs (e.g., CSE, CEE, and TERI) influence the discourse on the role of environment education for the formal educational system and need to be considered as valuable stakeholders and partners in revising curricula and developing teaching material.

In the following, the research questions are re-visited by reinterpreting the empirical results in regard to the theoretical framework. The theoretical and methodological approaches are reflected on in terms of their applicability to the context in Pune as well as their transferability to other contexts. Subsequently, the concept of transformative pedagogic practice is derived. The book ends with a conclusion and future directions of research as well as strategies for implementing ESD principles in Indian geography education.

9.1 How Does ESD Challenge Pedagogic Practice?

The empirical data shows how ESD as a transnational policy challenges and subverts existing pedagogic practice and social hierarchies within Indian classrooms. ESD as a democratizing teaching approach (UNESCO 2005b, 2009, 2011) stands in contrast to the traditional hierarchical structures that are reproduced in the country's myriad of educational contexts (Government of India 2004; Chauhan 1990). The code theory of Bernstein (1975a, b, 1990) helps to identify the fundamental challenges that ESD poses to teacher–student interactions. The demands of ESD and its current teaching methods and content can be described with competence-oriented and performance-based models of pedagogy, since the selection, pacing, sequence, and evaluation of knowledge and skills fundamentally differ in the two approaches. The analysis demonstrates that ESD objectives (de Haan 2008; UNESCO 2005b; Tilbury 2011) promote weak framing and classification (cf. Bernstein 1990) in classrooms, while existent pedagogic practice in the observed lessons upholds strong framing and classification. Strongly classified teaching contents and clearly defined roles and spaces between teachers and students, as well as educational authorities, contrast the weak classification as promoted by ESD, in which teaching contents are meant to be dealt with in an integrated manner, and open questions and tasks extend thinking beyond pre-defined categories. Weak framing refers to the communication style, e.g., to discussions as promoted by ESD, in which students have control over the selection of contents, whereas in observed pedagogic practice, the teacher strongly frames classroom communication, e.g., by taking most of the time to speak. This indicates that the understanding of power relations and cultural values of teaching and learning are fundamental for making ESD relevant to pedagogic practice. These empirical findings are discussed in detail by revisiting the three subordinate research questions.

9.1.1 How Do Academic Frameworks on the Topic of Water in Geography Education in India Relate to ESD Principles?

In the Indian educational system, two contrasting types of pedagogies are transmitted at different institutional levels. On the one hand, the National Curriculum Framework 2005 (NCF) presents an open curriculum and a competence model of pedagogy, which promotes pedagogic principles similar to those of ESD. The NCF 2005 promotes critical thinking and a student-centered teaching approach, which encourages weak framing of teacher–student interactions. On the other hand, principles in syllabi and textbooks represent a performance model of pedagogy, which aims at the reproduction of prescribed information in textbooks. This indicates that the instructional discourse around child-centered pedagogies is barely translated into syllabi and textbooks, and ultimately, pedagogic practice. That curriculum and geography textbooks are incongruent has also been found by Kuckuck (2014) who examined how spatial conflicts are addressed in 22 textbooks in the two German states of North Rhine-Westphalia and Brandenburg. Considering the principal role of the textbook for structuring subject knowledge (Rinschede 2005: 350, cf. Sect. 4.1.4), the lacunae between curriculum and textbooks are particularly relevant to consider in relation to improving educational quality (Crossley and Murby 1994; UNESCO 2005a). As earlier studies of the World Bank (1990) demonstrated, the provision of pedagogically sound and culturally relevant textbooks is cost-effective for improving the overall educational achievements in emerging countries. Particularly in India, where a "textbook culture" has already been identified by Kumar in 1988, the framing of the textbook strongly influences the manner of pedagogic practice. These observations are in line with other studies (Clarke 2003; Sriprakash 2012; Berndt 2010), which examined how textbook-centric pedagogic practice limits the space for student participation. The approach in syllabi and textbooks is more influential on authors and teachers than the NCF 2005: They prescribe geographical topics and objectives for geography lessons and guide teachers with respect to geographical teaching contents, methods, and student evaluation. Due to their foremost importance, a systematic textbook analysis examined how principles of ESD and argumentation skill development relate to national and Maharashtra state geography textbook contents and methods relating to the topic of water.

In the examined geography textbooks, a performance model of pedagogy is transferred, rather than a competence model as promoted by ESD. Textbooks are strongly classified as they contain a high number of chapters and follow a regional, and not a problem-oriented approach as suggested by ESD. Chapters on water in geography textbooks display a strongly fact-oriented and fragmented approach to teaching contents. Water is depicted as a fixed commodity, rather than a contested resource, which is constructed differently by groups. Hence, water is not presented as an area of conflict. This contrasts the argument of Stöber (2011) that conflicts should already be addressed in primary school, as they contribute to the formation of opinion and the ability to be critical. The author found that multi-perspectives of different stakeholders do not become clear in German geography textbooks, and that conflicts are presented as a factor hindering development, rather than conflicts being addressed inherently. A similar finding by Kuckuck (2014) confirms that a conflictual approach to resources is barely explored in German geography textbooks. The author found that on only 2.16% of 4445 German geography textbook pages, conflicts of any kind are mentioned. However, the resource of water is a prevalent topic in the few cases where conflicts are addressed, as 25% of all thematic approaches to conflicts tackle water (Kuckuck 2014: 72). Interestingly, the regional examples of water conflicts in German textbooks are mostly located in the Middle East to exemplify geopolitical conflicts. Other water conflicts are addressed in irrigation and the Three Gorge Dam on the Yangtze River in China. In line with the textbook study conducted, these results demonstrate both the need and opportunity to address multi-perspectives on natural resource conflicts for textbook development in India, Germany, and possibly elsewhere.

The textbook analysis in this study emphasizes that differential access to water is addressed in human geography textbooks without elaborations on the causes and consequences of such. In physical geography textbooks, technical explanations on the hydrological cycle and river locations are prevalent. These contents are repetitive over the sequence of school years, and more facts are added, while interlinkages and an increase in the complexity of cognitive skills to promote learning progression are lacking. This separate approach to human and physical geography hinders a critical perspective on water in terms of sustainable development. The need for critical perspectives on resources instead of homogenizing them can be supported by evidence from Stöber (2007). He examined the relevance of presenting heterogeneous group interests by analyzing the case of a violent outbreak on textbook politics in Gilgit (Northern area of Pakistan) in 2004. The case presents how political the presentations of homogenous religious practices in textbooks can become within a religiously heterogeneous population.

This is also something the ESD principle of network-thinking attempts, as it favors an integrated and problem-based approach to topics at the human–environment interface. However, the examined geography textbook chapters in this study include definitions and mainly one-dimensional statements. For example, in a geography textbook for class 6 in Maharashtra (M-Geo-6), types of rain and their distribution are tackled, but the relevance of water to different users and the environment is not addressed. Similarly, critical aspects such as the reasons for acid rain are not elaborated on, apart from stating it through a definition. The high numbers of separate chapters of textbooks promote a strong pacing within teaching, which do not allow a stronger focus on competence development as promoted by ESD. This contradicts the postulation for life skill education (WHO 2001) on the front page of several textbooks.

Controversial topics at the human–environment interface such as urban water conflicts are not elaborated in textbooks. The topic of water is relevant in all geography textbooks. However, the topic of water is not depicted in the form of a conflictual approach, in which differential perspectives on water access and control are displayed. Hence, the controversy of water in students' urban environment is not presented, and the development of students' own productive and receptive argumentation skills are not facilitated. The multiple-choice questions at the end of each chapter do not promote critical thinking, but the reproduction of knowledge in a limited number of words. This highlights the performance model of pedagogy in contrast to the competence model.

Teaching approaches which facilitate higher-order thinking skills are crucial for enhancing students' critical thinking and related characteristics, such as truth-seeking, open-mindedness, self-confidence, and maturity (McElvany et al. 2012; Barak et al. 2007). As Barak et al. (2007) demonstrated in a longitudinal case study, dealing with real-world problems, encouraging open-ended discussions, and fostering the asking of questions contribute to critical thinking abilities. Cognitively demanding learning materials are those in which pictures are integrated into texts, as combining two sources of information requires high cognitive abilities (McElvany et al. 2012). However, higher-order thinking skills promoted through productive and receptive argumentation skills are predominantly not encouraged in texts and instructional pictures in the examined geography textbooks. Instead, mostly the lowest level of cognitive learning is promoted through reproductive tasks, which do not deepen the understanding of a topic. Water pollution and differential access to water are stated, but causes, effects, and consequences not explained, and the questions of "why?" and "how?" are not addressed. Similarly, sustainable development is defined, but not further engaged with and transferred to students' environments. In one exercise in the NCERT geography textbook for class 12 (N-Geo-12_IndPpEc), students are instructed to elaborate on social water conflicts, but the textbook itself does not provide an example, methodological guidance, or more in-depth knowledge which could help students to understand the nature of social conflicts over natural resources. This indicates the need to review how higher-order thinking skills could be promoted in textbooks, as these also promote better text comprehension and memory (McElvany et al. 2012).

The textbooks do not always cover the wide diversity of students' backgrounds, as mostly urban middle-class students are depicted. Linkages to students' own experiences and the expression of their own perceptions are not sufficiently encouraged. This might have a negative impact on students' motivation; as for many students, these depictions may be meaningless. Students may feel helpless as information is too generalizing. Students expressed a "blaming attitude" toward the government and the "poor" with respect to water problems, which might indicate that student-oriented and action-oriented approaches to environmental challenges are not sufficiently promoted in textbooks. Similarly, the study on environmental concepts by Bharati Vidyapeeth Institute of Environment Education and Research (2002) revealed a lack of in-depth information, local references, and activities encouraging environmental action in textbooks of all subjects.

The knowledge and skills in textbooks demonstrate that the paradigmatic shift toward child-centered pedagogy of the NCF 2005 from earlier NCFs has mostly not been transmitted into selection criteria for syllabi and textbooks. This indicates that the translation of the academic frameworks in the NCF 2005 to syllabi and textbooks needs a comprehensive conceptual and methodological framework. Concrete didactic criteria could systematically guide syllabi designers and textbook authors for choosing the selection, sequence, pacing, evaluation of knowledge and skills. The gap between the postulations of the NCF 2005 and the syllabi and textbooks can also be attributed to the hierarchies and power relations in the form of strong classification between educational institutions, which hinder cooperation within and between committees. Because of the different responsibilities and context-specific challenges for each state and district, pedagogic principles are interpreted differently and do not trickle down.

Curriculum and textbook development is a lengthy process, which needs coordination between authors. Geography textbook authors are not always teachers themselves, and many stated that they were not trained in didactic principles, not to mention ESD. This may be the reason for inconsistent didactic approaches and a lack of continuity between topics and textbooks of different school years. Lacunae and fragmentation in textbooks could also be linked to the necessary translation of textbooks into English, which depend on the linguistic capacities of translators. Hence, one objective could be to strengthen the reciprocal interdependence of curricula and syllabi designers, textbooks authors, and teachers to develop texts, tasks, and capacities to work toward the aspired transformation of pedagogic practice.

In conclusion, the contents and methods regarding water in geography curricula, syllabi, and textbooks hardly relate to the principles of ESD and argumentation skill development. The lack of multi-perspectives in natural resource conflicts and a low amount of tasks which promote argumentation was, however, also observed in German textbook analysis (Stöber 2011; Kuckuck 2014; Budke 2011). This demonstrates the need to revise geography textbooks to promote opinion-formation (Stöber 2011) in both India and Germany and beyond. The examined syllabi and textbooks with prescribed contents and methods show a strong framing and classification of knowledge and skills which contradict a competence mode of pedagogic practice, which ESD promotes. The syllabi and textbook task analyses indicate the expectations of students' conformity through the exact repetition of textbook contents.

9.1.2 How Do Power Relations and Cultural Values Structure Pedagogic Practice on Water in Indian Geography Education, and How Are These Linked to ESD Principles?

Teaching methodology in observed pedagogic practice contrasts with the postulations for environmental education and Critical Pedagogy in the NCF 2005 and its numerous position papers (National Council of Educational Research and Training 2009). This demonstrates that the institutional regulations conveyed in textbooks and syllabi, and to a lesser extent the NCF 2005, influence pedagogic practice. Another factor-structuring pedagogic practice is teachers' perception of "good geography education," and their knowledge and vision in teaching (cf. Kennedy 2006). As in an observed geography lesson, a teacher praises a "good class" (geography lesson in S2, cf. Sect. 7.1.3), she does so as students nod and agree with her speech, and do not ask questions. The focus on reproduction transmits respect and authority, rather than questioning and critical thinking. ESD and the promotion of argumentation skills intrude with social order, that is, in the case of India, highly stratified relations by age, caste, class, gender, and other social categories (Mandelbaum 1975). Although India is a diverse society, the educational system focuses on competition based on those who best reproduce prescribed knowledge. This demonstrates how ESD's objectives for critical thinking challenge pedagogic practice in which affirmative and obedient behavior toward the teacher is valued.

Hence, teachers' role in classroom teaching corresponds to the textbook, which structures classroom interaction in a closed mode. Observed pedagogic practice centers around the teacher who transmits textbook contents to the students. The latter indicates a strong framing of classroom communication (Bernstein 1990) in which students' recognition and realization rules are limited to word-by-word reproductions of the teachers' words. Knowledge is evaluated on the basis of reproducing correct answers from textbooks, whereas students often utter single-word rather than elaborate explanations. The teacher speaks a great deal, whereas students' communication skills are hardly promoted. Students recite textbook content and what the teacher says in unison. The principles of selection, sequence, pacing, and evaluation of knowledge and skills in pedagogic practice are closely linked to principles found in textbooks.

The analyses of these classroom observations with the concepts of Bernstein (1990) reveal the mechanisms of social reproduction and indicate the limited space in this strongly framed context to develop skills such as critical thinking as per the objectives of ESD principles (UNESCO 2005b). On the one hand, the analysis confirms a culture of teacher-centered pedagogic practice which has been studied in India from as early as the 1980s (Kumar 1988) until more recent studies (Sriprakash 2010; Clarke 2003). On the other hand, this study provides a more in-depth understanding of these observations, particularly on the role and perceptions of the teacher in pedagogic practice. Although these are described in other studies, power relations in pedagogic practice have barely been investigated in detail, particularly in regards to transnational educational policies. For example, Berndt (2010: 244) observed primarily inquiry–response cycles and teachers who dictate texts, while these observations have not been theoretically grounded with an approach engaging with power relations and cultural values relayed in pedagogic practice. Therefore, this study further developed approaches of Sriprakash (2012) who analyze power relations with the terminology of Bernstein (1975b, 1990) by adding additional theoretical perspectives on the transformative potential of transnational educational reforms (Freire 1996; Thompson 2013; Vavrus and Barrett 2013; Manteaw 2012).

The identification of the explicit boundaries in pedagogic practice, e.g., between the standing teacher and the sitting students facing the teacher, depicts the expected respect and obedience of the learners toward the teacher. The teacher performs on stage what the state expects her or him to deliver to the students—manifested in the textbook (Goffman 1971). This performance mode of pedagogic practice is visualized by the use of physical space and resources in classrooms. The tables screwed to the floor represent the immovable and inflexible order of the classroom, which channels teacher–student communication. As mentioned in the studies by Thompson (2013) and Kanu (2005), educational reforms shall not only consider activities and curriculum revision, but also cultural values such as discursive practices and indigenous approaches to learning, which are, for example, expressed in classroom arrangements.

The performance mode in observed geography lessons promotes reproduction, rather than transformation through pedagogic practice. Bernstein's terms help to understand and differentiate these observations in regard to ESD: The type of knowledge and skills prescribed through the syllabus and the textbooks does not leave enough time and space for students to participate in class, and limit the teacher's agency and control over teaching contents. What Kumar (1988) labeled a "textbook culture" in Indian classrooms remains relevant today. The textbook is one political instrument for the state authorities to exercise political and cultural power over teachers and students (Kumar 1988: 453). Textbooks depict what counts as legitimized knowledge, while at the same time the curriculum is a means to access public thought (Bernstein 1975b: 87). The pre-structured contents and methods in syllabi and textbooks represent norms and values as well as constructions of learning and teaching.

The teacher is the transmitter of the pre-structured selection of knowledge as depicted in textbooks. Due to the strong textbook orientation, the teacher is also the timekeeper of the sequence of knowledge to ensure the coverage of the textbook within lessons and school years. Since the teachers' main task is to pace geography lessons with the aim of completing the dense syllabus, they often have to overlook the students' specific questions and interests. This leads to a strong framing of classroom interaction, as there is little space for students to develop own argumentations and to select and sequence knowledge themselves, as demanded by ESD. This contrasts multi-directional communication patterns through which ESD principles can be realized.

Despite the amount they speak, teachers are not given control over the selection, sequence, pacing, and evaluation of knowledge and skills in geography lessons as individual lesson planning and material preparation are not included. Hence, the textbook has a steering and authoritative role (Pal 1993: 8) in pedagogic practice as it governs teacher–student interaction. This implies that the role of the teacher as the central agent of cultural transmission is neglected (Batra 2005: 4348) and has to be strengthened before ESD can be introduced at the classroom level.

The student is the receiver of knowledge, as she or he has to reproduce the knowledge of the teacher and the information in the textbook. This position at the end of the educational chain does not enable students' empowerment, as students' skills and creative participation are not sufficiently promoted as demanded by ESD. Learning is perceived as a collective and not as an individual process, an observation which also Kanu (2005) states for South Asia, based on an ethnographic study on curriculum change through "Western" skill development attempts. Despite schooling processes which transmit collective approaches to learning, students diversely perceive, reproduce, and subvert these intentions of conformity under the influences of peer cultures, media, and marketing, as Thapan (2014) outlined in ethnographic studies of schooling in India (cf. Leder 2015). Because of these diverse perceptions within a heterogeneous society, educational stakeholders have to integrate principles of schooling, which empower students with skills and knowledge to become critical, transformative citizens.

The power relations and cultural values in pedagogic practice, as observed in geography lessons at English-medium schools in Pune, pose a challenge to the principles of ESD. The definition developed for this study, "ESD is concerned with pedagogic practice that promotes critical consciousness through argumentation to empower for debates and decision-making on sustainable natural resource use by facilitating learner-centered, problem-posing, and network-thinking teaching approaches" (Sect. 3.3), presents an ambitious objective. The term to "empower" indicates that ESD interferes with power relations, and that it is a multi-dimensional and relational process (cf. Leder 2016). Empowerment is defined as "the process by which those who have been denied the ability to make strategic life choices acquire such an ability" (Kabeer 1999: 346). This can be linked to the ESD objective (UNESCO 2005b) to promote strategic decision-making on social and environmental concerns. Within a society in which teachers' concepts of self are "constructed and enacted in terms of social relatedness, interdependence, and communality with others" (Kanu 2005: 502), group work as promoted by ESD can contribute to a collective notion of empowerment. This indicates that ESD principles can be integrated and linked to cultural values.

However, ESD cannot be reduced to classroom activities, as it implies a comprehensive approach building on educational research. This can identify how perceptions on the role and the authority of the teacher and the textbook can be integrated with the constructions of learning and teaching of ESD. ESD, based on the didactic framework developed for this study, favors an argumentative approach facilitating multiple perspectives, critical and network thinking, and learner-centered teaching methods. The findings indicate that observed teaching contents and methods in pedagogic practice hardly parallel ESD principles, and therefore, the implementation of ESD is an opportunity for a gradual transformation in which both teachers and students are introduced to translate ESD according to their understanding.

9.1.3 How Can ESD Principles Be Interpreted Through Argumentation on Water Conflicts, and How Can This Approach Be Applied to Pedagogic Practice in Indian Geography Education?

The implementation of ESD principles such as the promotion of critical thinking and the ability to make decisions on sustainable lifestyles (UNESCO 2005b; Tilbury 2011; de Haan 2008; Schockemöhle 2011) in pedagogic practice requires a didactic approach which promotes a critical awareness of multiple perspectives on resource access, as well as negotiation skills. Building upon the argument that a critical consciousness (Freire 1996) is a pre-condition for environmental action, this study demonstrated how teacher and student interaction shifted toward weaker framing by linking ESD to the didactic approach of argumentation (Budke 2012a, b; Budke et al. 2010a, b). The selection, sequence, pacing, and evaluation of knowledge were weakened with three developed ESD teaching modules focusing on differential access to water resources as well as addressing urban water conflicts in the students' own environment of Pune. Both teachers and students welcomed this change, and the debates, which the teaching modules encouraged, provided space for students to express their opinions. This emphasizes the valuable link of the developed didactic framework for ESD through argumentation on resource conflicts in geography education (cf. Sect. 3.3). Considering the societal relevance of argumentations (Habermas 1984; Kopperschmidt 2000; Andrews 2009; Wohlrapp 2006), the didactic approach of argumentation facilitates reasoning of diverse perspectives without prescribing the content of argumentations. Arguing with people of higher rank (caste, class, gender, age) is generally considered culturally undesirable, despite India's tradition of skeptical argument, the acceptance of heterodoxy, and divergent viewpoints (Sen 2005; cf. Mandelbaum 1975; Marrow 2008). However, the approach in this study can be adopted and interpreted for pedagogic practice in the Indian context, as it offers an open way to interpret the objectives of the NCF 2005. Since the students developed argumentations addressing marginalized perspectives such as the limited access to water for informal settlements, the ESD teaching modules promoted awareness on the "subaltern" (Spivak 2008). The ESD teaching approach could be extended to facilitate direct communication and exposure of students to multiple stakeholders, including marginalized communities, and another study could examine the degree to which this influences students' argumentation on the sustainable development of water resources, and whether this translates into a change in agency. The implementation of the three ESD teaching modules indicates how ESD principles can be facilitated through argumentation, but further research is necessary to understand if argumentation levels and types of argumentation skills (receptive, productive, interactive) as developed by Budke et al. (2010a, b), and Kuckuck (2014) for the German context are transferable to the Indian context, as this study indicated that ESD principles are reinterpreted by teachers to a great extent.

Teachers' interpretation of the ESD principles as defined for this study (cf. Sect. 3.3) is shaped by their cultural constructs of the teacher as transmitter and the student as receiver. The implemented ESD modules oriented to promote argumentation skills relax the classification and framing through the focus on group work, but underlying cultural constructs continue to influence and subvert the teaching modules' goals. The analyses of power relations throughout the implementation process show that the authoritative position of the teacher and the strong framing of teacher–student communication are only slightly altered through the ESD modules which were designed and introduced in a teacher workshop. This becomes apparent, for example, in the use of the speaking cards, which were meant to facilitate a spontaneous student debate and to encourage shy students to participate. However, they were only applied to indicate the teacher's prescribed order of students speaking, which subverted the intended meaning of them. Furthermore, the teacher does not take the facilitating role as intended, but strongly steers through the discussion ensuring the pre-arranged sequence of presenting students. Articulate students selected beforehand were presenters, while the other students only listened. The intended discussion turned into a presentation of short student statements, in which students did not impulsively interfere to contradict other students' statements. These results correspond to similar observations of hierarchically structured classroom interactions by Sriprakash (2012: 166) who observed how a teacher controls discussions with strong framing of the content, sequence, and pacing of classroom interaction. This control of the teacher continued throughout the introduction of child-centered educational reforms and "joyful learning" (Nali Kali) in rural government schools in Karnataka, India.

The application of the ESD teaching modules accented the focus on order, sequence, and form in classroom teaching. The intervention study further shows how culturally rooted norms, perceptions, and practices of teachers and students' interaction shape the adoption process of ESD principles (cf. Levin and He 2008). This is, however, not specific to pedagogic practice in India. In Germany, for example, ESD is considered as a school development program, which builds on the school culture, the organization of a particular school and pedagogic practice (Programm Transfer 21, 2007). This indicates that not "a single model of excellent teaching" (Vavrus 2009) is needed, but rather international educational research concerned with the cultural, economic, and political dimensions of learner-centered education in different contexts (Vavrus and Barrett 2013; Thompson 2013; cf. Sriprakash 2010; Mukhopadhyay and Sriprakash 2011a; Leder 2014).

The implementation of ESD teaching modules demonstrates how the construction of the teacher and the students' roles strongly persist, as the students' relationship of respect toward and distance to the teacher is maintained. The focus on a factual understanding and the perceived need for a "right answer" continue to exist within the weakly framed ESD teaching modules. The teacher reinterprets the continuum of the "Position Bar" along which statements on 24/7 water supply are meant to be arranged. Using binary categories of "pro" and "contra," the teacher introduces a third category "in between" to simplify the answer categories for the students. This reflects both teachers and students' difficulty in differentiating perspectives as tendencies along a continuum and not in clearly defined categories. It also reflects teachers and students' insecurity in terms of not knowing the "right answer."

The "Visual Network" initiated multi-directional communication within the classroom through group work and disrupted hierarchically structured communication patterns between the teacher and the students. Visual inputs further diverted students' attention from reproducing textbook contents to expressing their own perceptions. When the group discussions were meant to be transferred to the classroom level, this turned back into a formal and hierarchically structured presentation in which one student takes the teacher's position in front of the class to state the groups' opinion. This indicates that the use of classroom space as well as uni-directional communication formats continue as usual, with the only exception that one student takes the teacher's role. The other students maintain their passive roles. This indicates that the participatory approach envisioned by ESD only pertains to single students. The rest of the class unanimously agrees with their peer's presentation, a behavior observed in earlier lessons held by teachers. This demonstrates the difficulty of introducing discussions to classrooms.

Teachers only give students a limited amount of time to react to their peers, which indicates that they do not expect any objections from the students. Instead of discussing the students' presentation, the teacher is focused on asking the next student to present. This shows that discursive and engaging approaches to teaching contents are hardly promoted in geography teaching. The teacher's focus on maintaining order rather than paying attention to content has also been demonstrated by Kumar (2005). In the introduction of his book "Political Agenda of Education. A study of colonialist and nationalist ideas," he notes a prevailing focus on maintaining order, and teachers and students' lack of interest and curiosity in a museum's exhibition or teachers' indifference when students throw stones at animals in a zoo. He attributes this to "the concept of knowledge that underlies our system of education [which] stops teachers from perceiving "order" in its extended sense" (p. 15). This indicates that it is not only skill development which both teacher and students require, e.g., on how to facilitate discussions. Instead, a comprehensive approach to and discussion on weakening the strong framing and classification of pedagogic practice is needed. Educational stakeholders who were interviewed for this study at hand explain barriers to such an approach with the challenging conditions of high numbers of students with diverse backgrounds in one classroom, limited access to learning resources and teachers' limited exposure to teaching methodology training (NGO 10). Next to these structural challenges, reflections on students and teachers' skill development have to engage with the distance and communication patterns between the teacher and students which structure the culture of pedagogic practice.

Educational stakeholders, teachers, and students perceive ESD topics and related argumentation skill development as a critical issue in urban English-medium schools. Observations in other school forms (vernacular government schools, schools of different trusts and educational boards, etc.) as well as communicative validation with educational stakeholders at Maharashtra state and national level confirm the relevance of rethinking principles of pedagogic practice. This refers to a paradigm shift from a "textbook culture" (Kumar 1988), and "teaching the text" (Pal 1993: 19) to student-centered teaching methodologies and critical thinking. These are encouraged in the opening pages of textbooks, e.g., by introducing "life-skills education" (World Health Organization 2001) in Maharashtrian geography textbooks, however, the implementation of these postulations present a challenge. Nevertheless, this endorses the didactic relevance of ESD principles despite their amended intention through the translation by teachers. Teachers assured that the ESD modules demonstrated entry points for interpreting ESD for geography lessons, but found it challenging to maintain class discipline and to deal with insecurities of knowledge, as they are used to the clear answers given in textbooks. Teachers on the one hand feel bound to prepare students for standardized exams, and on the other hand criticize students' lack of interest beyond textbook answers. The students expressed that the introduced ESD teaching modules triggered their empathy, awareness, and enthusiasm for the topic of water conflicts. They stated that they do not like to study answers from textbooks. This indicates both teachers and students' interest in and appreciation of student-centered teaching methods.

The introduction of ESD principles demonstrates how existing power relations and constructions of teaching and learning are challenged. However, the results indicate that changes in pedagogic practice require the stepwise introduction of student-centered and argumentative teaching approaches, which address teachers and students' culturally shaped learning and teaching perceptions. The empirical study sheds light on the multiple factors, processes, and perspectives involved in translating ESD to pedagogic practice in urban India (Fig. 9.1). The example of English-medium schools in Pune demonstrates the multiple challenges of re-contextualizing ESD to the formal educational system in India and its local classroom realities. The analysis shows that there is a great translation gap between the fact-based textbooks and the aspirations of the open National Curriculum Framework 2005. The vision of the NCF 2005 stresses the importance of promoting ESD principles and a constructivist approach and moves beyond textbook-based learning. However, classroom observations in Pune and the syllabus and textbook analyses show that in practice, teaching is marked by strong classification and framing focusing on student performance. The results further point out the central role of the teacher in the implementation process of ESD, which was also the result of the meta-analysis of over 800 meta-analyses on the most influential indicators for students' achievements by Hattie (2009). As further studies have pointed out, teachers' specific professional knowledge and competences are necessary for implementing pedagogic principles in general and ESD principles in particular (Hellberg-Rode et al. 2014; Schrüfer et al. 2014; Baumert and Kunter 2006). This points out the relevance of effective teacher training through which teachers become aware of and can change their pedagogic practices, something which Morais (2002) examined by applying Bernstein's theory of pedagogic discourse. Furthermore, the content and sources of teacher's personal practical theories (Levin and He 2008) and the identity and practices of teachers influence pedagogic practice, e.g., whether characterizations such as inquiry-oriented teaching styles are favored (Eick and Reed 2002). Hence, ESD demands for a fundamental

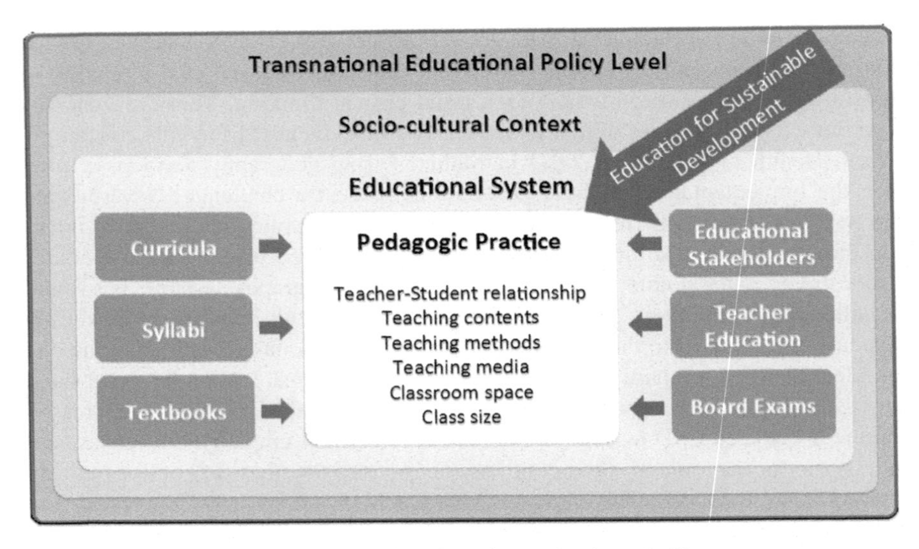

Fig. 9.1 Multi-level translation of ESD in pedagogic practice (own draft)

change of strongly socio-culturally shaped classroom interaction processes depend highly on the teacher, along with spatial classroom structures, as well as the pacing and sequence of learning content and methods in geography lessons prescribed in textbooks.

The translation of ESD as transnational educational policy encompasses engaging at different interlinked levels, i.e., the socio-cultural context, the educational system level, and the level of pedagogic practice (cf. Fig. 9.1). The adaptation of different pedagogical approaches such as ESD intervene with existing cultural values of learning and teaching, which are transmitted within the educational system. Pedagogic practice encompasses the complex teacher–student interaction relationship, teaching contents, methods, and infrastructure such as teaching media, as well as the classroom space and class size. Pedagogic practice is influenced by the selection, sequence, and pacing of knowledge and skills as prescribed in academic frameworks, that is, curricula, syllabi, and textbooks. Board examinations shape the evaluation of students' performance toward memorization and rote learning. Teacher education and educational stakeholders with a government and NGO background further influence pedagogic practice. Thus, such a systemic approach (Benedict 1999) is necessary to attribute a transformative character to ESD. This depicts how pedagogic practice is embedded in the wider societal framework and highly dependent on the rules and norms guiding the manner of enforcement through various stakeholders within the educational system. Thus, ESD requires a theory and process of change to be initiated and furthered through the actions of government institutions at the national level. Although this study focuses on pedagogic practice in particular, it includes a holistic view highlighting the institutional change and educational network support necessary for the transformation of pedagogic practice. Steiner (2011: 396) proved with a project on teacher training that changes in institutional conditions rather than single classroom introduction of ESD can be successful. Thus, comprehensive approaches are needed to attribute ESD with a truly transformative character.

9.2 Reflection on the Theoretical Framework for Transformative Pedagogic Practice

The combination of different approaches in a theoretical framework to examine transformative pedagogic practice links concepts beyond disciplinary boundaries. This book sought a transdisciplinary synthesis between geographical developmental research, geographical education research, and sociology of education research. This offered multi-level analyses of changes in pedagogic relations through transnational educational policies. The descriptive–analytical approach of Bernstein's Sociological Theory of Education (1975–1990) with the Critical Pedagogy perspective by Freire (1996) as a normative orientation theoretically grounds the empirical study on pedagogic practice in geography lessons in the socio-cultural context of India. Bernstein's approach was developed from the analysis of the role of language and communication in sustaining class relations in England. With his structuralist approach, he analyzed how power relations and control is exercised within pedagogic practice. This approach was also used in other studies in India and other developing contexts (2011, 2012; Barrett, 2007) and allowed an in-depth analysis of the status quo of pedagogic practice, which focuses on reproduction.

Linking Bernstein's concepts to the Brazilian educationist's vision for social transformation through pedagogic practice enabled the development of a grounded approach for dialogue learning and an approach for the promotion of a critical consciousness in highly stratified contexts (Fetherston and Kelly 2007; Freire 1996; Giroux 2004; Kumar 2006). The combination of these two approaches allowed for the studying of possibilities for a paradigm change toward transformative pedagogic practice, while engaging with the socio-cultural context and power relations shaping pedagogic practice.

Since Bernstein as a British sociologist of education and the Brazilian educationist Freire developed their theories and concepts in an abstract manner, they proved applicable to the analysis of pedagogic practice in India, as they left space for context-specific interpretations. However, Bernstein's sociology of pedagogy (1975a, b, 1990) has two shortcomings which have been addressed in this study. Firstly, Bernstein's concepts do not grasp the dynamic and powerful influences and the directing regulations of the educational institutions on pedagogic practice (cf. Archer 1995). Therefore, this study has examined institutional regulations through curricula, syllabi, and textbook analyses. The comparison of these analyses with observations of pedagogic practice demonstrates how they closely guide teachers and their communication with students. These results exemplify the mechanisms of social reproduction which relay societal structures, socio-cultural constructions, and power relations into pedagogic practice (cf. Sriprakash 2012).

The embedding of this approach in the educational system (Archer 1995) brings in institutions guiding the instructional discourse, as well as a multi-level reflection on the cultural relativity of concepts of learning and teaching. Between the conflicting priorities of reproduction and transformation of pedagogic practice, educational policies can be situated in the attempt to change existing patterns and forms of teaching contents and methods. The reinterpretation of educational policies is shaped by socio-cultural power relations on different levels pedagogic practice. Policies are not "borrowed," but undergo a "process of translation" (Mukhopadhyay and Sriprakash 2011b; Behrends et al. 2014; Thompson 2013; cf. Kanu 2005), which involves the contextualization and transformation of policies. The theoretical framework on transformative pedagogic practice illustrates how the transnational objective of ESD poses an opportunity to rethink the structuring elements of pedagogic practice. Through this, an isolated perspective on teaching methodologies could be preempted as the role of educational authorities, textbook development and teacher training, among others, could be reflected upon.

Secondly, Bernstein's concepts offer an in-depth analysis of power relations and control mechanisms in pedagogic practice and educational discourses, but do not directly address the question as to how concretely the existing relations could be changed. Similarly, Critical Pedagogy formulates normative postulates on the aims of education (cf. Freire 1996), however, equally fails to answer the question as to how to facilitate change in existing teaching methods and contents (cf. Bernstein 1990). Hence, this study combined skill development approaches from geographical teaching methodology research, which explicitly address these questions with the concepts of Bernstein (1975a, b, 1990) and Freire (1996).

To translate ESD objectives into pedagogic practice, the didactic approach of argumentation helps guide the wider perspective of endorsing critical thinking in classrooms. The selected didactic approach of argumentation skills developed by Budke et al. (2010a, b) promotes a participative and critical approach to teaching. Students are given space to voice their perceptions and to integrate multiple perspectives in meaningful discussions. By linking argumentation to ESD, the status quo and opportunities for implementing ESD in geography education could be analyzed in-depth with the application of the theoretical concepts of Bernstein (1975a, b, 1990) and Freire (1996).

The application of the abstract concepts developed by Bernstein (1975b, 1990) to the observed pedagogic practice in English-medium schools in Pune helped to identify constraints for the implementation of educational reforms and related teaching methodologies within the Indian context. Bernstein's concepts as well as the approach of Critical Pedagogy based on Freire (1996) help to understand the principles and norms of both ESD and pedagogic practice in India. This indicates that this theoretical framework could be valuable to apply to similar contexts as well, particularly in developing countries with highly stratified societies. The implementation of ESD can initiate a transformation of strong classification and framing principles of pedagogic practice into a weak classification and framing. This transition moves away from dichotomous progressive/traditional teaching categorizations to new forms of mixed pedagogic practice of weak and strong classifications and framings which was also observed by Morais (2002). Such foundational change to less restricted teaching structures emphasizes how pedagogic practice is deeply intertwined with social structures and cultural values (Sriprakash 2012; Fetherston and Kelly 2007). This perspective explains why educational stakeholders used terminologies similar to ESD principles, but interpreted them in various ways in practice. Throughout the empirical study, educational stakeholders, teachers, and students were motivated and enthusiastic about implementing principles of ESD in classroom teaching, syllabi, and textbooks. To an extent, the introduction of promoting argumentation skills subverted existing power relations between the authority of the teacher and the student as a perceiver of knowledge. However, political resistance against "Western" values and changes of the respected position of the teacher as "transmitter of knowledge" can be expected within India. As there has been resistance against the modernization attempt of the NCF 2005 (The Hindu 2005, Aug. 7), even greater opposition can be expected if ESD principles are to be integrated into curricula, syllabi, and textbooks.

9.2.1 Derivation of the Concept of Transformative Pedagogic Practice

Bernstein's code theory (Bernstein 1975b, 1990) helps to gain an in-depth understanding of how ESD relates to pedagogic practice: The implementation of ESD demonstrates how a transformation of strong classification and framing of pedagogic practice into weaker classification and framing takes place, and how a new form of mixed pedagogic practice emerges (Morais 2002). Using Bernstein's terminology, ESD represents a competence model of instruction—that is, one marked by weak framing and classification. ESD underlies a constructivist understanding of learning that promotes student-centered teaching methodologies and flat hierarchies with flexible evaluation criteria and a strong focus on the learning process and on skill development; it is an advocate of invisible pedagogic practices. The teacher is meant to have the role of facilitator, and thus classroom interaction is supposed to be weakly framed. Power relations are implicit, and consequently classification is weak. In contrast to ESD, framing and classification of the observed classroom interactions in English-medium schools in Pune are strong. The introduced ESD teaching modules are meant to bridge these two modes of pedagogy. However, they were interpreted differently than intended, and a mixed form of weak and strong classification and framing emerged. Similar reflections were stated in the study by Sriprakash (2010), who observed teachers' tensions when democratizing learner-centered performance models of teaching emphasized in national and global development goals were introduced in rural Indian classrooms. She further identified how the introduction of new activities reinforces social messages of control and hierarchy relayed to the students

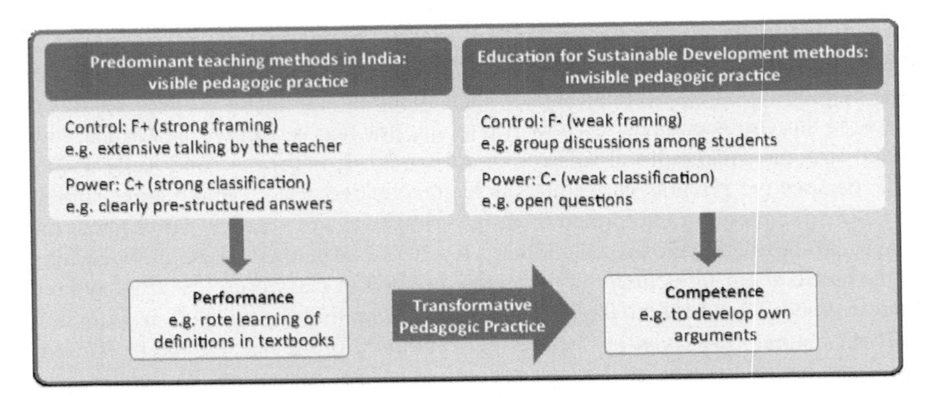

Fig. 9.2 Linking ESD and observed pedagogic practice as performance and competence models through a transformative pedagogy (own draft)

when learning cards and individual attention to students turned into a differentiation not based on interests, but on whether students were "dull" and needed such support (Sriprakash 2010: 302).

Thus, as a form of invisible pedagogy, ESD would need to be culturally adapted to fit with the prevalent visible pedagogy of the contemporary Indian context. This evokes the need for a transformative pedagogic practice: An intermediate, transitional form of pedagogy which links performance and competence models of pedagogy (Fig. 9.2). The aim of transformative pedagogy is to contribute to the transformation of current pedagogic practices of reproduction, into pedagogic practices of transformation. This integrates weak framing and classification as means of empowering students to become critical, responsible citizens.

By taking existing power relations and control mechanisms into account, a different methodological approach in which weakly framed and classified elements are embedded into existing structures of strong framing and classification seems more useful than the ambition to transform classroom teaching into a new, pre-defined type of pedagogic practice. Hence, the visionary principles of ESD as a "travelling model" (Behrends et al. 2014) are reinterpreted for the local context by linking them to existing principles and forms of pedagogic practice. For example, multi-perspectives on water access and the role of different stakeholders can be integrated as factual learning contents into textbooks to raise awareness about the topic. Teaching methodologies are accepted and integrated best when they only slightly change the degree of student–teacher interaction and link them to existing material, institutional, and (infra)structural constraints, as the broader educational approach "Contingent Constructivism" (Vavrus 2009) states. A "cultural translation" (Thompson 2013) of ESD implies the need for a gradual transformation from strong to weak framing, and likewise classification. As other studies have pointed out, learning in schools in India is culturally perceived as knowledge assimilation rather than knowledge construction (Sriprakash 2010). This assigns teachers with the role of being the transmitters of knowledge. Change in this regard will only be achieved through discussion and reflections on this in teacher training sessions, syllabi, textbooks, and other teaching material development committees. This cannot be done for ESD in isolation; other subjects will also have to be subject to this transition process.

9.2.2 Practicability of the Didactic Framework in Pune, India

The didactic framework for ESD through argumentation in geography education presents resource conflicts as a result of diverging social, environmental, and economic interests. ESD can be approached through teaching methods promoting networking thinking, argumentation, and student-orientation. The combination of these three didactic teaching approaches with the three themes of the triangle of sustainability—social, ecological, and economic dimensions—served as framework for the empirical study (cf. Sect. 3.3). These six dimensions of the didactic framework guided the analyses of the integration of ESD principles into educational policies and curricula, national school syllabi, and NCERT and MSCERT textbooks. Based on the dimensions of the didactic framework, pedagogic practice in Pune was observed, textbooks, syllabi, and curricula were analyzed, and three ESD teaching modules were developed, implemented, and evaluated.

Particularly, the didactic approach of argumentation is societally and didactically relevant for implementing ESD in geography lessons as it stimulates a critical consciousness through dialogue learning (Freire 1996). This approach translates the principles of the transnational educational policy of ESD into an approach for geographical teaching methodology. In this study, the approach was implemented with the locally relevant topic of urban water conflicts in Pune by addressing the interrelated factors causing differential access to water. This was done through the development of three teaching methodologies. The "Visual Network" encouraged network thinking and generated a discussion on differential water access through pictures, which had to be arranged—instead of textual input. The concept of argumentations being "correct" or "false" was challenged by asking students to arrange different perspectives on the sustainability of 24/7 water supply along a continuum, the "Position Bar." For the "Rainbow Discussion," students had to develop arguments for the social, economic, and ecological sustainability of saving water. The analysis of these three ESD teaching modules revealed how bi-directional teacher–student communication patterns changed toward multi-directional interaction. Both teachers and students perceived this change as enriching. Based on the teaching intervention, students identified a majority of argumentations as "in between" the two poles of pro and contra. They could identify that limited water access in slums is not a "given" problem, but a result of conflicting interests.

Through the approach of argumentation, students are given more space to develop and express their own opinions, while the reproduction of textbook contents and oral inputs are overcome through visual inputs and group work. This implies a shift from strong classification and framing to weak classification and framing, and hence a change toward more competence-oriented rather than performance-oriented teaching styles. However, several challenges became clear which reflect how socio-cultural constructions of learning and teaching are sustained. While the use of space and resources was changed in the classroom, the focus on sequence, order, and form in presenting results persisted. Alternatively, an open atmosphere that encourages spontaneous reactions in a debate would have strengthened a more multi-directional interaction. Furthermore, to align different perspectives along a continuum proved difficult in a learning environment in which usually the "right answer" in the textbook is learned. Teachers' uncertainties were also observed during the module implementation when they reinterpreted a fluid continuum in three categories to simplify the answer options. They took the lead in debates by steering students through a preselected order of speakers. Similar observations are noted by Sriprakash (2010), as activity cards meant to encourage child-centered learning animated series of question answers.

Despite these observations, the developed ESD teaching modules promoting argumentation on the topic of urban water conflicts prove practicable for the implementation in geography education in several regards. Critical perspectives in applied classroom methodologies promote multiple perspectives and depict natural resource use as a topic of struggle and conflict, which stands in stark contrast to current presentations of the resource of water in textbooks. Particularly, topics on human–environment interaction provide educational examples of how social relations of power persist over time and space, and structure the experienced reality of unequal access to resources. As these modules were tested successfully in classroom teaching, they could serve as a sample of how to integrate ESD principles into textbooks. Recognizing the complexity of classroom cultures and the power of argumentative approaches to interfere with the existing social order, contextualized and approaches open to various interpretations rather than prescriptive contents can be valuable to pedagogic practice. The intervention in the form of three teaching modules can be implemented in geography lessons and textbooks, when ESD principles and the promotion of argumentation skills have been previously outlined in a teacher workshop to ensure intentional classroom implementation.

9.2.3 Reflecting the Methodological Approach

The knowledge produced in this study on pedagogic practice does not reflect an objective reality, but is situated within Bernsteinian and Critical Pedagogy perspectives. My relationship to the research participants and my positionality shaped my interpretations in various ways. This raises ethical dilemmas about my authority to write on pedagogic practice in a different socio-economic context. I myself am a "broker or translator of knowledge" (Sriprakash and Mukhopadhyay 2015: 231) as my knowledge is situated, contingent, and relational. The methods I employ, as well as the choice of theoretical lenses and my interpretation of the principles of the transnational educational policy ESD define what I perceive as relevant to consider, examine, and analyze in this study. Hence, I would like to outline how I have produced a particular understanding of transformative pedagogic practice, and the limitations of such.

The methodological approach is derived from the theoretical framework and takes a social constructivist perspective to observe how social reality and knowledge is constituted in social interactions, subjective meanings, and shared artefacts relevant to pedagogic practice. Next to quality criteria of qualitative social research such as subject proximity, rule-governed, and communicative validity, I employed multilevel and multi-dimensional methodological approaches to examine pedagogic practice and its transformative potential. The qualitative research design balanced field research with document analysis and an intervention study. Ranging from qualitative interviews and focus group discussions, curricula, syllabi, and textbook analyses and extensive classroom observations, also with videography documentation, I precisely documented data and triangulated findings and perspectives. Through this, I was able to develop a grounded analysis embracing students, teachers, and educational stakeholders' perspectives, and reflecting on their power within the educational system. This methodological approach was chosen to reflect on the transformative potential of pedagogic practice.

Although unequal power relations in research cannot be removed, I applied measures to mitigate them and engaged in the research process reflexively (cf. Chap. 5). Based on the critical reflection of my role in the research process and the explication of the rules and norms guiding my analyses, I seek to identify contradictions in the educational structure and possibilities for educational reforms to engage with power relations in pedagogic practice. These are grounded in evidence drawn from a rich data set as a result of applying a variety of rigorous research methods (cf. Chap. 5).

The five selected schools in Pune in which the majority of classroom observations and the intervention study were conducted provide only case studies of the heterogeneous educational system of India. Since all schools were English-medium schools, in which both teachers and students speak English as a second language, I missed out on the enriched communication of teacher–student interactions in their mother tongue. The selected schools represented different middle and upper classes; the teaching approaches might not have worked as well in government schools, as these are less flexible in diverting from the curriculum. Further my role and perspective as a researcher with a different socio-cultural background influenced the processes of data collection and data analysis. To mitigate this, the applied abstract concepts provide an external language to precisely describe and analyze empirical observations. Thus, this study derived categories for data collection after explorative field investigations, observations, and interactions with Indian partners and the beliefs and values of students, teachers, and educational stakeholders. This study's strength is that it triangulates a variety of data sources to verify and integrate different perspectives on classroom observations and document analyses.

9.2.4 Transferability of Theoretical and Methodological Approaches to Other Contexts

ESD is meant to be introduced at all educational levels, and thus, the theoretical framework and the multi-level methodological approach might support collecting, processing, and analyzing empirical data for an in-depth analysis of pedagogic practice. The theoretical framework and the methodological approach could be applied and tested in combination with or separately to subjects other than geography, as well as other forms of schools, informal education, or education at the university level. The benefit of an analysis with the framework developed in this study is that it does not look at pedagogic practice and transnational educational policies in isolation, but relates them to the socio-cultural context as well as power relations within the educational system. The principles and codes with which Bernstein describes pedagogic practice hold true for any relationships of unequal power, in which there is a transmitter and a receiver (e.g., a doctor and a patient). The theoretical framework as well as the methodological approach can be transferred to other cultural contexts and used for cross-country comparative geographical education research. As earlier studies have shown, Bernstein's concepts (1975a, b, 1990) proved applicable to India and other countries (Sriprakash 2011; Clarke 2003; Neves et al. 2004; Morais 2002; Neves and Morais 2001). Similarly, the concepts of Freire (1996) are widely discussed internationally (Punch and Sugden 2013; Fetherston and Kelly 2007; Giroux 2004; Kumar 2006), whereas their combination and the link to the didactic approach of argumentation is novel in this study. However, as this study demonstrated by investigating the translation of the transnational policy ESD, pedagogic practice is a highly contextualized and complex social interaction. This builds on the existing research on the translation of educational policies (Mukhopadhyay and Sriprakash 2011b, 2013; Thompson 2013; Tikly 2004; Merry 2006; Behrends et al. 2014). Whereas, international comparisons are interesting, an in-depth analysis of power relations within pedagogic practice requires a long and culture-specific engagement with concepts and processes of learning and teaching. Culturally, specific argumentations (Sect. 3.2.1) exemplify how different systems of thoughts reflect various principles and categories of norms, beliefs, and practices. These are again entangled in complex intersectionalities of age, gender, caste, class, religion, ethnicity, and other social divides which constitute culture. One example of an international comparison of cultural perspectives on pedagogy is by Alexander (2001) on primary education in five countries. While such in-depth studies are rare, a greater lack of studies is found on pedagogy in developing and under-resourced contexts.

Relations between a transmitter and a receiver can be described with weak or strong modes of classification and framing and might provide useful insights into socio-cultural constructions of learning and teaching. These may help to understand why and how new forms of teaching methodology are interpreted and embedded within existing social interactions. In regard to Critical Pedagogy approaches, it is insightful to combine them with teaching methodology research to move beyond a language of critique toward identifying opportunities for implementing different principles of pedagogy.

References

Alexander, R. (2001). *Culture and pedagogy. International comparisons in primary education*. Singapore: Blackwell Publishing.

Andrews, R. (2009). *The importance of argument in education*. London: Institute of Education.

Archer, M. (1995). The neglect of the educational system by Bernstein. In A. R. Sadovnik (Ed.), *Knowledge and pedagogy: The sociology of Bernstein* (pp. 211–235). Norwood, New Jersey: Ablex Publishing Corporation. http://books.google.de/books?id=3tgXQ_ISJHYC&printsec=frontcover&hl=de&source=gbs_ge_summary_r&cad=0-v=onepage&q&f=false.

Barak, M., Ben–Chaim, D., & Zoller, U. (2007). Purposely teaching for the promotion of higher-order thinking skills: A case of critical thinking. *Research in Science Education, 37*(4), 353–369. http://link.springer.com/article/10.1007/s11165-006-9029-2.

Barrett, A. M. (2007). Beyond the polarization of pedagogy: Models of classroom practice in Tanzanian primary schools. *Comparative Education, 43*(2), 273–294. https://doi.org/10.1080/03050060701362623.

Batra, P. (2005). Voice and agency of teachers: The missing link in National Curriculum Framework 2005. *Economic and Political Weekly, 40*(1), 4347–4356.

Baumert, J., & Kunter, M. (2006). Stichwort: Professionelle Kompetenz von Lehrkräften. *Zeitschrift für Erziehungswissenschaften, 50*, 469–520.

Behrends, A., Park, S.-J., & Rottenburg, R. (2014). Travelling models: Introducing an analytical concept to globalisation studies. In A. Behrends, S.-J. Park, & R. Rottenburg (Eds.), *Travelling models in African conflict management: Translating technologies of social ordering* (pp. 1–40). Leiden: Brill.

Benedict, F. (1999). A systemic approach to sustainable environmental education. *Cambridge Journal of Education, 29*(3), 433.

Berndt, C. (2010). *Elementarbildung in Indien im Spannungsverhältnis von Macht und Kultur. Eine Mikrostudie in Andhra Pradesh und West Bengalen*. Berlin: Logos Verlag.

Bernstein, B. (1975a). *Class and pedagogies: Visible and invisible*. Paris: OECD.

Bernstein, B. (1975b). *Class, codes and control. Towards a theory of educational transmission*. London: Routledge.

Bernstein, B. (1990). *Class, codes and control. The structuring of pedagogic discourse*. London: Routledge.

Bharati Vidyapeeth Institute of Environment Education and Research. (2002). Study of status of infusion of environmental concepts in school curricula and the effectiveness of its delivery.

Budke, A. (2011). Förderung von Argumentationskompetenzen in aktuellen Geographieschulbüchern. In E. Matthes, Heinze, C. (Ed.), *Elementarisierung im Schulbuch* (pp. 253–264). Bad Heilbrunn: Verlag Julius Klinkhardt.

Budke, A. (2012a). Argumentationen im Geographieunterricht. *Geographie und ihre Didaktik, 1*, 23–34.

Budke, A. (2012b). "Ich argumentiere, also verstehe ich." - Über die Bedeutung von Kommunikation und Argumentation für den Geographieunterricht. In A. Budke (Ed.), *Kommunkation und Argumentation* (pp. 5–18). Braunschweig: Westermann Verlag. Geo Di 14 (KGF), Alexandras einführungsartikel ausgdruckt.

Budke, A., Schiefele, U., & Uhlenwinkel, A. (2010a). Entwicklung eines Argumentationskompetenzmodells für den Geographieunterricht. *Geographie und ihre Didaktik, 3*, 180–190.

Budke, A., Schiefele, Ulrich, & Uhlenwinkel, Anke. (2010b). Entwicklung eines Argumentationskompetenzmodells für den Geographieunterricht. *Geographie und ihre Didaktik, 3,* 180–190.

Chauhan, C. P. S. (1990). Education for all: The Indian scene. *International Journal of Lifelong Education, 9*(1), 3–14. https://doi.org/10.1080/0260137900090102.

Clarke, P. (2003). Culture and classroom reform: The case of the district primary education project, India. *Comparative Education, 39*(1), 27–44. https://doi.org/10.1080/0305006032000044922.

Crossley, M., & Murby, M. (1994). Textbook provision and the quality of the school curriculum in developing countries: Issues and policy options. *Comparative Education, 30*(2), 99–114. http://www.jstor.org/stable/3099059.

de Haan, G. (2008). Gestaltungskompetenz als Kompetenzkonzept für Bildung für nachhaltige Entwicklung. In I. Bormann, & G. de Haan (Eds.), *Kompetenzen der Bildung für nachhaltige Entwicklung. Operationalisierung, Messung, Rahmenbedingungen, Befunde.* (pp. 23–43). Wiesbaden: VS Verlag für Sozialwissenschaften. http://link.springer.com/book/10.1007/978-3-531-90832-8, http://www.amazon.de/Kompetenzen-Bildung-nachhaltige-Entwicklung-Operationalisierung-ebook/dp/B001BS64SC/ref=dp_return_2?ie=UTF8&n=530484031&s=digital-text-reader_3531155296.

Eick, C. J., & Reed, C. J. (2002). What makes an inquiry-oriented science teacher? The influence of learning histories on student teacher role identity and practice. *Science Education, 86*(3), 401–416.

Fetherston, B., & Kelly, R. (2007). Conflict resolution and transformative pedagogy: A grounded theory research project on learning in higher education. *Journal of Transformative Education, 5*(3), 262–285. https://doi.org/10.1177/1541344607308899.

Freire, P. (1996). *Pedagogy of the oppressed*. London: Penguin Books Ltd.

Giroux, H. A. (2004). Critical pedagogy and the postmodern/modern divide: Towards a pedagogy of democratization. *Teacher Education Quarterly, 31*(1), 31–47.

Goffman, E. (1971). *Relations in Public. Microstudies of the Public Order*. New York: Harper and Row.

Government of India. (2004). *Education for all. India Marches Ahead*. New Delhi: Ministry of Human Resource Development.

Habermas, J. (1984). *The theory of communicative action*. Boston, Mass.: Beacon Press.

Hattie, J. (2009). *Visible learning. A synthesis of over 800 meta-analyses relating to achievement*. Oxon: Routledge.

Hellberg-Rode, G., Schrüfer, G., & Hemmer, M. (2014). Brauchen Lehrkräfte für die Umsetzung von Bildung für nachhaltige Entwicklung (BNE) spezifische professionelle Handlungskompetenzen? *Zeitschrift für Geographiedidaktik, 4,* 257–281.

Kabeer, N. (1999). Resources, agency, achievements: Reflections on the measurement of women's empowerment. *Development and Change, 30,* 435–464.

Kanu, Y. (2005). Tensions and dilemmas of cross-cultural transfer of knowledge: Post-structural/postcolonial reflections on an innovative teacher education in Pakistan. *International Journal of Educational Development, 25,* 493–513.

Kennedy, M. M. (2006). Knowledge and vision in teaching. *Journal of Teacher Education, 57*(3), 205–211.

Kopperschmidt, J. (2000). *Argumentationstheorie. Zur Einführung*. Hamburg: Junius.

Kuckuck, M. (2014). *Konflikte im Raum - Verständnis von gesellschaftlichen Diskursen durch Argumentation im Geographieunterricht*. Münster: MV-Verlag.

Kumar, K. (1988). Origins of India's "Textbook culture". *Comparative Education Review, 32*(4), 452–464. http://www.jstor.org/stable/1188251.

Kumar, K. (2005). *Political agenda of education. A study of colonialist and nationalist ideas*. New Delhi: Sage Publications.

Kumar, V. A. (2006). Gramsci and Freire: Bridging the divide in Indian context: An exploratory essay.

Leder, S. (2014). Das indische Bildungssystem im Wandel: Zwischen traditionellen Unterrichtspraktiken und dem Anspruch einer Bildung für nachhaltige Entwicklung. *Geographien Südasiens, 2*, 18–21.

Leder, S. (2015). Uncovering schooling ideals and student culture: the case of India. Book Review of Thapan, Meenakshi (Ed.), *Ethnographies of schooling in contemporary India*. http://www.booksandideas.net/Schooling-Ideals-and-Student-Culture-the-Case-of-India.html. Accessed May 25, 2015.

Leder, S. (2016). *Linking women's empowerment and their resilience*. Colombo, Sri Lanka: CGIAR Research Program on Water, Land and Ecosystems (WLE).

Levin, B., & He, Y. (2008). Investigating the content and sources of teacher candidates' personal practical theories. *Journal of Teacher Education, 59*(1), 55–68.

Mandelbaum, D. G. (1975). *Society in India*. Noida: Popular Prakashan.

Manteaw, O. O. (2012). Education for sustainable development in Africa: The search for pedagogical logic. *International Journal of Educational Development, 32*(3), 376–383. <Go to ISI> ://000301698300003http://ac.els-cdn.com/S0738059311001301/1-s2.0-S0738059311001301-main.pdf?_tid=e247c694-1df6-11e2-a864-00000aab0f6b&acdnat=1351095877_200e77592eca28a76d7de4643351e023.

Marrow, J. (2008). *Psychiatry, Modernity and family values: Clenched teeth illness in North India*. Chicago: ProQuest.

McElvany, N., Schroeder, S., Baumert, J., Schnotz, W., Horz, H., & Ullrich, M. (2012). Cognitively demanding learning materials with texts ans instructional pictures: Teacher's diagnostic skills, pedagogical beliefs and motivation. *European Journal of Psychology of Education, 27*(3), 403–420.

Merry, S. E. (2006). Transnational human rights and local activism: Mapping the middle. *American Anthropologist, 108*(1), 38–51.

Morais, A. M. (2002). Basil Bernstein at the micro level of the classroom. *British Journal of Sociology of Education, 23*(4), 559–569. https://doi.org/10.2307/1393312.

Mukhopadhyay, R., & Sriprakash, A. (2011a). Global frameworks, local contingencies: Policy translations and education development in India. *Compare—A Journal of Comparative and International Education, 41*(3), 311–326. https://doi.org/10.1080/03057925.2010.534668.

Mukhopadhyay, R., & Sriprakash, A. (2011b). Global frameworks, local contingencies: Policy translations and education development in India. *Compare—A Journal of Comparative and International Education, 41*(3), 311–326. https://doi.org/10.1080/03057925.2010.534668.

Mukhopadhyay, R., & Sriprakash, A. (2013). Target-driven reforms: Education for All and the translations of equity and inclusion in India. *Journal of Education Policy, 28*(3), 306–321. https://doi.org/10.1080/02680939.2012.718362.

National Council of Educational Research and Training. (2009). *National Curriculum Framework 2005. Position papers on national focus groups on systemic reform* (Vol. II). New Delhi: NCERT.

Neves, I., & Morais, A. M. (2001). Texts and contexts in educational systems: studies of recontextualising spaces. In A. M. Morais, I. Neves, B. Davies, & H. Daniels (Eds.), *Towards a sociology of pedagogy. The contribution of Basil Bernstein to research* (pp. 223–249). New York: Peter Lang.

Neves, I., Morais, A. M., & Afonso, M. (2004). Teacher training contexts. Study of specific sociological characteristics. In J. Muller, B. Davies, & A. M. Morais (Eds.), *Reading Bernstein, researching Bernstein* (pp. 168–186). London: Routledge.

Pal, Y. (1993). *Learning without burden*. New Delhi: Ministry of Human Ressource Development, Government of India.

Programm Transfer 21. (2007). Schulprogramm Bildung für nachhaltige Entwicklung. Grundlagen, Bausteine, Beispiele. Berlin: BLK-Programm Transfer 21.

Punch, S., & Sugden, F. (2013). Work, education and out-migration among children and youth in upland Asia: Changing patterns of labour and ecological knowledge in an era of globalisation. *Local Environment, 18*(3), 255–270. https://doi.org/10.1080/13549839.2012.716410.

Rinschede, G. (2005). *Geographiedidaktik*. Paderborn: Schöningh.

Schockemöhle, J. (2011). Regionales Lernen - Kompetenzen fördern und Partizipation stärken. Zur Wirksamkeit des außerschulischen Lernens in der Region. In H. Bayrhuber, U. Harms, B. Muszynski, B. Ralle, M. Rothgangel, L.-H. Schön, et al. (Eds.), *Empirische Fundierung in den Fachdidaktiken* (pp. 201–216). Münster: Waxmann Verlag.

Schrüfer, G., Hellberg-Rode, G., & Hemmer, M. (2014). Which practical professional competencies should teachers possess in the context of education for sustainable development? Theoretical foundations and research design. In D. Schmeinck, & J. Lidstone (Eds.), *Standards and research in geography education—Current trends and international issues* (pp. 135–143). Berlin: MBV.

Sen, A. (2005). *The argumentative Indian*. Noida: Penguin.

Spivak, G. (2008). *Can the subaltern speak?*. Wien: Turia + Kan.

Sriprakash, A. (2010). Child-centered education and the promise of democratic learning: Pedagogic messages in rural Indian primary schools. *International Journal of Educational Development, 30*(3), 297–304.

Sriprakash, A. (2011). The contributions of Bernstein's sociology to education development research. *British Journal of Sociology of Education, 32*(4), 521–539.

Sriprakash, A. (2012). *Pedagogies for development: The politics and practice of child-centred education in India*. New York: Springer.

Sriprakash, A., & Mukhopadhyay, R. (2015). Reflexivity and the politics of knowledge: researchers as 'brokers' and 'translators' of educational development. *Comparative Education, 51*(2), 231–246.

Steiner, R. (2011). Kompetenzorientierte Lehrer/innenbildung für Bildung für Nachhaltige Entwicklung. Kompetenzmodell, Fallstudien und Empfehlungen. Münster: MV-Verlag.

Stöber, G. (2007). Religious identities provoked: The Gilgit 'Textbook controversy' and its conflictual context. *Internationale Schulbuchforschung, 29*, 389–411.

Stöber, G. (2011). Zwischen Wissen, Urteilen und Hndeln - "Konflikt" als Thema im Geographieschulbuch. In C. H. Meyer, R., & Stöber, G. (Eds.), *Geographische Bildung* (pp. 68–81). Braunschweig: Westermann.

Thapan, M. (2014). *Ethnographies of schooling in contemporary India*. Delhi: Sage.

The Hindu. (2005, Aug. 7). NCERT draft curriculum framework criticised. Retrieved from http://www.thehindu.com/2005/08/07/stories/2005080705361000.htm.

Thompson, P. (2013). Learner-centred education and 'cultural translation'. *International Journal of Educational Development, 33*(1), 48–58. https://doi.org/10.1016/j.ijedudev.2012.02.009.

Tikly, L. (2004). Education and the new imperialism. *Comparative Education, 40*(2), 173–198.

Tilbury, D. (2011). *Education for sustainable development. An expert review of processes and learning*. Paris: UNESCO.

UNESCO. (2005a). *EFA global monitoring report 2005*. Paris: UNESCO.

UNESCO. (2005b). United Nations decade of education for sustainable development (2005–2014): International implementation scheme, Paris.

UNESCO (2009). UNESCO World Conference on ESD: Bonn Declaration.

UNESCO. (2011). *Education for sustainable development. An expert review of processes and learning*. Paris: UNESCO.

Vavrus, F. (2009). The cultural politics of constructivist pedagogies: Teacher education reform in the United Republic of Tanzania. *International Journal of Educational Development, 29*(3), 303–311. https://doi.org/10.1016/j.ijedudev.2008.05.002.

Vavrus, F., & Barrett, L. (2013). Teaching in tension. International pedagogies, national policies, and teachers' practices in Tanzania. Rotterdam: Sense Publishers.

Wohlrapp, H. (2006). Was heißt und zu welchem Ende sollte Argumentationsforschung betrieben werden? In E. Grundler, & R. Vogt (Eds.), *Argumentieren in der Schule und Hochschule. Interdisziplinäre Studien* (pp. 29–40). Tübingen: Stauffenburg Verlag Brigitte Narrr.

World Bank. (1990). *Papua New Guinea, Primary Education Project Completion Report*. Washington D.C: World Bank.

World Health Organization (2001). *Skills for health*. http://www.who.int/school_youth_health/media/en/sch_skills4health_03.pdf. Accessed Mar 10, 2014.

Chapter 10
Research Prospects and Strategies for Translating ESD in Pedagogic Practice

Abstract This chapter synthesizes the key empirical findings of the book and outlines research prospects and policy implications for translating Education for Sustainable Development (ESD) into pedagogic practice. I argue that the educational discourse of ESD, if implemented, has the potential to fundamentally challenge the reproductive mode of pedagogic practice in the case of geography education in India over time, as it subverts cultural values, norms, and constructions of teaching and learning. Despite this, if ESD is framed as transformative pedagogic practice, it can contribute to gradually revising current geography teaching contents and methods toward promoting learner-centered teaching, critical thinking, and argumentation skill development. However, the discrepancy between institutional objectives, structural conditions and pedagogic practices needs to be addressed. Currently, there is little space for teachers' agency to allow the recontextualization of ESD principles, for example, to adapt natural resource governance topics to students' perspectives and to include an argumentative approach to local water conflicts. Syllabi, textbooks, and examinations need new selection and evaluation criteria that emphasize knowledge which promotes argumentation skills. This would allow to integrate ESD through critical thinking on natural resource use into existing forms of pedagogic practice. To encourage innovation in curriculum design, syllabi, and textbook development, cooperation between educational institutes, teaching methodology research, and, most of all, teachers' agency need to be strengthened. Discretion, opportunities, and support are necessary so that teachers can respond to the unique needs of their students.

This study set out to explore the challenges for translating the transnational educational policy of Education for Sustainable Development (ESD) into pedagogic practice in geography education in India. Within the setting of a heterogeneous educational system in India, this study investigated how global educational objectives are reenacted in pedagogic practice at a local level. While educational reforms in India have focused on increasing student enrollment and access to schools (Berndt 2010; cf. Kingdon 2007), recent endeavors address the quality of teaching (National Council of Educational Research and Training 2005). In the context of environmental challenges such as the pressure and conflicts on natural resource use, this study

S. Leder, *Transformative Pedagogic Practice*, Education for Sustainability,
https://doi.org/10.1007/978-981-13-2369-0_10

builds on the argument that developing a critical consciousness (Freire 1996) in relation to controversial human–environment relations is relevant to participating in decision-making. This locates the translation of ESD objectives (UNESCO 2005) between the priorities of social reproduction (Bernstein 1975, 1990) and transformation (Freire 1996) through pedagogic practice and entails a multi-level analysis to identify challenges and opportunities for educational reforms.

So far, geographical educational research in developing contexts with grounded theoretical frameworks remains inconclusive, and empirical studies that investigate the translation of the transnational educational policy of ESD lack theoretical grounding and comprehensive methodological approaches (cf. Tilbury 2011; Manteaw 2012). This study sought to answer the questions how pedagogic practice and academic frameworks in geography education in India relate to ESD objectives (UNESCO 2005), and how the interpretation of ESD principles through argumentation on water conflicts in geography education (Budke et al. 2010; Budke 2012) challenges power relations in pedagogic practice. Based on the dialectic relationship of theoretical considerations and empirical data analyses consisting of fieldwork, document analyses, and an intervention study at English-medium schools in Pune, this book analyzed relayed power structures and cultural values in pedagogic practice for their transformative potential.

This study demonstrated how educational reforms promoting learner-centered education are engaging with foundational change within society: Pedagogic practice is intertwined with cultural values of teaching and learning and relays deeply entrenched social structures in the form of age, gender, class, caste, and other divides (Bernstein 1990; Clark 2005; Fetherston and Kelly 2007; Giroux 2004). On the one hand, the teacher is perceived as the transmitter of knowledge, while the students are the reproducers of knowledge. Values attributed to a "good class" are those of respect and affirmation toward the authority of the teacher, as observed in geography lessons and also noted by Thapan (2014). On the other hand, Indian syllabi and public textbooks perpetuate a strong focus on the reproduction of knowledge. As syllabi and textbooks carry the message of what is perceived as legitimate knowledge (Kumar 1988), their strong classification and framing of knowledge and skill development in pedagogic practice control communication in pedagogic relations to a great extent (cf. Bernstein 1975). Despite the amount of speaking done by teachers, teachers only have limited control over the selection, sequence and pacing of knowledge and skills. It is rather the textbook that governs teacher–student communication in geography lessons (cf. Kumar 1988). The textbook analysis revealed explicit categories of knowledge and the strongly framed sequence, pacing, and evaluation of knowledge leave little space for teachers and students' agency, e.g., the choice to develop their own argumentations on teaching contents. Water resources, for example, are presented as a fixed commodity, and the access to water is not depicted as socially constructed. Information on water in national and state textbooks is primarily fact- and definition-oriented, while local examples and causes and consequences of socio-economically differentiated access to water are barely explained. Most textbooks contain a great number of chapters, which indicates a strong classification of knowledge, and the majority of textbook tasks focus on reproduction through single-word answers or definitions or lists stated in the textbooks. This also results in narrowly framed classroom communication driven by lectures or inquiry–response cycles led by the teacher, which was observed in geography lessons at English-medium schools in Pune and noted in the literature (Sriprakash 2010, 2012; Clarke 2003). Students had barely any space to express their own opinions and to ask questions extending beyond the prescribed information in the textbook, which indicates a performance mode of pedagogic practice (Bernstein 1990).

These principles in observed pedagogic practice and geography textbooks stand in stark contrast to the educational objectives promoted by the transnational educational policy ESD (UNESCO 2005), and, in a similar way, the principles emphasized in the National Curriculum Framework (NCF 2005). The attempt to influence instructional discourse by the national apex body of education, the National Council of Educational Research and Training (NCERT), faces difficulties establishing new principles in pedagogic practice. Although the NCF 2005 promotes critical thinking and skill development, these are hardly implemented in the observed geography lessons. This is demonstrated in two contrasting types of pedagogy at different institutional levels: The NCF 2005 lays out an open curriculum and promotes an integrated approach to content, while geography syllabi and textbooks represent a collection and have a closed mode. Conflicting views on natural resource use from social, environmental, and economic perspectives, and teaching methodologies relating to *how* skills can be promoted are not sufficiently addressed in educational policies, curricula, and syllabi analyzed in this study. Furthermore, critical perspectives and the representation of voices other than the urban middle class are not comprehensively legitimized through syllabi and textbooks. Hence, the transformative aspirations of ESD and NCF 2005 objectives contrast such a pedagogic practice marked by principles of reproduction and performance, which is structured through strong classification and framing. This indicates that ESD's competence model of pedagogy with a weak classification and framing interferes with power relations and cultural constructs of teaching and learning relayed in pedagogic practice.

Hence, the dominant educational discourse of ESD turns out to fundamentally challenge pedagogic practice in the case of geography education in India, as it attempts to subvert cultural values and constructions of teaching and learning. The intervention study demonstrated how power relations and communication patterns in pedagogic practice sustain when teaching methodologies that facilitate critical thinking and argumentation skills are introduced. The interpretation of new activities is linked to existing concepts of clearly defined answers, presentations, and a strong focus on form and sequence. This became obvious when, for example, a teacher interpreted the ESD teaching module "Position Bar" not as a continuum along which opinions of different stakeholders on 24/7 water supply in Pune could be arranged, but through the introduction of a third category "in between" pro and contra arguments. This demonstrated how a new way of interpreting the method emerged while drawing on well-known categories of thought. Similarly, the speaking cards in the ESD teaching module 3 "Rainbow Discussion" were interpreted to indicate the pre-arranged order of students through the teacher instead of facilitating spontaneous discussion and ensuring that all students participate. Both examples indicate how teachers recontextualize and translate activities within their patterns of pedagogic practice, as similarly observed in other studies in India, Nigeria, and Tanzania (Thompson 2013; Sriprakash 2010; Manteaw 2012; Vavrus 2009; Vavrus and Barrett 2013). Teachers and students experience uncertainties as to how to deal with activities and classroom communication that deviate from the predominant interaction patterns in which the teacher has an authoritative position. These aspects need to be considered as part of the interpretation of ESD principles that aim to encourage critical and creative thinking and dealing with uncertainties, as well as promoting action competence for people to shape their own environments (UNESCO 2005; Tilbury 2011; de Haan 2008; Schockemöhle 2011).

This implies that transnational educational policies can only partially be embedded in existing pedagogic practice. Novel teaching methodology approaches, which indicate a paradigmatic shift of principles in the teacher–student communication, need a gradual translation into the cultural context (Thompson 2013; Behrends et al. 2014; Vavrus and Barrett 2013; Mukhopadhyay and Sriprakash 2011) with the intense cooperation and training of teachers, syllabi designers and textbook authors. Hence ESD, if interpreted as transformative pedagogic practice as in this study at hand, can contribute to discussions and exemplify approaches on how critical teaching methodologies and multi-perspectivity can let "the Subaltern speak" (Spivak 2007). Although arguing with people of higher status is culturally not desirable in India (Mandelbaum 1975; Sen 2005; Marrow 2008), education shall raise students' awareness of natural resource conflicts and promote skills for decision-making on lifestyles in today's democratic societies (Bharati Vidyapeeth Institute of Environment Education and Research 2002; National Council of Educational Research and Training 2005, 2009; National Green Tribunal 2014). This study reveals large gaps between the demands of citizens in modern urban India and the selection of knowledge and skills to be promoted in schooling through textbooks. Diverting textbook contents toward promoting critical thinking and argumentation skills can prepare students for reflecting and participating in decision-making on social justice in water resource conflicts, e.g., by addressing marginalized perspectives such as the limited access to water for informal settlements.

The study's theoretical approach implies that analyses drawing on conceptual insights from different disciplines generate detailed knowledge on pedagogic practice and its transformative potential. The transdisciplinary synthesis between geographical developmental research, geographical education research and sociology of education research offers multi-level analyses of changes in pedagogic relations through transnational educational policies. The conceptual contribution of this study is the integration of Bernstein's descriptive–analytical approach of the *Sociological Theory of Education* (1975, 1990) with the Critical Pedagogy of Freire (1996), an educational system perspective (Archer 1995), and the didactic approach of argumentation skill development (Budke 2012). So far, teaching methodology research has not been linked with Bernstein's sociology of education and Critical Pedagogy to allow an in-depth analysis of structural and cultural constraints on the implementation of educational reforms. In-depth research, however, can inform the implementation and translation process of educational policies. This study exemplified how

ESD approaches to teacher trainings and textbook writing can be interpreted through the didactic approach of argumentation to bridge the lacunae between global policy rhetoric and classroom realities. The developed didactic framework for ESD through argumentation initiated meaningful debates with educational stakeholders and teachers about whether and how the predominant reproductive mode of pedagogic practice can and should be altered toward the promotion of critical and network thinking and multi-perspectivity.

The empirical analysis that has been presented here offers a range of challenges and opportunities accompanying ESD. These are first and foremost cultural values and power relations in pedagogic practice such as collective, affirmative, and reproductive norms in teaching, as well as respect for authorities and persons in higher positions within a stratified Indian society. These values are closely related to the strong framing and classification of knowledge in textbooks and overloaded syllabus, which barely promote critical thinking on local examples, communicative approaches, and multi-perspectivity. Additionally, there are a number of structural challenges such as high number of students per class, insufficient teacher training, lack of teaching material, tables which are screwed to the classroom floor and thus hinder group work, a highly competitive educational system, a great diversity in students' socio-cultural backgrounds and parents and principals' pressure on teachers and students to perform well in standardized examinations which mostly promote only the reproduction of knowledge. Particularly, the structural challenges have been mentioned in interviews with educational stakeholders and teachers and in a range of different studies on the quality of education in India (Govinda 2002; Kumar 2003; Kumar and Oesterheld 2007; Lall 2005; Government of India 2004). As this study has demonstrated, educational reforms have to engage with existing cultural values and power relations in pedagogic practice to create meaningful change in pedagogic principles.

Despite the foundational gap that currently exists between ESD and contemporary pedagogic practices, the former—with a comprehensive, argumentative approach—does at least offer an opportunity for national contexts to rethink their traditional ways of imparting knowledge in school. The postulations for a paradigm change in teaching by the NCF 2005, as well as the great interest of educational stakeholders, teachers and students make evident the opportunities for introducing ESD through argumentative teaching methods to India's classrooms. The transnational educational reforms posited by ESD do allow sufficient room for local interpretation, but concrete objectives, policies, and financial as well as human resources for their implementation in national, regional, and local contexts need to be developed and better allocated if we are to see more widespread global citizen participation in sustainable development practices. For policy and curricula developers, textbook authors, and most importantly teachers, a national ESD implementation strategy included in the next NCF (the due date has not yet been determined) could open up new perspectives on how and why to teach certain contents. This suggests that mechanisms are needed to strengthen the link between educational institutions and teacher education as well as textbook production in order to weaken strong classification and framing in pedagogic practice (Batra 2005; cf. Giroux 2004; Schweisfurth 2011).

Water and other natural resources are and will continue to be sites of struggle and conflict (UN World Water Assessment Programme 2016). In the dawn of the twenty-first century, argumentation skills are pivotal in a democratic society to identify different interests, perspectives, and access to resources and to solve resource conflicts through negotiations (Budke and Meyer 2015; Kuckuck 2014). This is all the more important in light of the important role of education for sustainable social, environmental, and economic development, as also highlighted in the Sustainable Development Goals (SDGs). Thus, ESD could function as a directive, and thereby orient existing educational reforms and approaches toward the further promotion of argumentation skills and teaching methods that facilitate critical and network thinking in classroom learning. These principles of ESD contribute to a transformative pedagogic practice approach, one that is aimed specifically at greater critical student participation. Debates between researchers and policy makers on meaningfully interpreting ESD objectives for national educational policies and pedagogic practice can contribute to the process of gradually promoting a fundamental change in Indian teaching methods and content—especially relating to environmental education in schools. Through changes on state and national educational levels, pedagogic practice can move toward context-specific realization of the ESD objective of better-equipping students to be the decision-makers of tomorrow—thereby identifying approaches to promote argumentation skills, values, and knowledge to become the vanguards of Sustainable Development.

10.1 Research Prospects

Further comparative cross-cultural educational research is needed to understand how transnational educational policies like ESD are and can be implemented in other socio-economic contexts, as well as how underlying structures and processes influence pedagogic practice. As Bernstein's concepts proved useful to analyzing the messages and models of pedagogy, it still needs to be complemented by approaches in order to actually change current pedagogic practices. This study combined Bernstein's concepts with normative postulations of Critical Pedagogy, and teaching methodology research. For socio-cultural contexts different from India, this conceptual framework can be altered or combined with other approaches to analyze the transformative potential of pedagogic practice.

Further research is also necessary to analyze how ESD can be concretely translated into local Indian contexts. There is a great need for research on context-relevant ESD teaching methodological approaches for geography and other subjects, such as history, languages, natural, and social sciences in India. Critical questions to be asked in all these subjects are: How do students become empowered through teaching methods and contents in the particular subject to participate in societal decision-making processes? How can students acquire the knowledge and skills needed to participate in discussions and form their own opinion, as suggested by ESD and the NCF 2005?

Most importantly, research on the challenges and opportunities to strengthen the agency of the teacher in the formal educational system of India is crucial to changing pedagogic practice in the long term. Similarly, subject-related and overarching pedagogic capacity development concepts for textbook writers and teachers need to be developed and tested in formal in-service and pre-service training and interdisciplinary workshops. The success of the concrete recommendations mentioned in this book needs to be verified, e.g., for textbook development by comparing skill development through conventional textbook chapters with chapters with an ESD approach. Another focus could be set on whether traditional knowledge systems of students are addressed in textbooks. This could be relevant to linking new contents to student perceptions and hence increase their interest and motivation, and the effectiveness of learning.

Comprehensive educational research should also address how the role of state bodies of education at the national, state and district level, such as the NCERT, the SCERT, and the DIETs, is constituted and could be strengthened within the educational system. A broader approach promoted by ESD is the whole school approach (Henderson and Tilbury 2004), which not only aims at promoting ESD in all subjects taught, but also aims at implementing ESD thinking into the school structure. How this approach could be implemented in the Indian educational system needs to be further investigated using broader concepts and empirical studies to develop approaches. Overall, educational research needs to be strengthened to understand how pedagogic principles can be implemented in effective teaching methodologies and capacity development.

This study has offered an evaluative perspective on the challenges of translating ESD into pedagogic practice in a developing context. The theoretical perspective of Bernstein and my interpretation of ESD as a Critical Pedagogy approach limit the study's insights. Other theorists and concepts used to examine educational reforms could result in a different language of analysis and hence produce alternative knowledge to this study. As this study was conduced in geography lessons at urban English-medium schools, other subjects, school forms, and settings might provide different outcomes. Nevertheless, this focused approach strengthened the coherence of the study's empirical results and enabled an in-depth analysis through the applied theoretical and didactic frameworks.

10.2 Strategies for Translating ESD in Geography Teaching

The insights provided by the study at hand have broad implications for educational policy in India as well as for strategies of educational stakeholders affected by these policies. This is particularly relevant in the current changing educational landscape in India, where the public interest in providing access to schools is partly shifting toward a discussion on the quality of education. To ensure that quality education is accessible for the masses and not only for those who can afford to attend private schools, there needs to be a gradual change by revising current teaching contents and methods toward student-oriented, creative and critical-thinking skill development for government schools.

The transnational educational policy ESD offers a new framework of thought and exercises public and communicative power despite being a not legally binding soft approach. National, state, and local educational bodies within the particular socio-cultural context develop the particular approaches, contents, and methods to translate these objectives into practice. Hence, these institutions need to be strengthened to redirect social reproduction in pedagogic practice toward transformative pedagogic practice.

The results of the empirical study raise the question as to how current teaching practices can better prepare urban youth for the challenges of a rapidly changing environment, in which conflicts about progressively degrading natural resources are everyday experiences. Skills to develop and express opinions and to understand others' argumentations are necessary to recognize and adequately deal with other perspectives from various backgrounds and with different interests in the context of urbanization, population pressure, and natural resource conflicts. The analysis of pedagogic practice from a Bernsteinian lens identifies how power relations and cultural values of teaching and learning at the macro- and micro-level challenge the emergence of a transformed pedagogic practice in India. The study of how power relations in educational contexts are exercised and reproduced helps to better formulate policy strategies for transforming pedagogic practice. By doing so, multiple contradictions are revealed against which educational reforms must be redirected. To tackle the challenges examined in this book, both broader and concrete strategies are necessary to overcome the rhetoric of student-oriented educational reforms and to implement teaching methods with meaningful principles in local contexts in line with the transnational educational policy of ESD and the National Curriculum Framework 2005.[1]

The implementation of ESD principles could be brought forward through the MOEF and MHRD[2] issuing a directive on the need to holistically integrate ESD principles into curricula, textbooks, and pedagogic practice. The NCERT could develop concrete strategies and guiding models for an ESD approach to learning contents and allot more time in school to ESD. These national stipulations can then be rolled out on the state level. The directives and strategies would ideally be modified by SCERTs and passed on to those state government organizations and DIETS who devise both textbooks for schools and the pre-teacher and post-teacher training programs. Concrete strategies could be implemented at various levels in the following ways:

(1) In order to integrate ESD through critical thinking on natural resource use into existing forms of Indian geography education, syllabi, textbooks, and examinations need *new selection and evaluation criteria* vis-à-vis knowledge. Textbooks

[1] For a detailed analysis of the overlaps and differences between ESD and the NCF 2005, see Sect. 6.1.1.

[2] MoEF = Ministry of Environment and Forests; MHRD = Ministry of Human Resource Development of the Government of India.

need to be designed to encourage and facilitate classroom discussions rather than transmitting facts to students.

The focal role of textbooks, which guides both students and in particular teachers, needs to be reconceptualized into a tool for orientation with strong methodological guidance. The textbook analysis highlighted the need to reformulate texts, tasks, and questions to facilitate a shift from a descriptive, factual and monocausal to a critical and argumentative conflict approach as envisioned by ESD. Concerning the current factual approach to the topic of water in geography education, multi-perspectives on the causes and effects of local water conflicts should be added to the current declarative regional geographical knowledge in textbooks. For example, this requires taking an argumentative approach to urban water conflicts integrating multiple perspectives to be strengthened by thorough pre-service and in-service teacher training. Such a renewal of the teaching approach from a factual toward an argumentative approach in syllabi and textbooks may better meet students' interests and motivation, and most importantly, students' skill requirements to become critically thinking citizens for social, economic, and environmental sustainable development.

(2) *Teachers' agency* needs to be strengthened by giving them more ownership, choice, and responsibility to structure teaching contents and methods according to the needs of the students. Teacher training should facilitate teaching methodologies, which promote discussions, critical thinking and argumentation skills on natural resource use conflicts. Teachers need to reflect on their own argumentations before they are capable of identifying methods to facilitate students' argumentation skills. Teacher capacity development should tackle the insecurity teachers feel when not knowing the "right answer" to redefine their established role as "transmitter of knowledge." As the analysis with Bernstein's concepts shows, structural conditions and cultural values inhibit teacher's agency. Teacher's agency needs to be strengthened not only through providing tools and training for lesson planning, but also by encouraging the rethinking of the principles of pedagogic practice: "Training of teachers that makes them aware of the meaning and effects of their actions and gives them the opportunity to change their practices" (Morais 2002: 561). Teachers should be trained not to limit their pedagogic practice to the "transmission" of textbooks, but to use them as one source to select knowledge and pace their teaching according to the students' needs. Next to textbook texts, teachers should have time to address students' particular interests. This indicates that textbooks need to have a reduced number of facts to be learnt and rather focus on a comprehensive skill development approach with easily transferable examples. Thus, a new space for teachers' agency is developed where perception, values, attitudes, and skills can be reworked and reinterpreted for a particular context.

(3) *Teacher training concepts* need to be developed and implemented in revised teaching methodologies for argumentation skill development, as well as greater grounding in the psychological and didactic principles of learning. ESD can ultimately only be imparted through trained teachers. Textbook authors and teachers should be sensitized as to how to develop contents, methods, and tasks that focus

explicitly on sustainability in the students' own local contexts. When teachers are sensitized on their own knowledge, skills and practice of ESD and Critical Pedagogy approaches through reconceptualized teacher training, they can implement these and function as multipliers. Didactic research on different skills and their respective levels contributes to the development of a deeper knowledge of how to transform pedagogic practice. Adequate teacher training based on scientifically grounded didactic knowledge improves pedagogic practice in which teachers have greater autonomy. Teaching methodologies focusing on argumentation skill development need to be implemented systematically in lessons. Further, teachers need to receive greater access to a multitude of teaching materials apart from textbooks. Teacher training at the district level and teaching material offered in different print formats need to be revised and widely disseminated. Teachers should be empowered to question textbook contents and to choose the material appropriate to the specific class they are teaching. If teachers have more freedom to choose contents and methods, which they find appropriate for their class, they can create space for student participation and encourage argumentation in class. This might work in line with a stronger decentralization of the educational system in which greater structural support for teacher training allows more contextualized teaching material and teacher training and increase the agency of the teacher. Financial resources need to be invested into institutionalized teacher capacity development and strengthening the cooperation with resource centers of teaching research and training. Developing teacher capacities—specifically through training on the meaning and effects of explaining complex interdisciplinary contents—will further increase the ability of students to appreciate and internalize the spirit of ESD while strengthening critical thinking and argumentation skills, as well as the will to personally take action for a better future through lifestyle choices.

(4) *Educational institutes*, particularly the NCERT and SCERTs, need to be strengthened to encourage innovation in curriculum design, syllabi, and textbook development. The political responsibility of educational stakeholders should be increased at the state and national level. Currently, senior officials are most influential in decision-making, but young professors and government officials, who are highly motivated to contribute, become reluctant to push for change due to the hierarchical structure between, and, most importantly, within these institutions. Therefore, space needs to be created to develop and integrate new ideas to reform the existing processes of curriculum design and textbook development. One measure would be to include more qualified teachers in the planning process of new curricula. Another option would be to develop formal training for teacher trainers, textbook authors, and curriculum designers, which builds on new teaching methodologies. Hence, NCERT and SCERT need to be strengthened for training and research toward the implementation pathways and teaching methods for ESD, to improve the process of translating policies into practice through consistent terminology, methodology, and conceptualization at all levels during implementation.

(5) *Close cooperation between educational researchers and educational stakeholders*, such as curriculum designers, policy makers, textbook authors, and teachers, need to be developed to closely link scientifically evidenced strategies, teaching contents and methods promoted in educational policies, curricula, textbooks, and pedagogic practice. A greater interaction between the two relevant ministries MHRD and MoEF, the NCERT, the SCERTs, the DIETS and most importantly, teachers, may strengthen their ownership and increase their impact in influencing pedagogic practice. Concrete concepts of teaching methods and contents which are empirically and theoretically grounded and contextualized may become more powerful in their implementation if they are developed and reviewed from multiple perspectives.

(6) *Teaching methodology research* needs to be strengthened for geography and other subjects with the aim of developing evidence-based approaches to competence models for each subject. Teaching methodology research centers as well as teacher education centers could be developed in which junior researchers and teachers' skills are promoted. One issue inhibiting the development of a strong teaching methodology research network is that Ph.D. theses and studies are often only found in libraries of the university where the degree was obtained, and transparency through access to studies online would facilitate the visibility of teaching methodology research.

(7) *Textbook development* at the national and state level needs to be more competitive to enrich the approaches, structures, contents, texts, and methods used in current textbooks. Currently, textbook development is marked by a process strongly structured through existing hierarchies, in which it often depends on senior officials to include new approaches and methods in textbooks. A transparent, participatory process of systematic textbook development across all grades needs to be developed and monitored. Although discussions within textbook committees take place, they follow a strong framing. As stated in the NCF 2005, capacities of textbook writers and teacher trainers for critical approaches to human–environment topics need to be promoted. There is a great need for textbook designers to be trained in psychological principles, and in diverse sets of teaching methods and writing styles to be included in the textbook to promote skills beyond the reproduction of information. These skills include critical-thinking skills, argumentation skills, network thinking and a focus on students and their interests and skills. Concrete formulations in textbooks and additional teacher handbooks are necessary to guide this change in pedagogic approach. This calls for a complete revision of textbooks as well as monitoring by external reviewers to ensure the quality of contents and teaching methods in textbooks. The orientation on the urban middle-class student could be revised by providing separate textbooks for urban and rural contexts. Textbooks should integrate an approach sensitive to gender, caste, age, and other relevant social divisions.

(8) Multiple *teaching resources* apart from the blackboard and the textbooks should be provided to enrich the learning experience of students. Material available from NGOs, newspapers, etc., may broaden students' horizons on different per-

spectives. Teachers should be trained to teach students skills to question contents and perspectives, which they are not used to doing for textbooks. Another policy to create more flexible use of classroom space and multi-dimensional communication, which is not only symbolically powerful, is to unscrew tables from the classroom floors to enable group work. This was a major logistic barrier to implementing the ESD teaching methods, as some geography lessons had to be relocated to courtyards or libraries to conduct group work.

The listed strategies work toward "systematic reforms to strengthen process [sic] of democratization of all existing educational institutions" (NCF 2009: 13). This is necessary to create an educational system in which national democratic values such as "social justice, liberty, equality, and fraternity" are realized through life skill education at the classroom level. New perspectives of teaching contents and methods which encourage critical and creative thinking can transform pedagogic practice toward promoting more inclusive forms of citizenship.

References

Archer, M. (1995). The neglect of the educational system by Bernstein. In A. R. Sadovnik (Ed.), *Knowledge and pedagogy: The sociology of Bernstein* (pp. 211–235). Norwood, NJ: Ablex Publishing Corporation. http://books.google.de/books?id=3tgXQ_ISJHYC&printsec=frontcover&hl=de&source=gbs_ge_summary_r&cad=0-v=onepage&q&f=false.

Batra, P. (2005). Voice and agency of teachers: The missing link in National Curriculum Framework 2005. *Economic and Political Weekly, 40*(1), 4347–4356.

Behrends, A., Park, S.-J., & Rottenburg, R. (2014). Travelling models: Introducing an analytical concept to globalisation studies. In A. Behrends, S.-J. Park, & R. Rottenburg (Eds.), *Travelling models in African conflict management: Translating technologies of social ordering* (pp. 1–40). Leiden: Brill.

Berndt, C. (2010). *Elementarbildung in Indien im Spannungsverhältnis von Macht und Kultur. Eine Mikrostudie in Andhra Pradesh und West Bengalen*. Berlin: Logos Verlag.

Bernstein, B. (1975). *Class, codes and control. Towards a theory of educational transmission*. London: Routledge.

Bernstein, B. (1990). *Class, codes and control. The structuring of pedagogic discourse*. London: Routledge.

Bharati Vidyapeeth Institute of Environment Education and Research. (2002). *Study of status of infusion of environmental concepts in school curricula and the effectiveness of its delivery*.

Budke, A. (2012). Argumentationen im Geographieunterricht. *Geographie und ihre Didaktik, 1*, 23–34.

Budke, A., & Meyer, M. (2015). Fachlich argumentieren lernen - Die Bedeutung der Argumentation in den unterschiedlichen Schulfächern. In A. Budke, M. Kuckuck, M. Meyer, F. Schäbitz, K. Schlüter, & G. Weiss (Eds.), *Fachlich argumentieren lernen*. Münster: Waxmann.

Budke, A., Schiefele, U., & Uhlenwinkel, A. (2010). Entwicklung eines Argumentationskompetenzmodells für den Geographieunterricht. *Geographie und ihre Didaktik, 3*, 180–190.

Clark, U. (2005). Bernstein's theory of pedagogic discourse. *English Teaching: Practice and Critique, 4*(3), 32–47.

Clarke, P. (2003). Culture and classroom reform: The case of the district primary education project, India. *Comparative Education, 39*(1), 27–44. https://doi.org/10.1080/0305006032000044922.

de Haan, G. (2008). Gestaltungskompetenz als Kompetenzkonzept für Bildung für nachhaltige Entwicklung. In I. Bormann, & G. de Haan (Eds.), *Kompetenzen der Bildung für nachhaltige Entwicklung. Operationalisierung, Messung, Rahmenbedingungen, Befunde* (pp. 23–43). Wiesbaden: VS Verlag für Sozialwissenschaften. http://link.springer.com/book/10.1007/978-3-531-90832-8, http://www.amazon.de/Kompetenzen-Bildung-nachhaltige-Entwicklung-Operationalisierung-ebook/dp/B001BS64SC/ref=dp_return_2?ie=UTF8&n=530484031&s=digital-text-reader_3531155296.

Fetherston, B., & Kelly, R. (2007). Conflict resolution and transformative pedagogy: A grounded theory research project on learning in higher education. *Journal of Transformative Education, 5*(3), 262–285. https://doi.org/10.1177/1541344607308899.

Freire, P. (1996). *Pedagogy of the oppressed*. London: Penguin Books Ltd.

Giroux, H. A. (2004). Critical pedagogy and the postmodern/modern divide: Towards a pedagogy of democratization. *Teacher Education Quarterly, 31*(1), 31–47.

Government of India. (2004). *Education for all. India marches ahead*. New Delhi: Ministry of Human Resource Development.

Govinda, R. (2002). *India education report*. New Delhi: Oxford University Press.

Henderson, K., & Tilbury, D. (2004). *Whole-school approaches to sustainability: An international review of sustainable school programs*. Report prepared by the Australian Research Institute in Education for Sustainability (ARIES) for The Department of Environment and Heritage, Australian Government.

Kingdon, G. G. (2007). *The progress of school education in India*. Oxford: ESRC Global Poverty Research Group.

Kuckuck, M. (2014). *Konflikte im Raum - Verständnis von gesellschaftlichen Diskursen durch Argumentation im Geographieunterricht*. Münster: MV-Verlag.

Kumar, K. (1988). Origins of India's "textbook culture". *Comparative Education Review, 32*(4), 452–464. http://www.jstor.org/stable/1188251.

Kumar, K. (2003). *Quality of education at the beginning of the 21st century. Lessons from India*. http://portal.unesco.org/education/en/ev.php-URL_ID=37798&URL_DO=DO_TOPIC&URL_SECTION=201.html. Accessed March 10, 2014.

Kumar, K., & Oesterheld, J. (2007). *Education and social change in South Asia*. Orient Longman.

Lall, M. (2005). *The challenges for India's education system* (pp. 1–10). London: Chatham House.

Mandelbaum, D. G. (1975). *Society in India*. Noida: Popular Prakashan.

Manteaw, O. O. (2012). Education for sustainable development in Africa: The search for pedagogical logic. *International Journal of Educational Development, 32*(3), 376–383. <Go to ISI>://000301698300003. http://ac.els-cdn.com/S0738059311001301/1-s2.0-S0738059311001301-main.pdf?_tid=e247c694-1df6-11e2-a864-00000aab0f6b&acdnat=1351095877_200e77592eca28a76d7de4643351e023.

Marrow, J. (2008). *Psychiatry, modernity and family values: Clenched teeth illness in North India*. Chicago: ProQuest.

Morais, A. M. (2002). Basil Bernstein at the micro level of the classroom. *British Journal of Sociology of Education, 23*(4), 559–569. https://doi.org/10.2307/1393312.

Mukhopadhyay, R., & Sriprakash, A. (2011). Global frameworks, local contingencies: Policy translations and education development in India. *Compare—A Journal of Comparative and International Education, 41*(3), 311–326. https://doi.org/10.1080/03057925.2010.534668.

National Council of Educational Research and Training. (2005). *National Curriculum Framework 2005*. New Delhi: NCERT.

National Council of Educational Research and Training. (2009). *National Curriculum Framework 2005. Position Papers on National Focus Groups on Systemic Reform* (Vol. II). New Delhi: NCERT.

National Green Tribunal (2014). *M. C. Mehta vs University Grants Commission Ors on 17 July, 2014*. https://indiankanoon.org/doc/155218083/. Accessed March 29, 2016.

Schockemöhle, J. (2011). Regionales Lernen - Kompetenzen fördern und Partizipation stärken. Zur Wirksamkeit des außerschulischen Lernens in der Region. In H. Bayrhuber, U. Harms, B.

Muszynski, B. Ralle, M. Rothgangel, L.-H. Schön, et al. (Eds.), *Empirische Fundierung in den Fachdidaktiken* (pp. 201–216). Münster: Waxmann Verlag.

Schweisfurth, M. (2011). Learner-centred education in developing country contexts: From solution to problem? *International Journal of Educational Development, 31*(5), 425–432. https://doi.org/10.1016/j.ijedudev.2011.03.005.

Sen, A. (2005). *The argumentative Indian*. Noida: Penguin.

Spivak, G. (2007). *Can the subaltern speak?* Berlin: Turia + Kant

Sriprakash, A. (2010). Child-centered education and the promise of democratic learning: Pedagogic messages in rural Indian primary schools. *International Journal of Educational Development, 30*(3), 297–304.

Sriprakash, A. (2012). *Pedagogies for development: The politics and practice of child-centred education in India*. New York: Springer.

Thapan, M. (2014). *Ethnographies of schooling in contemporary India*. Delhi: Sage.

Thompson, P. (2013). Learner-centred education and 'cultural translation'. *International Journal of Educational Development, 33*(1), 48–58. https://doi.org/10.1016/j.ijedudev.2012.02.009.

Tilbury, D. (2011). *Education for sustainable development. An expert review of processes and learning*. Paris: UNESCO.

UN World Water Assessment Programme. (2016). *The United Nations world water development report 2016: Water and jobs*. Paris: UNESCO.

UNESCO. (2005). *United Nations decade of education for sustainable development (2005–2014): International implementation scheme*. Paris.

Vavrus, F. (2009). The cultural politics of constructivist pedagogies: Teacher education reform in the United Republic of Tanzania. *International Journal of Educational Development, 29*(3), 303–311. https://doi.org/10.1016/j.ijedudev.2008.05.002.

Vavrus, F., & Barrett, L. (2013). *Teaching in tension. International pedagogies, national policies, and teachers' practices in Tanzania*. Rotterdam: Sense Publishers.

Appendix

Empirical Data Sources and Questionnaires

See Tables A.1, A.2, A.3, A.4, A.5, A.6, A.7, A.8, A.9, A.10, A.11 and A.12.

Table A.1 Interviews with educational and water stakeholders

Number	Code	Category	Position	Gender
1	NCERT1	NCERT	Assistant professor for geography	Female
2	NCERT2	NCERT	Professor for science and mathematics	Female
3	NCERT3	NCERT	Professor for environment education	Female
4	NCERT4	NCERT	Professor of sociology	Male
5	NCERT5	NCERT	Professor of biology	Female
6	NCERT6	NCERT	Director and professor for mathematics	Male
7	NCERT7	NCERT	Professor of geography	Female
8	NCERT8	NCERT	Professor of geography	Female
9	NCERT9	NCERT	Head of international relations division	Male
10	MSCERT 1	MSCERT	Committee of authors for environment education textbook designing	Group interview (mixed gender)
11	MSCERT 2	MSCERT	Geography textbook author	Male
12	MSCERT 3	MSCERT	Geography textbook author	Male
13	PMC 1	PMC	Executive engineer for water supply and sewage	Male
14	PMC 2	PMC	Executive engineer for water works	Male
15	PMC 3	PMC	Electrical engineer for water works	Male
16	PMC 4	PMC	Executive director	Male

(continued)

S. Leder, *Transformative Pedagogic Practice*, Education for Sustainability,
https://doi.org/10.1007/978-981-13-2369-0

Table A.1 (continued)

Number	Code	Category	Position	Gender
17	PMC 5	PMC	Environment officer	Male
18	PMC 6	PMC	Office deputy director of education	Male
19	PMC 7	PMC	Education officer at the zoo	Female
20	Prof 1	Professor	Director of education trust	Female
21	Prof 2	Professor	Assistant professor for geography education	Female
22	Prof 3	Professor	Ph.D. for environment education and geography	Female
23	Prof 4	Professor	Professor for geography	Female
24	Prof 5	Professor	Professor for sociology	Female
25	Prof 6	Professor	Professor for water economics	Female
26	Prof 7	Professor	Professor for environment education	Male
27	NGO 1	NGO	School program director of an environment organization	Male
28	NGO 2	NGO	Environment education and training program coordinator of an environment organization	Female
29	NGO 3	NGO	Director of environment education and priest	Male
30	NGO 4	NGO	Director of NGO promoting urban farming in schools	Male
31	NGO 5	NGO	Consultant for eco-spirituality	Female
32	NGO 6	NGO	Director of a school for alternative education	Male
33	NGO 7	NGO	Director of ecological service and education organization	Male
34	NGO 8	NGO	Director of environment education organization	Female
35	NGO 9	NGO	Zonal director of environment education organization	Female
36	NGO 10	NGO	Faculty coordinator and teacher trainer at slum education organization	Female
37	NGO 11	NGO	Intern for environment education organization	Female
38	NGO 12	NGO	Director of environment education organization	Male
39	NGO 13	NGO	Green school program coordinator of an environment organization	Male
40	NGO 14	NGO	Director of sustainability division of an environment organization	Female
41	Law 1	Lawyers	Supreme court judges	Group interview (mixed gender)
42	Corp 1	Corporate	Director of corporate sustainability	Male

Table A.2 Interviews with teachers, principals, and students

Number	Code	Category	Position	Gender	Participation in intervention study
1	T1_S2	Teacher	Senior geography teacher at S2	Female	Yes
2	T2_S2	Teacher	Junior geography teacher at S2	Female	Yes
3	T3_S2	Teacher	Junior geography teacher at S2	Female	Yes
4	T4_S1	Teacher	Junior geography teacher at S1	Female	Yes
5	T5_S3	Teacher	Junior geography teacher at S3	Female	Yes
6	T6_S4	Teacher	Senior geography teacher at S4	Female	Yes
7	T7_S4	Teacher	Junior geography teacher at S4	Female	Yes
8	T8_S5	Teacher	Junior geography teacher at S5	Female	Yes
9	T9_S5	Teacher	Senior geography teacher at S5	Female	Yes
10	T10_S4	Teacher	Senior geography teacher at S4	Female	No
11	T11_S4	Teacher	Senior geography teacher at S4	Female	No
12	T12_S6	Teacher	Senior geography teacher at S6	Male	No
13	T13_S7	Teacher	Senior geography teacher at S7	Female	No
14	T14_S8	Teacher	Senior geography teacher at S8	Female	No
15	P1_S3	Principal	Principal and geography teacher	Group interview (mixed gender)	No
16	P2_S2	Principal	Principal	Female	No
17	St1_S1	Student	Class 9 geography student	Female	Yes
18	St2_S2	Student	Class 9 geography students	Group interview (mixed gender)	Yes
19	St3_S2	Student	Class 9 geography students	Group interview (mixed gender)	Yes
20	St4_S2	Student	Class 9 geography student	Female	Yes
21	St5_S6	Student	Class 7 geography student	Male	No
22	St6_S5	Student	Class 9 geography students	Group interview (mixed gender)	Yes

(continued)

Table A.2 (continued)

Number	Code	Category	Position	Gender	Participation in intervention study
23	St7_S1	Student	Class 9 geography students	Group interview (mixed gender)	Yes
24	St8_S2	Student	Class 9 geography student	Male	Yes
25	St9_S5	Student	Class 9 geography student	Male	Yes
26	St10_S5	Student	Class 9 geography student	Female	Yes
27	St11_S2	Student	Class 9 geography student	Female	Yes
28	St12_S2	Student	Class 9 geography student	Male	Yes
29	St13_S1	Student	Class 9 geography student	Male	Yes
30	St14_S2	Student	Class 9 geography student	Female	Yes
31	St15_S3	Student	Class 9 geography students	Group interview (mixed gender)	Yes
32	St16_S4	Student	Class 9 geography students	Group interview (mixed gender)	Yes
33	St17_S3	Student	Class 9 geography student	Female	Yes
34	St18_S2	Student	Class 9 geography student	Male	Yes
35	St19_S4	Student	Class 9 geography student	Female	Yes
36	St20_S4	Student	Class 9 geography student	Female	Yes
37	St21_S2	Student	Class 9 geography student	Female	Yes
38	St22_S2	Student	Class 9 geography student	Female	Yes
39	St23_S2	Student	Class 9 geography student	Male	Yes
40	St24_S2	Student	Class 9 geography student	Male	Yes
41	St25_S1	Student	Class 9 geography student	Female	Yes
42	St26_S1	Student	Class 9 geography student	Male	Yes
43	St27_S4	Student	Class 9 geography student	Male	Yes

Table A.3 Visited schools

School code	School type	Location	Purpose of visit
S1	English-medium school	Pune	Classroom observation, Visual Network in group work, M3
S2	English-medium school	Pune	Classroom observation, Visual Network in group work, M2
S3	English-medium school	Pune	Classroom observation, Visual Network in group work, M1
S4	English-medium school	Pune	Classroom observation, Visual Network in group work, M1
S5	English-medium school	Pune	Classroom observation, Visual Network in group work, M1
EMS6	English-medium school	Pune	Classroom observation, Visual Network in group work
MBS7	Marathi Boys School	Pune	Visual Network in group work, Shashwat Green Day
MGS8	Marathi Girls School	Pune	Classroom observation, classroom discussion
HES9	Hindi elementary school	Pune	Classroom observation, teacher discussion
MRS10	Marathi rural school	Pune district	Exploration, interview with principal
EMS11	English-medium school	Mumbai	Exploration and student discussion, Green School Campaign
EMS12	English-medium school	Mumbai	Exploration and student discussion, Green School Campaign
EMS13	English-medium school	Mumbai	Exploration and student discussion, Green School Campaign
EMS14	English-medium school	Mumbai	Exploration and student discussion, Green School Campaign
EMS15	English-medium school	Mumbai	Exploration and student discussion, Green School Campaign
EMS16	English-medium school	Mumbai	Exploration and student discussion, Green School Campaign
MIS17	International School in Mumbai	Mumbai	Classroom observation, teaching
MSS18	Slum School in Mumbai	Mumbai	Classroom observation, teaching
ARS19	Alternate rural school	Kudal, Maharashtra	Exploration, Alternate School, NGO
URS20	Hindi rural school	Uttarakhand	Health and Education Camp, observation, and discussion
URS21	Hindi rural school	Uttarakhand	Health and Education Camp, observation, and discussion
URS22	Hindi rural school	Uttarakhand	Health and Education Camp, observation, and discussion

(continued)

Table A.3 (continued)

School code	School type	Location	Purpose of visit
URS23	Hindi rural school	Uttarakhand	Health and Education Camp, observation, and discussion
URS24	Hindi rural school	Uttarakhand	Health and Education Camp, observation, and discussion
URS25	Hindi rural school	Uttarakhand	Health and Education Camp, observation, and discussion

Table A.4 National and Maharashtra state geography textbooks analyzed in this study

Textbook code	Textbook title	Board	Standard	Year/edition
M-Geo-3P	Geography Standard Three, Pune District	MSBSHSE	III	2012/1
M-Geo-3M	Geography Standard Three, Brihan Mumbai	MSBSHSE	III	2008/1
M-Geo-4	Geography Standard Four	MSBSHSE	IV	2009/1
M-Geo-5	Geography Standard Five	MSBSHSE	V	2010/1
M-Geo-6	Geography Standard Six	MSBSHSE	VI	2011/1
M-Geo-7	Geography Standard Seven	MSBSHSE	VII	2011/1
M-Geo-8	Geography Standard Eight	MSBSHSE	VIII	2010/1
M-Geo-9	India: Physical Environment, Geography Standard IX	MSBSHSE	IX	2011/1
M-Geo-10	India: Human Environment, Geography Standard X	MSBSHSE	X	2012/1
M-Geo-11	Principles of Physical Geography	MSBSHSE	XI	2009/4
M-Geo-12	Principles of Human Geography	MSBSHSE	XII	2011/7
M-Geo-9 Ec	Geography and Economics Standard IX	MSBSHSE	IX	2012/1
M-Geo-11	Geography Standard: XI	MSBSHSE	XI	2012/1
N-EVS-3	Environmental Studies: Looking around, Textbook for Class III	NCERT	III	2013/1
N-EVS-4	Environmental Studies: Looking around, Textbook for Class IV	NCERT	IV	2012/1
N-EVS-5	Environmental Studies: Looking around, Textbook for Class V	NCERT	V	2013/1
N-Geo-6	Social Science: The Earth our Habitat, Textbook in Geography for Class VI	NCERT	VI	2012/1
N-Geo-7	Social Science: Our Environment, Textbook in Geography for Class VII	NCERT	VII	2010/1
N-Geo-8	Social Science: Resources and Development, Textbook in Geography for Class VIII	NCERT	VIII	2010/1

(continued)

Table A.4 (continued)

Textbook code	Textbook title	Board	Standard	Year/edition
N-Geo-9	Social Science: Contemporary India I, Textbook in Geography for Class IX	NCERT	IX	2006/1
N-Geo-10	Social Science: Contemporary India II, Textbook in Geography for Class X	NCERT	X	2012/1
N-Geo-11 PhyEnv	India: Physical Environment, Textbook in Geography for Class XI	NCERT	XI	2011/1
N-Geo-12 IndPpEc	India People and Economy, Textbook in Geography for Class XII	NCERT	XII	2011/1
N-Geo-11 FPhyGeo	Fundamentals of Physical Geography, Textbook for Class XI	NCERT	XI	2011/1
N-Geo-12 FHumGeo	Fundamentals of Human Geography, Textbook for Class XII	NCERT	XII	2007/1
N-Geo-11 PWGeo 1	Practical Work in Geography Part I, Textbook for Class XI	NCERT	XI	2008/1
N-Geo-12 PWGeo 2	Practical Work in Geography Part II, Textbook for Class XII	NCERT	XII	2008/1

Table A.5 National educational policy documents, curricula, and syllabi analyzed in this study

Year of publication	National Educational Policy Documents	Publishing institution
1986	National Policy on Education (NPE) 1986	MHRD
1988	National Curriculum for Elementary and Secondary Education. A Framework 1988	NCERT
1993	Learning without Burden. Report of the National Advisory Committee appointed by the Ministry of Human Resource Development	MHRD
2000	National Curriculum Framework for School Education 2000	NCERT
2004	Syllabus for Environmental Education in Schools 2004	NCERT
2004	Curriculum Framework for Teacher Education 2004	NCERT
2005	National Curriculum Framework (NCF) 2005	NCERT
2005	Position Papers of National Focus Groups for the National Curriculum Framework 2005	NCERT
2005	Syllabus for Classes at the Elementary Level, Volume 1, National Curriculum Framework 2005	NCERT
2005	Syllabus for Secondary and Higher Secondary Classes, Volume 2, National Curriculum Framework 2005	NCERT
2006	The Reflective Teacher. Organization of In-Service Training of the Teachers of Elementary Schools under SSA	NCERT
2009	National Curriculum Framework–A Historical Perspective	NCERT
2009	National Curriculum Framework for Teacher Education 2009	NCTE

Table A.6 Structure of the teacher workshop

Time	Program	Facilitator
10.00–10.20	Introduction to ESD and the institute BVIEER	Prof. Dr. Bharucha, Dr. Shamita Kumar
11.20–11.30	Introduction of participating teachers	Plenum
10.30–11.00	Introduction of research project, preliminary results and objectives of intervention study with teachers, teacher group work on opportunities, and barriers of ESD in geography lessons	Stephanie Leder
11.00–11.10	Coffee break	
11.10–11.30	Introduction to ESD teaching modules and workshop procedure	Stephanie Leder
11.30–12.30	Working groups: Three groups of teachers test one ESD teaching module	Stephanie Leder, Dr. Shamita Kumar, Dr. Kranti Yardi
12.30–13.30	Lunch break	
13.30–14.30	Evaluation, modification, and reflection of ESD teaching modules in three groups of teachers	Stephanie Leder, Dr. Shamita Kumar, Dr. Kranti Yardi
14.30–14.40	Coffee break	
14.40–15.30	Discussion in plenum on implementation of teaching module in own geography lessons, identification of barriers and opportunities	Stephanie Leder
15.30–16.30	Distribution of teaching material (teaching modules, laminated pictures, worksheets) and detailed arrangement of lesson plan	Stephanie Leder

Table A.7 Student questionnaire and interview questions

Student Questionnaire and Interview Questions	
Geography Students at English-medium Secondary Schools	
Name:	School:
Gender:	Standard/Class:
Caste:	Place of living:
Religion:	Number of people in your household:

	Water in Pune
1	Which water do you drink (bottled, filtered, tap)? Why?
2	From where do you get the water?
3	Who provides you with water? How much does it cost?
4	How does water get to your neighborhood?
5	Have you been sick because of water? Which disease did you have? Why?
6	Do you think there are problems with water in Pune? If so, please list them and explain.
7	What do you think how much water do you/your household need every day? What is water used for?
8	Do you think you have sufficient water? Why?
9	Do you think you could save water? Why would this be important to you?
10	What do you think: why do some people drink better water in Pune than others?
11	How could this problem be solved?
12	Do you think there is enough drinking water for everyone in India? Why? To which extent?
13	Do you know about drinking water in other countries? Which problems do they face? Do they have sufficient water? Why?
	Studying Experience
14	From a scale from 1 (very much) to 5 (not at all), how much do you like geography?
15	What do you like about geography?
16	What do you NOT like about geography?
17	Do you like the teaching style in geography? Why (not)?
18	Would you like to have a different teaching style in geography?
19	What have you learned about water in school?
20	Do you think it is important to learn more about water?
21	What would you like to learn about water?
22	Have you ever seen or visited a dam, water works or a treatment plant? If so, what did you do there and what did you learn?
23	Do you have discussions in geography class? What do you discuss about?
24	Do you like discussions? Why (not)?
25	Do you think it is important to discuss in geography class? Why (not)?

Table A.8 Teacher interview questions

Teacher Interview Questions	
Teachers and Stakeholders of Water Education (NGOs)	
Name:	Standards taught:
Gender:	Mobile No/ Email:
Age:	Date/ Place:

1	Could you please introduce yourself?
2	Which subjects and standards are you teaching?
3	How long have you been teaching (at this school)?
4	How many teachers are teaching at this school?
5	Are there any projects related to water at this school? (green school projects, competitions, seminars, workshops, documentations, books, DVD in library)
6	Are there any cooperation with NGOs, government programs, environment clubs & colleges?
7	Have you heard of the Green Teachers program?
8	Is your teaching or your school influenced by these programs?
	Water Supply
9	Are there any problems for the people in Pune regarding water? Why?
10	How could it be solved?
11	Who is responsible for the drinking water supply?
12	What do you understand in regard to sustainable water supply?
	Teaching on Sustainable Development
13	Have you heard about sustainable development? Is it important to teach about it? If so, why and how?
14	To which extent is it the task of schools to integrate sustainability aspects in school?
15	To which extent is geography class important for teaching sustainability?
16	Do you integrate sustainability aspects (content wise) in your teaching?
17	How are different grades sensibilized for sustainability/water conflicts?
18	What should be done to improve teaching on sustainability?
19	Which difficulties do you see in making students understand the sustainable development concept?
	Teaching on Water
20	Do you teach about water use?
21	Which difficulties do you see in making students understand aspects of water use?
22	Does anything hinder you to promote sustainable development and/in water education (institutional, experiential)?
23	Which information is included in the curriculum? What is the development of different grades, class material, teaching methods?
	Student Focus
24	Which understanding do students have of drinking water supply (individually every day, local, municipal, national, global)?
25	Do students know where they get their water from and how water reaches to them? Do students know whether water is clean?
26	Which understanding do students have of sustainable development (local, regional, global)?

Table A.8 (continued)

27	In which way is drinking water supply a problem in the students' living environment?
28	Which problems should be addressed to students concerning drinking water/tap water?
29	How could these problems be tackled in geography class?
30	How can the understanding of complex man-environment relations be enhanced in class?
31	Which strategies would students help to understand sustainability?
32	How can students learn to transfer their environmental knowledge into actions?
	Communication in Classrooms
33	How do your students communicate in the classroom?
34	Do you have any experiences with discussions/argumentation in class?
35	How could this be promoted?
36	How can students improve their communication skills?
37	(How) do you promote communication skills?
38	(How) do your students learn to argue in class?
39	Are there ways for students to participate on a political level?
40	How could students be prepared to argue with politicians?
41	Is there a way to promote the capacity to act?
42	Which teaching material do you use in class to promote discussions?
43	Do you have any teaching material on water I could copy?
44	Could you recommend anyone else I could talk to?

Table A.9 Water stakeholder interview questions

	Water Stakeholder Interview Questions
	Stakeholders of Municipal and Private Water Supply; Academics
	Name: / Profession/ Position:
	Gender: / Mobile No/ Email:
	Age: / Date/ Place:
	Introduction
1	Could you please introduce yourself?
2	Could you tell me about your working area?
	Current Situation of Water Supply in Pune
3	Could you tell me about the drinking water supply in Pune? How does it work?
4	Which factors determine if people have sufficient and good quay drinking water in Pune? How? Why? (economic/ social/ political/ infrastructural/ natural)
5	How much water is available to different residential districts?
6	How does the water quality differ in different residential districts?
7	What different kinds of water do people drink (filtered, bottled, tap)?
8	Who is responsible for the water supply in Pune (stakeholders) and which power do they have?
9	What is the role of private water providers? Where are they efficient?
	Water Complexities
10	What are strength and weaknesses of the water infrastructure in Pune (and why)?
11	Do you see the need to change anything about the drinking water supply in Pune for the future (leakages/ storage/ metering/ intermittent supply/ sewage/ administrative)
	Water Solutions
12	How would you improve about the water supply in Pune? Is it realistic?
13	Do you think providing only bottled water would be a solution?
14	(How) can sustainable water supply (economic, environmental, social) be achieved in Pune?
15	What are schemes related to drinking water supply sustainable?
16	What are your takes of JNNURM in regards to water supply in Pune (positive or negative impact)?
	Role of Education
17	Do you think education on water is important? Why?
18	Which aspects of water in Pune should be addressed for students?
19	Using which teaching methods and which teaching material should students be taught on water?
20	How is teaching on water supply integrated in education system?
21	Are there any programs you can tell me about, e.g. by NGOs?
	Data, Maps & Contacts
22	Do you have any data or maps of the water supply (quantity) and quality in different residential areas?
23	Could you recommend anyone else I could talk to? (NGOs, people working for PMC water department, private providers)?

Table A.10 Teacher workshop questionnaire and rating results

Teacher Workshop Questionnaire

Education for Sustainable Development in Geography Teaching: Water in Pune

Name:	**Standards taught:**
Gender:	**Mobile No./ Email:**
Age:	**Date/ Place:**

Please rate the following questions from 1 = totally agree to 5 = do not agree at all.
Please explain your answer on the back of this page. Thank you!

1	NCERT/ MSCERT curricula and textbooks focus on content rather than methods. Rote learning is most important.	3,2 (n=9)
2	Pre- and in-service teacher training focus on learning the content of the syllabus, rather than practicing teaching strategies.	2,8 (n=9)
3	I lack the support of the school management and the parents to focus on other topics rather than exam preparations.	4,2 (n=9)
4	For exams, too many facts and details need to be learned rather than that concepts are understood.	1,9 (n=9)
5	I face the problem of large class sizes, which hinders me to teach more learner-centered methods.	2,2 (n=9)
6	I do not feel that I have enough supportive teaching material. I also have only limited access to reference material and teaching aids.	3,9 (n=9)
7	I feel that students have a rather fragmented knowledge. I guess this is because exercises focus on definitions and listing facts in textbooks and exams.	1,9 (n=9)
8	I think that the content in textbooks is not locally specific and examples given are very general.	1,4 (n=9)
9	The language used in textbooks is not appropriate for children.	4,8 (n=9)
10	I do only little partner or group work in class. Projects take place rather outside of the classroom.	2,2 (n=9)

Thank you for answering the following questions in detail on the back of this sheet.

11	Which difficulties do you face trying to teach on topics other than the curriculum suggests or exams focus on?
12	Which difficulties do you face trying to teach with student-centered teaching methods?
13	Which opportunities do you see in teaching on water with a focus on Pune and sustainability?
14	Which opportunities do you see in teaching with communicative, student-centered methods?
15	What did you like about the workshop?
16	What would you like to improve about the workshop?

Table A.11 Textbook authors' evaluation survey of ESD in textbooks and rating results

Textbook Authors' Evaluation of ESD in Textbooks

Name of textbook/ subject/ class/ chapter: ____________________

Please rate each question: 1= I fully agree, 2= I agree, 3= I partly agree, 4 = I disagree, 5 = I fully disagree

Please explain relevant statements on the back of this sheet.

No	Content Statement	NCERT textbooks (n=8)	MSCERT textbooks (n=10)	NCERT and MSCERT textbooks (n=18)
1	The curiosity of the student is raised.	3,0	2,2	2,4
2	The structure of the chapter is logic in content and visualized clearly.	2,5	2,2	2,3
3	Relevant key terms are bold printed.	3,0	1,7	2,1
4	Key terms are defined well.	2,8	2,3	2,4
5	Different sources (quotes or diagrams from books) are given.	2,3	2,7	2,5
6	Sustainability aspects are mentioned sufficiently.	3,0	3,0	3,0
7	Social aspects are covered sufficiently.	3,0	3,1	3,1
8	Economic aspects are covered sufficiently.	4,0	2,8	3,2
9	Environmental aspects are covered sufficiently.	2,5	2,4	2,5
10	Causes and effects are explained properly.	2,8	2,8	2,8
11	The amount of information given in the text is too much.	3,8	2,9	3,2
12	Multi-perspectives or different views are stated.	3,3	3,0	3,1
13	Controversies are stated to encourage critical thinking.	3,0	4,0	3,7
14	The language used is age appropriate for the students.	2,0	2,1	2,1
15	The content is related to the students' local environment.	2,3	3,1	2,8
16	Diagrams, tables and figures are sufficient and good in quality.	2,8	3,1	3,0
17	Pictures and drawings are sufficient and explanatory.	3,3	3,1	3,2
18	Material additional to the text is numbered and sources are mentioned.	3,3	3,3	3,3
	Competences Addressed in Tasks			
19	Reproduction skills (define, list....) are enhanced.	2,5	2,9	2,7
20	Reorganization/Transfer skills (explain...) are enhanced.	2,3	2,7	2,6
21	Problem-solving skills (evaluate...) are enhanced.	2,8	3,4	3,2
22	Discussions/communication between students are encouraged.	2,5	3,5	3,2
23	Method competence is encouraged.	2,5	3,1	2,9
24	Action competence is encouraged.	3,8	3,1	3,3
25	Evaluation competence is encouraged.	2,3	2,9	2,7
26	Group work is encouraged.	2,5	3,5	3,2
27	Students have to make interlinkages.	3,0	3,3	3,2
28	The answers of the questions are given in the text.	2,8	2,2	2,4
29	Meaningful revision tasks are given.	3,0	3	3,0
30	A comprehensive overview for each chapter is given.	4,0	2,8	3,1
31	Clear learning objectives are given.	3,5	2,6	2,9
32	Complex thinking is encouraged.	3,8	2,8	3,1
33	Methodological advice of how to develop competencies is given.	3,3	3,2	3,2

Table A.12 Student evaluation survey of the ESD teaching methods and rating results

Student Evaluation of the ESD Teaching Methods

Please rate each question on the teaching module (M1, M2 or M3) from:
1 = I fully agree, 2 = I agree, 3 = I partly agree, 4 = I disagree, 5 = I fully disagree
Please explain relevant statements on the back. Thank you!

No	Statements	M1 (n=79)	M2 (n=44)	M3 (n=25)	M1, M2, M3 (n=148)
1	I think the lesson was very interesting.	1,3	1,3	1,0	1,2
2	It was difficult to understand what I have to do.	4,2	3,7	4,6	4,1
3	I have learned something new.	1,6	1,3	1,4	1,5
4	I found it difficult to arrange the picture with arrows.	3,9	3,8	4,2	3,9
5	I had difficulty to think of explanations.	3,7	3,5	4,2	3,7
6	I felt that I learnt about different perspectives.	1,9	1,8	1,5	1,8
7	I felt that there was a conflict in the discussion.	3,7	3,1	4,4	3,6
8	It was difficult to react to what the other said.	3,9	4,2	4,3	4,1
9	I could relate the topic to my environment or myself.	2,1	1,8	1,8	1,9
10	I felt I did not have enough information to solve the assignment.	3,9	3,6	4,1	3,8
11	The material given was difficult to understand.	4,3	4,1	4,6	4,3
12	I liked the lesson.	1,4	1,2	1,0	1,3
13	I could use the knowledge from some other lesson.	2,2	2,4	2,7	2,3
14	I explained something to my classmates during the activity.	2,1	2,0	1,8	2,0
15	I felt I could solve the problem.	2,1	1,8	1,9	2,0
16	I discussed more than usual in the class with my classmates.	2,3	2,1	1,8	2,2
17	I do not like group work.	4,6	4,4	4,9	4,6
18	I feel now that I can judge on the topic better than before.	1,8	2,0	1,8	1,9
19	I felt I could get a better overview of the topic than in normal classes.	2,2	2,0	1,4	2,0
20	I felt disturbed because of my classmates.	4,0	3,7	4,5	4,0
21	I like to speak in the group.	1,8	1,9	1,6	1,8
22	I would like to have a lesson like that more often.	1,6	1,8	1,5	1,7
23	I feel confused about the topic now.	4,5	4,0	4,5	4,3
24	I feel I can use the knowledge from the lesson at home.	1,7	2,1	1,2	1,7
25	I do not understand why I have to do this assignment.	4,2	4,0	4,8	4,2
26	It was too complicated for me. Normal classes are easier.	4,4	4,1	4,7	4,4
27	I would need more help and better instructions by the teacher.	4,1	3,4	3,2	3,7
28	If I do a similar thing again I will do much better next time.	1,4	1,6	1,4	1,4

Photographs

See Photos A.1, A.2, A.3, A.4, A.5 and A.6.

Photo A.1 Arrangement of regular classroom teaching during an inquiry–response cycle: uni-directional communication

Photo A.2 Students discuss pictures in groups before arranging them in a "Visual Network" in a relocated lesson

Photo A.3 Students arrange a "Visual Network" in the library

Photo A.4 Drawing of a teacher to explain students that river water is polluted by bathing and by washing clothes, dishes, and buffalos, and at the same time used for drinking water

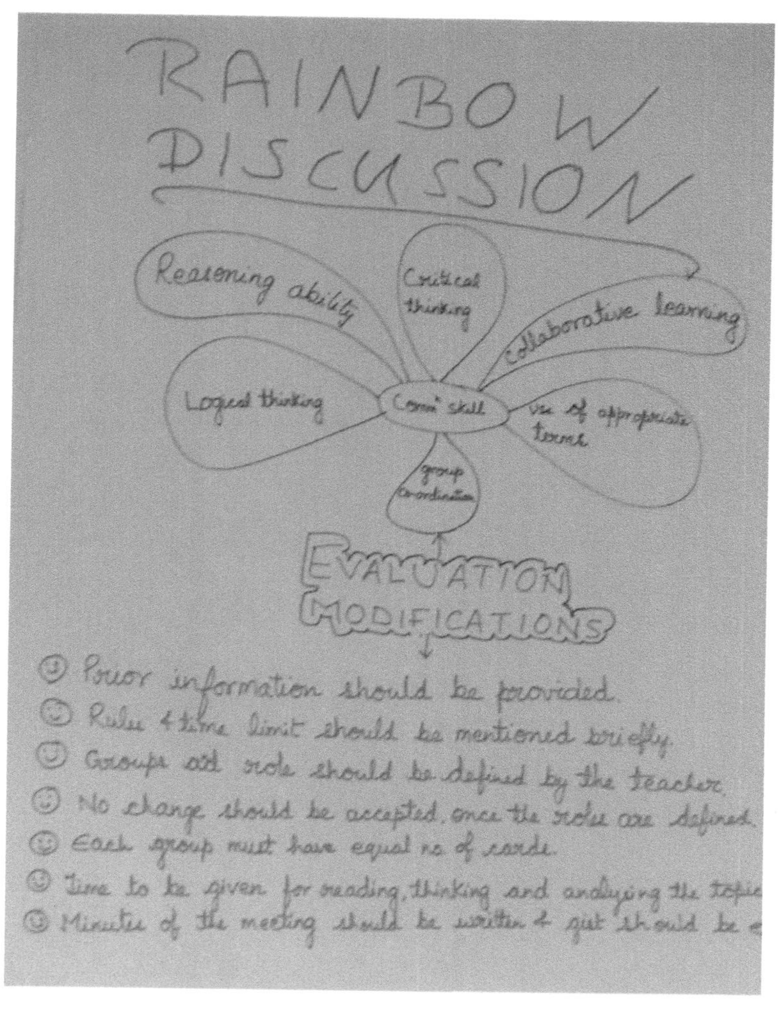

Photo A.5 Teacher evaluation of the ESD teaching module "Rainbow Discussion"

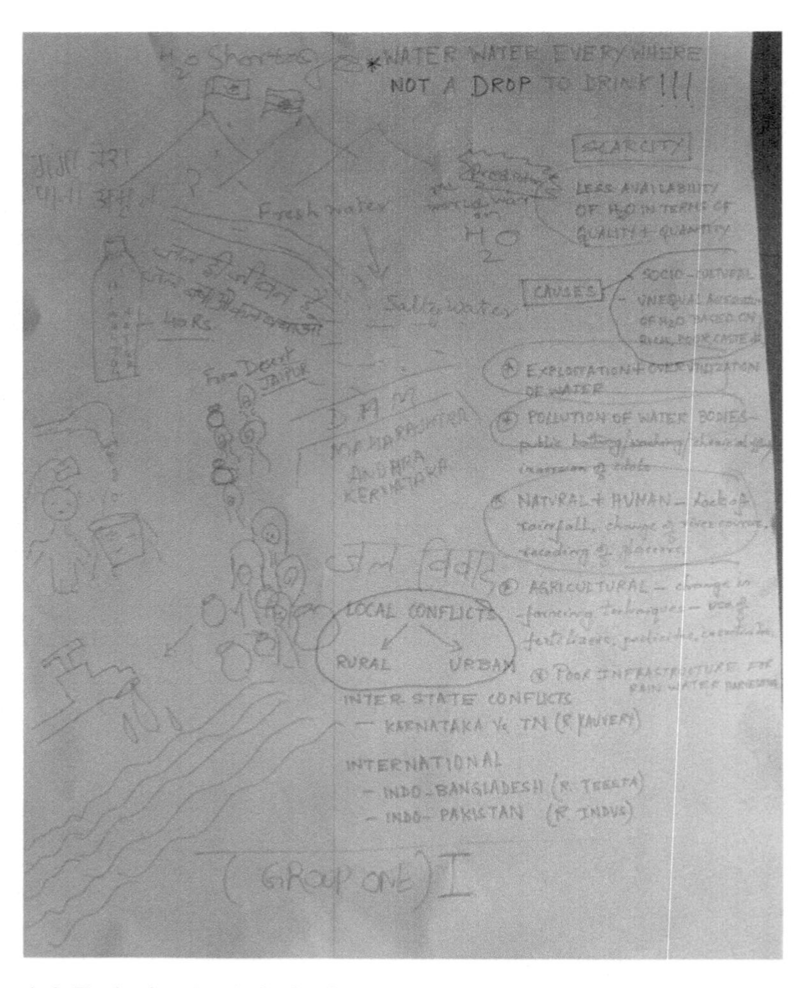

Photo A.6 Textbook author draft of an ESD approach to a water chapter in a geography textbook

Textbook Excerpts

See A.7, A.8 and A.9

Photo A.7 MSCERT geography textbook for Pune district for class 3 (M-Geo-3P)

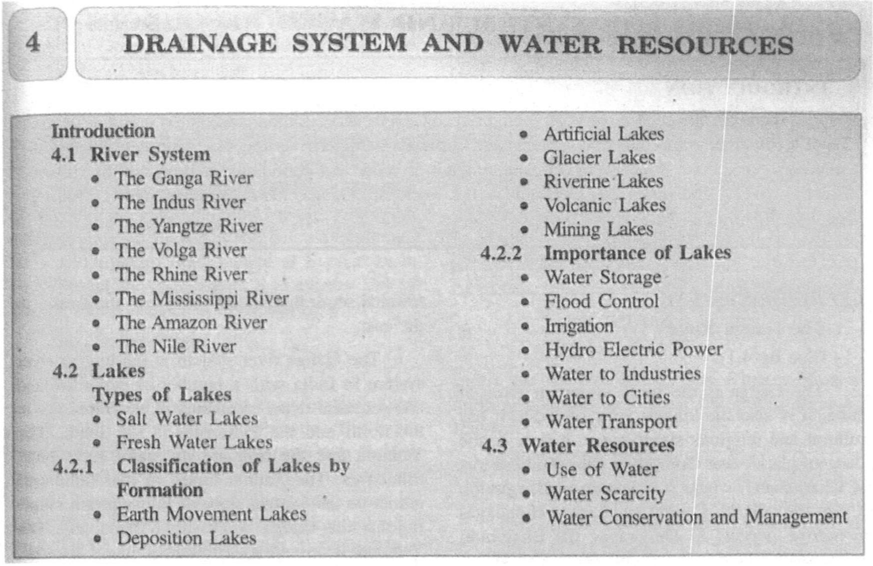

4 DRAINAGE SYSTEM AND WATER RESOURCES

Introduction
4.1 River System
- The Ganga River
- The Indus River
- The Yangtze River
- The Volga River
- The Rhine River
- The Mississippi River
- The Amazon River
- The Nile River

4.2 Lakes
Types of Lakes
- Salt Water Lakes
- Fresh Water Lakes

4.2.1 Classification of Lakes by Formation
- Earth Movement Lakes
- Deposition Lakes
- Artificial Lakes
- Glacier Lakes
- Riverine Lakes
- Volcanic Lakes
- Mining Lakes

4.2.2 Importance of Lakes
- Water Storage
- Flood Control
- Irrigation
- Hydro Electric Power
- Water to Industries
- Water to Cities
- Water Transport

4.3 Water Resources
- Use of Water
- Water Scarcity
- Water Conservation and Management

Photo A.8 Content of Chap. 4 "Drainage System and Water Resources" in MSCERT textbook for class 11 (M-Geo-11)

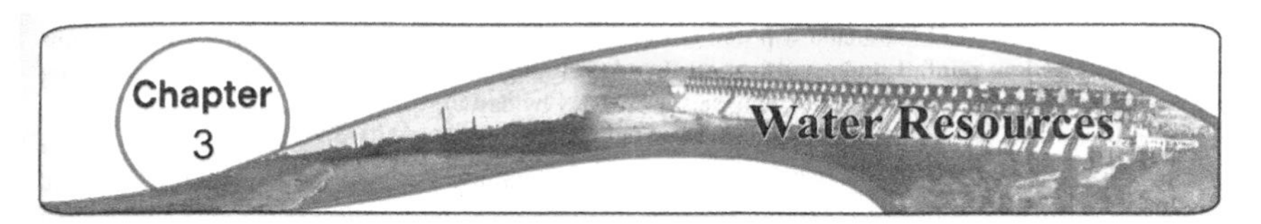

Importance of Water Resources :

• House use • Agriculture • Industries • Hydro electricity • Water Transport • Fishing.

Water is an important and valuable resource in India. **Monsoon** rainfall is the primary source of water resources in India. Monsoon rainfall influences availability and use of water resource. Rivers, glaciers, lakes, springs and wells are the secondary sources of water. Rivers and lakes are important for flow, storage and regulation of the currents of rain water.

Surface Water :

Rivers, tanks, lakes etc. are the main forms of surface water resources. In India, rivers are the main source of surface water. There are 24 major river basins in India. The annual flow of water in a river valley forms the water resource of that river basin. About 60 per cent of the total water resources in India are concentrated in the Ganga and Brahmaputra basins. These basins cover 33 per cent of the total geographical area of India. Besides, the water resource in Godavari basin is the largest among the river basins in Southern India. The west flowing rivers have only 11 per cent of water resources. Krishna, Narmada and Indus river systems are also significant water resources in India.

Sub-surface Water :

Sub - surface water is known as 'ground water'. Since ancient times ground water has been used as a resource in India. The Northern Plains in India have a great potential for ground water resource development. States of Uttar Pradesh, Maharashtra, Madhya Pradesh, Tamil Nadu etc. are leading in the development of ground water resources. Out of the total water resources usable ground water accounts for 31 per cent. In India, ground water resources have been developed in the regions where surface water resources are limited. The states of Punjab, Haryana, Uttarakhand, Rajasthan, Gujarat, Tamil Nadu, Andhra Pradesh, Madhya Pradesh, Karnataka, Maharashtra etc. have developed ground water resources in the regions of low rainfall.

Water resources play an important role in human life. In addition to domestic uses water is largely used for agricultural and industrial purposes.

Agriculture in India is dependant mainly on rain water. For the proper growth of crops and for their large scale production, regular and adequate supply of water is essential. This supply of water to the crops becomes available by rainfall and irrigation.

Irrigation :

Due to seasonal nature of rainfall, its irregularity and variability in quantity, water resources are not uniformly available in all parts of the nation. It is found that merely 30 per cent of the total cropped area receives adequate rainfall and 70 per cent area gets low rainfall. Therefore irrigation is required for proper growth of crops. Requirement of water for crops like rice and sugarcane can be met only by irrigation.

While considering the relation between statewise distribution of amount of rainfall and cultivated area we come to know the areas which need irrigation. States like Kerala, Orissa, West Bengal, Manipur, Assam, Tripura, Nagaland, Meghalaya, Sikkim etc. have almost all cropped areas in the regions of heavy rainfall. However, states like Haryana, Rajasthan, Gujarat, Maharashtra and Karnataka have more than 75 per cent of cropped area in regions of inadequate rainfall. In spite of inadequate rainfall some regions have made enough progress in irrigation and development of agriculture.

9

Photo A.9 Introductory page of MSCERT geography textbook for class 10, Chapter "Water Resources" (M-Geo-10)

ESD Teaching Modules

ESD Teaching Module I: Visual Network on Water Conflicts in Pune

Learning Objectives

1. Students should be able to arrange pictures in a network, which reflects interconnected factors of water supply in their own city.
2. Students should be able to explain each interlinkage between two pictures
3. Students should be able to explain different causes (water pollution, precipitation…) and effects (unequal access to water, diseases…) of water supply problems in their own city.
4. Students should be able to write a well-structured essay explaining water problems and their interlinkages in Pune.

Teaching phase	Content and method	Medium
Phase 1: Warm-up	Describe your pictures and place it in front of you	Pictures
Phase 2: Pre-activity I	In partner work, write titles for the pictures on the work sheet	Worksheet A: Water in Pune–Picture Titles
Phase 3: Pre-activity II	Exchange your titles with the other groups and agree on titles	Worksheet A: Water in Pune–Picture Titles
Phase 4: Activity I	Arrange pictures with arrows in a network	Arrows
Phase 5: Activity II	Explain the meaning of the arrows on the worksheet with your partner	Worksheet B: Water in Pune–Arrow explanations
Phase 6: Reinvestment	Write an essay on water problems in Pune	Notebook

Material needed per group of 4 students:

- Ca. 12 pictures and 20 arrows
- 2× Worksheet A: Water in Pune—Picture Titles
- 2× Worksheet B: Water in Pune—Arrow explanations

Worksheet A: Water in Pune—Picture Titles

Picture A	
Picture B	
Picture C	
Picture D	
Picture E	
Picture F	
Picture G	
Picture H	
Picture I	
Picture J	
Picture K	
Picture L	
Picture M	
Picture N	
Picture O	
Picture P	
Picture Q	
Picture R	
Picture S	
Picture T	

Worksheet B: Water in Pune—Arrow Explanations

Arrow between ______ and _______	
Arrow between ______ and _______	
Arrow between ______ and _______	
Arrow between ______ and _______	
Arrow between ______ and _______	
Arrow between ______ and _______	
Arrow between ______ and _______	
Arrow between ______ and _______	
Arrow between ______ and _______	
Arrow between ______ and _______	
Arrow between ______ and _______	
Arrow between ______ and _______	
Arrow between ______ and _______	
Arrow between ______ and _______	
Arrow between ______ and _______	

Selected photographs and symbols for the picture network

ESD Teaching Module II: Position Bar[1] on 24/7 Water Supply in Pune

Learning Objectives

1. Students should be able to arrange statements on the sustainability of 24/7 water supply in their own city on a position bar between pro and contra, so they realize that there are not only extreme positions but also balanced once.
2. Students should be able to identify various stakeholders who could have made each statement.
3. Students should be able to take over the role of a stakeholder and give reasons for her or his position.
4. Students should be able to write their opinion grounded with firm arguments.

Teaching phase	Content and method	Medium
Phase 1: Warm-up	1. Compare the statements on 24/7 water supply. How do they relate to each other? 2. Classify the statements in pro/contra arguments. Is it easy to classify them? Are they all contradictive? How does the way of argumentation differ?	Argument cards
Phase 2: Pre-activity I	Arrange the arguments on the position bar with your partner	Argument cards, position bar, worksheet
Phase 3: Pre-activity II	Identify possible stakeholders who could have made this statement and explain what their interest might be	Stakeholder cards
Phase 4: Activity	Take over the role of a stakeholder who is interviewed by your partner and give grounded reasons for your position	Worksheet
Phase 5: Reinvestment	Write an essay from the view of a stakeholder using firm arguments to convince the reader	Notebook

Material needed for 2 students:

- Worksheet "Position bar"
- Argumentation cards
- Worksheet "Starting sentences for an argumentative essay"

[1]**This teaching module is based on the "Meinungsstrahl" by** Mayenfels and Lücke (2012)

Argumentation Cards

Everyone would have equal, continuous and sufficient access to water with 24/7 water supply.	For 24/7 water supply, water needs to be metered. Then people would use less water because they have to pay. At the end, we would save water and stop wasting it.	How can this possibly be sustainable? Pune area has only seasonal rainfall- therefore we naturally have to live with minimal water and cannot afford 24/7 water supply as other regions and countries with continuous rainfall.
24/7 will prevent conflicts between the irrigation department and the municipality. There would be enough water then for everyone as per their needs.	People would overuse water instead of saving water. Maybe they simply manipulate the water meters and use 24/7 water supply for free.	Pune's water supply is not enough, and especially poor people do not have enough water. 24/7 is only a political strategy to win votes as it sounds like a wonderful idea.
24/7 water supply will only benefit the rich who can afford to pay for water. Poor people will never benefit.	Infrastructural investments should be done for recycling water in treatment plants before talking about supplying more water.	It is too expensive, because for 24/7, we need technical improvements on the pipelines which we cannot afford.
With 24/7 we will have less water cuts in summer. The result is finally more water for everyone.	Water pipelines need to be renewed and maintained, because there is a 40% of water loss while transmission. If leakages are fixed, we can talk about 24/7 water supply.	It is a new politician's idea to let people think they are improving the city's water supply- but in fact, they never will.
The water supply to Pune is not only sufficient, but much higher than the per-head water supply in other cities. Therefore 24/7 would prevent further disparity in distribution as everyone pays for their use.	If a basic amount of water is supplied for free and people have to pay for overuse, then 24/7 water supply is fair for the poor.	The continuous availability of water will lead to the exploitation of our water resources in summer. We will have even less water than now and water costs will be so high, that only the rich can afford it.

Starting sentences for an argumentative essay

In my opinion…
I (do not) agree that…
First of all…
Furthermore…
Another reason is that….
I would like to stress that…
I (do not) think that….
On the contrary, …
I am convinced that….
It should be mentioned that…
Last but not least, …
In conclusion, …

Further Conjunctions:

Because of…	Although…	As well as…
This shows that…	Responsible for this is…	This is contrary to…
One reason is that…	This results in…	This is because of…
Since…	Therefore…	If… then…
Most importantly is that…	…But…	So….

ESD Teaching Module III: Rainbow Discussion[2] on Sustainable Water Supply in Pune

Learning Objectives

1. Students should be able to come up with a list of pro and contra arguments on sustainable water supply in Pune with the help of material.
2. Students should be able to present arguments for both positions, pro and contra.
3. Students should be able to participate in a discussion with relevant, objective, and well-grounded arguments.
4. Students should be able to refer to counterarguments.
5. Students should be able to evaluate arguments on whether they are relevant, objective, and well grounded.

[2]This teaching method is based on the "Regenbogen-Vierer" by Kreuzberger (2012)

Teaching phase	Content and method	Medium
Phase 1: Warm-up	Form groups of six students each and read and discuss the worksheet	1 worksheet on sustainability
Phase 2: Pre-activity I	Fill in pro and contra arguments for one dimension of sustainability on the evaluation sheet with your partner, based on the worksheet. (2 students get one dimension and one evaluation worksheet)	Evaluation worksheet
Phase 3: Pre-activity II	Present your results to your group and fill in their answers for the other two dimensions in your dimension worksheet	Evaluation worksheet and 3 worksheets on social, economic, or ecological sustainability
Phase 4: Pre-activity	Draw a role card and five speech cards and prepare yourself for the discussion	Role cards and colorful speech cards
Phase 5: Activity	Discuss in your group and put down one colorful speech card whenever it is you wish to speak. Finish all of them	Colorful speech cards
Phase 5: Reinvestment	Feedback of observer and minute keeper. Discuss in the group what you have learned through this method	Observation worksheet

Material needed per group of 6 students:

- 4 × 5 colorful speech cards (5× red, 5× blue, 5× yellow, 5× green)
- 6 role cards
- 2 × 3 sustainability worksheets (2× economic sustainability, 2× environmental sustainability, 2× social sustainability)
- 6 evaluation worksheets
- 1 observation worksheet for minute keeper.

Social Sustainability

Access to water in Pune is very diverse. On average, the available amount of water is 229 liters per capita daily (lpcd) which is more than in most other Indian cities (ESR 2010). But while some people may use 350 lpcd or more, others have less than 80 lpcd for domestic purposes. Excessive overuse and equal access to water needs to be regulated through the municipality. People should pay for water overuse. Furthermore, people need to become aware of saving water. Water is a basic right for all. Clean and sufficient water will ensure hygiene and prevent diseases. Therefore, equal access to drinking water will decrease the gap between the rich and the poor.

"How could I possibly save water? I need to shower and clean my car every day and I think it is my right to keep things clean. I also have a garden which needs to be watered daily. This is greenery, for the environment, na?
For industries and farming, far more water is used than my family ever uses in their life for individual purposes. I am not responsible for any water wastage. So make them save water first, then I might start thinking about saving water!"
Resident (49), Koregaon Park

Ecological Sustainability

Water Pollution in Pune

Water is polluted through domestic, industrial, and agricultural wastewater which is directly led into Mula and Mutha rivers without any treatment to eliminate garbage or harmful chemicals, bacteria, etc. This has a strong impact on the ecosystems: Many species such as birds and fishes die and the rivers equal rather a sewage channel than a greenery. The health of people using river water for their needs as they do not have access to municipal water is in danger as many diseases can be transmitted through polluted water.

To prevent pollution, garbage needs to be minimized and wastewater needs to be treated in treatment plants or even better, at the source. Treating wastewater is a very complex and expensive process. In Pune, there are only eight treatment plants, which are treating 67 % of wastewaters. Therefore, the Municipal Corporation is planning for two more treatment plants to reach 100% treatment of wastewater (Environmental Status Report Pune 2010, PMC).

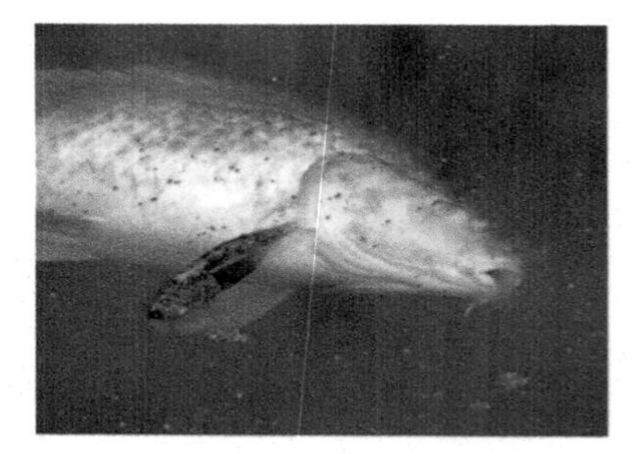

"With the Eco-Club at our school, we did a water audit of our whole school and calculated how much water the whole school uses every day. Then we developed projects of how to save water. We started rainwater harvesting in that we collected rainwater from rooftops in containers for gardening. We repaired all dropping taps we could find and started the project "Cut the flush" in our neighborhoods: We filled one-liter bottles with sand and convinced our neighbors to put them into the water tank of the toilet. In that way, one liter per flush will be saved. We collected signatures of our neighbors and checked whether they were actually doing it. All of us students started to realize how important it is to save water every day, especially in summer time."

Priyanka, 9th Standard, English Medium School

Ecological Sustainability

„Water is money. The Municipality and the citizens of Pune waste a lot of water because they don't realize how much water is worth. 40% of water is lost due to leakages and mismanagement. Why should we farmers be blamed? Pune has sufficient water and above average in comparison to other Indian cities. Therefore, water shortage in the city is not our fault. It is easy to blame us as we have quite a big share of the water in the dams. But we at least use it efficiently and turn crops into money. "

Farmer, 45

"Due to the industrial and population growth in Pune, the need for water in the city increases. At the same time, the irrigation department needs water for agriculture. Water is a basic need for everyone. Therefore it should be free of cost and sufficient for every citizen. Agriculture, especially cash crops, should come after the population is provided with enough water. We face water shortages in summer, because irrigation is demanding too much water."

Pune Resident, 38

Role Cards

PRO
In your opinion, water supply needs to be sustainable.

Prepare for the discussion:
Note down on your evaluation worksheet a few arguments which you think are most convincing.

PRO
In your opinion, water supply needs to be sustainable.

Prepare for the discussion:
Note down on your evaluation worksheet a few arguments which you think are most convincing.

CONTRA
In your opinion, there is no need for sustainable water supply.

Prepare for the discussion:
Note down on your evaluation worksheet a few arguments which you think are most convincing.

CONTRA
In your opinion, there is no need for sustainable water supply.

Prepare for the discussion:
Note down on your evaluation worksheet a few arguments which you think are most convincing.

You are the minute keeper.
Prepare for your role:
You have to explain the discussion rules to the group. It is your task to ensure that everyone is following these rules:

1. It is only allowed to talk if a colored speech talk is put in the middle of the table.
2. All speech cards have to be put in the middle during the discussion.
3. Two cards of the same color are not allowed to be put after each other.
4. Before one brings an argument, one has to counter the prior arguments.

You are the observer.

Prepare for your role:

- Fill in the observation sheet.
- Are the arguments objective, relevant and well explained?
- Do the participants refer to counterarguments?
- After the discussion, give each participant feedback on their argumentation skills.

Observation Worksheet

Discussion Participant 1: ______________________

	−−	−	+	++
The arguments were understandable and relevant				
The arguments were objective				
She or he gave strong reasons for the arguments				
She or he reacted to counterarguments				

Discussion Participant 2: ______________________

	−−	−	+	++
The arguments were understandable and relevant				
The arguments were objective				
She or he gave strong reasons for the arguments				
She or he reacted to counterarguments				

Discussion Participant 3: ______________________

	−−	−	+	++
The arguments were understandable and relevant				
The arguments were objective				
She or he gave strong reasons for the arguments				
She or he reacted to counterarguments				

Discussion Participant 4: ______________________

	−−	−	+	++
The arguments were understandable and relevant				
The arguments were objective				
She or he gave strong reasons for the arguments				
She or he reacted to counterarguments				

Evaluation Worksheet

Social Sustainability

Arguments against sustainability	Arguments for sustainability

Economic Sustainability

Arguments against sustainability	Arguments for sustainability

Ecological Sustainability

Arguments against sustainability	Arguments for sustainability

Curriculum Vitae

Dr. Johanna Stephanie Leder
Born 1985 in Bonn, Germany

Education

2011–2016	**Ph.D. (Dr. rer. nat.) in Cultural Geography** University of Cologne, Germany Ph.D. Thesis: "Pedagogic practice and the transformative potential of Education for Sustainable Development. Argumentation on water conflicts in geography teaching in Pune/India" ("Magna cum Laude"), Supervisors: Prof. Dr. F. Kraas/Prof. Dr. A. Budke
2006–2010	**State Examination in Geography, English, and Educational Sciences** (equivalent to M.Sc./M.A./ M.Ed.) University of Cologne, Germany Master Thesis: "Access to health care services for children in Mumbai/India" (Supervisor: Prof. Dr. F. Kraas)
2006-2011	**State Examination in Biology (additional subject)** University of Cologne, Germany
2009–2011	**Certificate for Further Studies on Bilingual Education** (English/German), University of Cologne, Germany
2008–2009	**Certificate for Teaching English as a Second Language (CTESL)**, University of New Brunswick Saint John, Canada; studies in Canadian literature, biology, and sociology
1996–2005	**A-Level at St. Ursula Gymnasium Brühl, Germany**
2002–2003	Clarendon High School, Arkansas, USA Duluth Central High School, Minnesota, USA

Appointments

since August 2017	Postdoctoral Researcher for Social and Environmental Justice Department of Urban and Rural Development Swedish University of Agricultural Sciences (SLU), Uppsala
November 2014–July 2017	Postdoctoral Fellow for Gender, Youth, and Inclusive Development, CGIAR Research Program on “Water, Land and Ecosystems” (WLE), International Water Management Institute (IWMI) Kathmandu, Nepal
June–September 2017	Lecturer, Institute of Geography Education University of Cologne, Germany
June 2014–October 2014	Consultant for Gender, Poverty, and Institutions CGIAR Research Program on “Water, Land and Ecosystems” International Water Management Institute (IWMI), Sri Lanka
June 2011–October 2014	Ph.D. Student at the Cologne Graduate School of Education University of Cologne, Germany Full three-year scholarship funded by the State Ministry of Innovation, Science and Research (MIWF) NRW
September 2011–June 2014	Research Affiliate at the Bharati Vidyapeeth Institute of Environment Education and Research (BVIEER), Pune, India
October–December 2013	Research Affiliate at the Centre de Sciences Humaines (CSH), New Delhi, India

Printed by Printforce, the Netherlands